Free-ranging Cats

Free-ranging Cats

Behavior, Ecology, Management

Stephen Spotte

WILEY Blackwell

This edition first published 2014 © 2014 by John Wiley & Sons, Ltd

Registered office: John Wiley & Sons, Ltd, The Atrium, Southern Gate, Chichester, West Sussex, PO19 8SQ, UK

Editorial offices: 9600 Garsington Road, Oxford, OX4 2DQ, UK
The Atrium, Southern Gate, Chichester, West Sussex, PO19 8SQ, UK
111 River Street, Hoboken, NJ 07030-5774, USA

For details of our global editorial offices, for customer services and for information about how to apply for permission to reuse the copyright material in this book please see our website at www.wiley.com/wiley-blackwell.

The right of the author to be identified as the author of this work has been asserted in accordance with the UK Copyright, Designs and Patents Act 1988.

All rights reserved. No part of this publication may be reproduced, stored in a retrieval system, or transmitted, in any form or by any means, electronic, mechanical, photocopying, recording or otherwise, except as permitted by the UK Copyright, Designs and Patents Act 1988, without the prior permission of the publisher.

Designations used by companies to distinguish their products are often claimed as trademarks. All brand names and product names used in this book are trade names, service marks, trademarks or registered trademarks of their respective owners. The publisher is not associated with any product or vendor mentioned in this book.

Limit of Liability/Disclaimer of Warranty: While the publisher and author(s) have used their best efforts in preparing this book, they make no representations or warranties with respect to the accuracy or completeness of the contents of this book and specifically disclaim any implied warranties of merchantability or fitness for a particular purpose. It is sold on the understanding that the publisher is not engaged in rendering professional services and neither the publisher nor the author shall be liable for damages arising herefrom. If professional advice or other expert assistance is required, the services of a competent professional should be sought.

Library of Congress Cataloging-in-Publication Data

Spotte, Stephen.
　Free-ranging cats : behavior, ecology, management / Stephen Spotte.
　　　pages cm
　Includes bibliographical references and index.
　ISBN 978-1-118-88401-0 (cloth)
　1. Feral cats. I. Title.
　SF450.S66 2014
　636.8–dc23
　　　　　　　　　　　　　　　　　　　　　　　　2014013795

A catalogue record for this book is available from the British Library.

Wiley also publishes its books in a variety of electronic formats. Some content that appears in print may not be available in electronic books.

Cover image: Two Stray Cats on Garbage Bins. © Vicspacewalker | Dreamstime.com

Set in 9.5/12pt BerkeleyStd by Laserwords Private Limited, Chennai, India
Printed and bound in Singapore by Markono Print Media Pte Ltd

1　2014

To Puddy, Tigger, Miss Sniff, Wilkins, Beavis, and Jinx

                                                              You enriched my life

Contents

Preface — xi
Abbreviations and symbols — xvii
About the companion website — xix

1 Dominance — 1
  1.1 Introduction — 1
  1.2 Dominance defined — 1
  1.3 Dominance status and dominance hierarchies — 6
  1.4 Dominance–submissive behavior — 10
  1.5 Dominance in free-ranging cats — 15

2 Space — 19
  2.1 Introduction — 19
  2.2 Space defined — 20
  2.3 Diel activity — 23
  2.4 Dispersal — 26
  2.5 Inbreeding avoidance — 27
  2.6 Home-range boundaries — 31
  2.7 Determinants of home-range size — 33
  2.8 Habitat selection — 41
  2.9 Scent-marking — 43

3 Interaction — 49
  3.1 Introduction — 49
  3.2 The asocial domestic cat — 49
  3.3 Solitary or social? — 52
  3.4 Cooperative or not? — 58
  3.5 The kinship dilemma — 61
  3.6 What it takes to be social — 66

4 Reproduction — 72
  4.1 Introduction — 72
  4.2 Female reproductive biology — 72
  4.3 Male reproductive biology — 84
  4.4 The cat mating system: promiscuity or polygyny? — 88
  4.5 Female mating behavior — 91

4.6	Male mating behavior	93
4.7	Female choice	96

5 Development 98
    5.1 Introduction 98
    5.2 Intrauterine development 98
    5.3 Dens 100
    5.4 Parturition 100
    5.5 Early maturation 104
    5.6 Nursing 108
    5.7 Weaning 109
    5.8 Survival 111
    5.9 Effect of early weaning and separation 113
    5.10 Early predatory behavior 114

6 Emulative learning and play 116
    6.1 Introduction 116
    6.2 Emulative learning 116
    6.3 Play 121
    6.4 Ontogenesis of play 125
    6.5 What is play? 130

7 Nutrition 137
    7.1 Introduction 137
    7.2 Proximate composition 138
    7.3 Proteins 139
    7.4 Fats 148
    7.5 Carbohydrates 150
    7.6 Fiber 155
    7.7 Vitamins 156

8 Water balance and energy 158
    8.1 Introduction 158
    8.2 Water balance 158
    8.3 Energy 162
    8.4 Energy needs of free-ranging cats 166
    8.5 Energy costs of pregnancy and lactation 172
    8.6 Obesity 178

9 Foraging 181
    9.1 Introduction 181
    9.2 Cats as predators 182
    9.3 Scavenging 185
    9.4 When cats hunt 189
    9.5 Food intake of feral cats 189
    9.6 How cats detect prey 190
    9.7 How cats hunt 200

	9.8	What cats hunt	205
	9.9	Prey selection	207
	9.10	The motivation to hunt	210
10	Management		214
	10.1	Introduction	214
	10.2	Effect of free-ranging cats on wildlife	215
	10.3	Trap–neuter–release (TNR)	224
	10.4	Biological control	233
	10.5	Poisoning and other eradication methods	237
	10.6	Integrated control	241
	10.7	Preparation for eradication programs	245
	10.8	"Secondary" prey management	247

References	251
Index	293

Preface

The dog is humankind's obsequious, slavering companion ever sensitive to its master's moods and desires. The cat is ambiguous, irresolute, indifferent to its owner, if indeed any human who co-habits with a cat can be called that. Many of my cats have been memorable, perhaps none moreso than Miss Sniff, who adopted me when I lived on a Connecticut farm. It happened like this. One night in late autumn I heard a noise outside and opened the door. In walked an ugly, leggy, calico cat. She had the triangular head and blank stare of a praying mantis, and her nose was in the air mimicking a sort of feline royalty. With startling arrogance she jumped onto the couch and made one end of it hers. And so I named her Miss Sniff.

For months my barn had been plagued by rats. Their excavations were everywhere, around the perimeter of the building and even deep into the clay floors of the horse stalls. Nothing I tried could eradicate them. They ignored traps, snickered at poisoned grain, shouldered aside the barn cats and ate the food from their bowl. Some, bored with the furtive life, lounged brazenly outside their burrows in full sunlight.

That first night I fed Miss Sniff and eased her out the door. She greeted me the next morning with a freshly killed rat, a large shaggy beast of frightening proportions. Female cats without kittens to raise often bring their prey home, laying it out in a convenient place and giving little churring calls to their humans. Paul Leyhausen (1979: 88–89) wrote: "The important thing for the cat is ... not the praise but the fact that the human serving as 'deputy kitten' actually goes to the prey it has brought home, just as a kitten thus coaxed does." I have no idea if Leyhausen's interpretation is true, but I nonetheless congratulated Miss Sniff, gave her a pat, and every morning thereafter she presented me with a dead rat. Within a few weeks she had caught them all. In retrospect I realize how mere praise was a paltry reward, and to express proper gratitude I should have sat down on the porch steps and eaten the rats in front of her. At least one or two simply to be polite.

The common cat is the most widespread terrestrial carnivoran on Earth, occupying locations from 55°N to 52°S and climatic zones ranging from subantarctic islands to deserts and equatorial rainforests (Konecny 1987a). This is possible because few carnivorans except possibly the red fox (*Vulpes vulpes*) can match its ecological flexibility and the capacity to find food and reproduce almost anywhere. As further evidence of protean adaptability, the cat has become the most common mammalian pet with an estimated 142 million having owners worldwide (Turner and Bateson 2000). Domestic cats are now the most popular house pet in the United States (Adkins 1997). According to the Pet Food Institute (2012) the estimated number of pet cats in the United States is >84 million, well in excess of the number of pet dogs (>75 million). Castillo and

Clarke (2003) set the total number of US cats at 100 million, including those without owners.

At the same time, free-ranging cats – many of them house pets – exact a devastating toll on wildlife around the world. May (1988) estimated that there were ~6 million free-ranging house cats in Britain. Although well fed, they killed an average of 14 prey items each per day, which extrapolates to ~100 million birds and small mammals annually. In the final chapter I present evidence that killing unowned cats is the only sensible method of controlling their depredation on wildlife. Eradication programs are unpopular with those bent on saving cats at all costs. However, the pressure placed on wild creatures should be alleviated whenever possible, and subtracting alien predators from terrestrial ecosystems is one way of reducing the carnage.

The underlying thesis throughout is that effective management of free-ranging cats is best achieved if based on understanding their behavior, biology, and ecology. In this respect I take issue with experts who claim cats to be social, occupy rank-order positions in dominance hierarchies, disperse under pressure from inbreeding avoidance, are territorial, have a polygynous mating system, and live in functioning kinship groups in which cooperation is common. The data do not support any of these positions, and failure to discard them stands in the way of real progress toward our understanding of why cats behave as they do. More important, casual disregard of the cat's reproductive biology and unusual nutritional requirements has hampered the search for novel methods of population control, limiting current choices to biological agents (e.g. feline panleucopenia virus) and nonselective poisons, augmented by trapping and shooting.

We should take a closer look at the domestic cat for other reasons too. The family Felidae is thought to contain ~40 species (Wildt *et al.* (1998: 505, Table 1), and all except the domestic cat are under threat of extinction (Bristol-Gould and Woodruff 2006, França and Godinho 2003, Goodrowe *et al.* 1989, Neubauer *et al.* 2004, Nowell and Jackson 1996, Pukazhenthi *et al.* 2001). The ordinary cat has therefore become a model for conserving other felids through study of its reproductive and sensory biology, genetics, behavior, use of habitat, and nutritional needs.

Cat biology is highly context-dependent. Laboratory studies have taught us much, and knowledge of free-ranging cats is paltry in comparison. My discussion focuses on the latter, but where lacunas exist I fill them with what we know from cats kept in confinement and presume that the differences are not too great. This is a reasonable approach, at least from a physiological standpoint. Cat genetics are well conserved (Plantinga *et al.* 2011), meaning the metabolic adaptations of cats are not likely to vary whether they occupy a laboratory cage, alley, or sofa cushion. Endocrine factors driving reproduction, for example, are difficult to monitor except in a lab, but differences compared with free-ranging cats are matters of degree, not kind.

I consider free-ranging cats classifiable into three categories: feral, stray, and house. *Feral cats* survive and reproduce without human assistance and often despite human interference (Berkeley 1982). *Stray cats* occupy urban, suburban, and rural areas where humans assist indirectly by making garbage available to scavenge and by offering shelter underneath houses and in abandoned buildings. Garbage represents a concentrated food source and also attracts rodents and birds, still other sources of food. Although strays are sometimes fed by sympathetic people, they are less likely to be offered shelter and veterinary care. *Free-ranging house cats* are those allowed outdoors unsupervised by their owners, who provide consistent shelter, food, and usually veterinary care.

Never take for granted a cat's understated ability to influence our own behavior. During an election year a while back in the village of Talkeetna, Alaska, the populace grew unhappy with its mayoral candidates. Someone started a write-in campaign for a yellow tabby named Stubbs, who hung out in the General Store. Stubbs won, and is now the mayor. Like politicians everywhere he spends much of his time asleep on the job, refusing to let the responsibilities of elected office become a distraction.

Stephen Spotte
Longboat Key, Florida

For cats, indeed, are for cats. And should you wish to learn about cats, only a cat can tell you.

Sōseki Natsume, *I Am a Cat*

Abbreviations and symbols

$\bar{x}$	mean	kcal	kilocalorie(s)
µmol	micromole	kg	kilogram(s)
a	scaling constant (power law)	kHz	kilohertz
		kJ	kilojoule(s)
ATP	adenosine triphosphate	L	liter(s)
BCFA	branch-chained fatty acid	LH	luteinizing hormone
BMR	basal metabolic rate	M	body mass
BSA	body surface area	MAF	minimum auditory field
cd	candela	mg	milligram(s)
CL	corpus (corpora) lutea	min	minute(s)
CM	center of mass	mmol	millimole(s)
CSF	contrast sensitivity function	ms	millisecond(s)
d	day(s)	MUP	major urinary protein
dB	decibel(s)	NFE	nitrogen-free extract
DHA	docosahexaenoic acid	ONL	obligatory nitrogen loss
DM	dry matter	PAPP	*p*-aminopropiophenone
DMI	density-mediated interaction	PUFA	polyunsaturated fatty acid
		RDH	resource dispersion hypothesis
E	energy		
EAA	essential amino acid	s	second(s)
EFA	essential fatty acid	SCFA	short-chained fatty acid
EPA	eicosapentaenoic acid	SD (or σ)	standard deviation of the mean
EUNL	endogenous urinary nitrogen loss		
		SEM	standard error of the mean
FC	food consumption	TMI	trait-mediated interaction
FPL	feline panleucopenia	TRSN	tecto-reticulo-spinal tract
FUNL	fasting urinary nitrogen loss	TS	total solids
g	gram(s)	UV	ultraviolet
GnRH	gonadotropin-releasing hormone	VNO	vomeronasal organ
		VR	vomeronasal receptor
ha	hectare(s)	W	watt(s)
k	scaling exponent (power law)	y	year(s)

About the companion website

This book is accompanied by a companion website:

> www.wiley.com/go/spotte/cats

The website includes:
- Powerpoints of all figures from the book for downloading
- PDFs of tables from the book and online appendices

1 Dominance

1.1 Introduction

The concept of dominance appears often in the animal behavior literature. When defined at all its meaning and usage are often inconsistent, making any comparison of results among experiments ambiguous. How we think of dominance necessarily influences findings obtained by observation (Syme 1974). Perhaps because domestic cats are asocial (Chapter 3), their expressions of dominance seem strongly situation-specific (Bernstein 1981, Richards 1974, Tufto *et al.* 1998) rather than manifestations of a societal mandate, making dominance–subordinate relationships less predictive of reproductive success and other fitness measures.

My objectives here are to define and describe dominance behavior and try to evaluate its relevance in the lives of free-ranging cats. Much experimental work on dominance and subordination in laboratory settings has only peripheral application to cats living outdoors. Consequently, I seriously doubt that watching cats crowded together in cages yields anything except measures of aberrant behavior, not at all unusual when circumstances keep animals from dispersing (Spotte 2012: 221–227).

The dominance concept has done little to enlighten our understanding of how free-ranging cats interact, its utility seemingly more applicable to animals demonstrating true sociality. As I hope to make clear, agonistic interactions between free-ranging cats are mostly fleeting, situational, and the consequences seldom permanent because neither participant has much to gain or lose. Baron *et al.* (1957) and Leyhausen (1965) used *relative dominance* when referring to how vigorously an individual dominates subordinates, meaning that some cats are more dominant than others in *relative* terms, perhaps by not allowing subordinates to usurp them even momentarily at the food bowl if a subordinate growls or by refusing to share food. That measurements of relative dominance, situational dominance, or dominance by any category have utility in assessing the interactions of free-ranging cats is doubtful. Food is not highly motivating. Small groups of cats, whether captive (Mugford 1977), feral (Apps 1986b), or stray (Izawa *et al.* 1982), seldom fight over food or anything else, raising the question of whether the "dominance" observed during arena tests and based on food motivation is not mostly an artifact of experimental conditions. As Mugford (1977: 33) wrote of laboratory cats fed *ad libitum*, "Less than 1% of total available time was accounted for by feeding, so it would be difficult for any single dominant animal to retain exclusive possession of the food pan. ... "

1.2 Dominance defined

The most useful definition of any scientific term consists of a simple falsifiable statement devised to reveal some causal effect in nature beyond mere description and data

---

Free-ranging Cats: Behavior, Ecology, Management, First Edition. Stephen Spotte.
© 2014 John Wiley & Sons, Ltd. Published 2014 by John Wiley & Sons, Ltd.
Companion Website: www.wiley.com/go/spotte/cats

analysis. Flannelly and Blanchard (1981: 440) made clear that "dominance is not an entity, but an attempt to describe in a single word the complex interactions of neurology and behavior." This is important to remember and useful conceptually, although difficult to wrestle into falsifiable hypotheses if the only available method of testing involves observation without manipulation of the subjects or conditions.

Any definition necessarily encompasses *agonism* (Drews 1993), which some consider a synonym of aggression, but properly interpreted and applied includes both dominance and submission (Spotte 2012: 40–42). Drews employed the terms dominant and subordinate to indicate relative rank in either a *dyad* (a group of two individuals) or more complex hierarchy (i.e. triad or higher). It follows logically that *dominance behavior* and *submissive behavior* denote specific responses (e.g. striking with a forepaw, sibilance, aggression, fleeing). Thus a subordinate owes its rank – as perceived by us – to behaving submissively when encountering a dominant conspecific.

Gage (1981) proposed studying dominance in either of two ways. One approach starts by proposing a theory that not only identifies the concept but encompasses conditions necessary to realize its application (*functional definition*). This step is followed by derivation of a testable hypothesis derived from theory that includes a definition. Empirical results then force acceptance or rejection of the null hypothesis of no difference along with the definition. The free-ranging cat literature largely ignores functional definitions. However, to qualify as scientific the design of an experiment is obliged to take a functional approach because all testable hypotheses must be grounded in theory. Descriptions not based on this principle leave no means of explaining the observations.

In the second approach (*structural definition*), observable states of dominance are tacitly assumed to exist outside theory, an operational definition is proposed, and tests are conducted to determine whether the term as defined has merit. The most complete structural definition is from Drews (1993: 308), who did not offer a functional counterpart: "*Dominance* [italics added] is an attribute of the pattern of repeated, agonistic interactions between two individuals, characterized by a consistent outcome in favour of the same dyad member and a default yielding response of its opponent rather than escalation." A consistent winner is therefore dominant, the consistent loser subordinate. This winner–loser format describes how agonistic encounters are resolved and assessed observationally by an investigator.

Drews' definition, along with the majority of others he reviewed, demonstrates that the animal behavior literature (including that portion dealing with free-ranging cats) is almost entirely data-driven, descriptive, and relies on structural definitions. In the absence of hypothesis testing, the causal basis of dyadic asymmetry and dominance hierarchies (see later) can only be inferred. To make inductive inferences is to step outside the boundaries of structurally-based experimentation and attempt to explain function, an impossible undertaking. When induction takes precedence, accounts of structurally based experiments morph into general, or universal, statements (Popper 1968: 27), none of which can ever be valid.

Some combination of signals is necessary before dominance ranks or hierarchies can assemble in sustainable configurations. *Communication* can be defined as "an association between the sender's signal and the receiver's behavior as a consequence of the signal" (Spotte 2012: 33). Assuming agonism is a form of communication – that is, measurable in terms of signal and response – then *dominance* considered within communication's restricted context is one animal's attempt to influence another's behavior (also see Krebs and Dawkins 1984, Maxim 1981, Smuts 1981). My purpose here is to ascertain how this is possible and attempt to assess the different manifestations.

Operationally, the individual signaling first (i.e. the cat attempting to influence how the other responds) can be either the dominant or subordinate member of a dyad. For example, crouching is considered submissive male behavior. If so, a male that crouches on encountering another male signals submission, announcing his subordinate status. The dominant male then has two choices: ignore the signal or respond by signaling his dominance. The latter behavior acknowledges respective status, although in either case the dominant-subordinate relationship likely has been established even between cats meeting for the first time (Cole and Shafer 1966), and any chances of aggression are diminished. The dominant male's first option (passive disregard) is evidence that "Subordinance-acknowledging ... is not always prompted by dominance-confirming, and either of them can serve as a signal or response" (Spotte 2012: 41).

As mentioned, an agonistic encounter produces a so-called "winner" and "loser," one animal emerging dominant, the other subordinate. A fight might serve to establish a dominant-subordinate relationship initially. However, mutual acknowledgement of status is what sustains the relationship over time, and perpetuation without change is based on recognition and familiarity. Fighting is rare afterward, and a stable relationship from both sides of the agonistic divide has been established. Dominant-subordinate status can be established quickly in dyadic contests. Cole and Shafer (1966) tested eight cats in 10 round-robin trials (28 combinations) and noted that in 82% of dyads the relationship became apparent during the first trial.

Dominance is conceptually fuzzy like "stress" and "species." As Hinde and Datta (1981: 442) emphasized, "If dominance is used to describe the directionality of interactions, it explains that directionality no more than the 'migratory instinct' explains migration." Familiarity makes dominance especially difficult to assess (de Boer 1977b). Landau's (1951: 1) rigorous mathematical analysis led to this conclusion: "The hierarchy is the prevalent structure only if unreasonably small differences in ability are decisive for dominance." Thus, "If all members are of equal ability, so that dominance probability is $1/2$, then any sizable society is much more likely to be near the equality than the hierarchy; and, as the size of the society increases, the probability that it will be near the hierarchy becomes vanishingly small." In Landau's view, what really controls dominance relationships are factors like the histories between individuals.

By age 8 weeks, cats are threatened by an unfamiliar conspecific or even a cut-out cardboard model of one, responding with *piloerection* (hair erect, or "standing on end") and arched back (Kolb and Nonneman 1975). Can two male cats recognize each other as individuals outside the context of dominant-subordinate or is familiarity predicated on signaling alone and subsequently learned through experience? Not presuming to know the answer raises another question: can dominance-submission be separated from learning and take place before mutual recognition has been established? Maybe the subordinate recognizes some feature of the dominant individual associated with a *prior attribute* (also called *supraindividual characteristic*), or individual trait that bestows rank, like greater body mass, a high-quality display, kinship, or a behavioral sign that induces submission without confrontation (Gauthreaux 1981, Winslow 1938). If so, it might predict the outcome of such meetings between strangers, but dominance *per se* would not be involved (Vessey 1981). This is not the case if the subordinate recognizes in the stranger a prior attribute associated with dominance that had previously consigned it (the subordinate) to its current status. As a result of that encounter the subordinate now defers and assumes the postures of submission (Bernstein 1981). In this hypothetical situation the *attribute* has prompted the dominant-subordinate relationship, not the individuals.

Dominance is presumably about conflict resolution and supposedly functions by dampening aggression (Hinde 1978). The capacity to prevent dominance from escalating into aggression might hold true in nature where subordinates can disperse. Captive animals are denied this option, and a subordinate is unable to escape the dominant's aggression (Spotte 2012: 221–227). Encounters between strangers require that both individuals recognize and correctly interpret certain properties possessed by the other. Encounters between two familiar animals, if unidirectional over time, are founded on learning, memory, and recognition, three factors that reinforce the agonistic status quo, repress aggression, and reduce the possibility of injury to either party. The expression of threat might be even more important than aggression in establishing a dominance relationship between cats (Cole and Shafer 1966).

As mentioned, dominance has been linked to prior attributes and patterned relationships between individuals, two incompatible concepts. The distinction requires understanding that dominance between animals as assessed by humans is a construct, in practical terms a relative measure rather than some inherent property possessed by certain individuals and not others. Dominance as a result of a prior attribute seems unlikely unless the physical feature (e.g. greater body mass) or trait conveyed (e.g. heightened aggression) exists in recognizable form in the absence of submission. Baron *et al.* (1957) found no consistent association between dominance status and prior attributes like differences in sex, body mass, passivity, and problem-solving ability. They wrote (Baron *et al.* 1957: 65): "Descriptive and correlational investigations such as this will not contribute greatly to our understanding of the determinants of social behavior in animals."

Dominance by definition must be relative, a dominant individual comprising one-half a dyad. It seems doubtful that physical attributes alone are reliable predictors of dominance, despite our sometimes explicit presumptions (e.g. piloerection makes body size appear larger, standing straight gives an appearance of being taller). I offer three reasons. First, examples abound of smaller, weaker individuals dominating stronger, more physically imposing opponents in dyadic situations. Second, because a prior attribute can be associated with putative rank (e.g. body mass, head volume, age in males) as claimed by Bonanni *et al.* (2007) raises the possibility of secondary associations that might be more meaningful (i.e. that one or more of these variables is merely a secondary expression of a behavioral trait and irrelevant in isolation). Third, if dominance can be recognized simply on the basis of prior attributes we should expect rank-order to mirror a continuum of the attributes themselves (e.g. heaviest is dominant followed by next heaviest, and so forth), but direct correlation of such factors is not consistently predictable (Hinde 1978).

Laboratory cats that had been dominant in both dyads and group hierarchies became timid and submissive to their former subordinates after psychological manipulation rendered them neurotic, yet nothing about their physical appearance had changed (Masserman and Siever 1944). In fact, the rank-order could be turned upside-down (*dominance reversal*) and then re-established by psychological manipulation of the test cats while keeping physical prior attributes constant. These last experiments indicate that dominance in cats emerges from a behavioral trait and not a physical attribute.

Dominance has sometimes been defined as "priority of access to resources" (Drews 1993: 299). As Drews made clear, for this to be predictable "implies *a priori* that dominance influences the pattern of access to resources or else that priority of access to resources be part of the definition of dominance." As a useful measure of behavior it presents two problems. First, if the premise states that dominance directly affects access to resources then measuring its impact based on access to resources involves circular reasoning (Richards 1974, Syme 1974). Second, limiting observations to any specific factor inevitably obscures interpretation: dominance envelopes all instances of conflict resolution, but in this example not every conflict is about resources (Drews 1993, Hand 1986).

Prediction is a necessary feature of dominance, but insufficient to define it (Vessey 1981), and description alone is obviously limited by a lack of both predictive and explanatory power, leaving underlying cause, or function, indeterminate. As pointed out by Drews (1993), definitions based on observation instead of theory are closed to empirical investigation, leaving no way of comparing them. Each such asserted definition stands isolated, untestable against any others. Distinctions devoid of theory are relegated to semantics (Gage 1981), and a definition that incorporates presumed synonyms lacks even descriptive value. The literature on free-ranging cats is notable in this respect, commonly identifying dominant males as "aggressive" or "winners." Making subjective evaluations in ways that dominance equates with aggressiveness and "winning" a dyadic encounter classifies one cat as dominant and the other subordinate (e.g. Bonanni *et al.* 2007, Cafazzo and Natoli 2009, Natoli *et al.* 2001). This method meets a basic statistical definition (Tufto *et al.* 1998: 1489) that "dominance is defined as a parameter characterizing the relationship between two individuals, determining the expected number of successes of the first individual in disputes with the other." In the end, however, attempts at explanation devolve inevitably into conjecture because cause has been omitted from the statement of hypothesis.

Tufto *et al.* (1998) pointed out that assessing dominance relationships in dyadic terms provides a parameter p_{ij} in which individual *i* dominates *j*. Dominance is therefore a parameter describing a relative relationship between two individuals along an infinite series of values spanning 0 and 1. Thus *i* dominates *j* if $p_{ij} > 0.5$. If the value of p_{ij} is exactly 0.5 then neither individual in the dyad is dominant, but the requirement is always

$$p_{ij} = 1 - p_{ji} \tag{1.1}$$

It should come as no surprise that the process of devising and then sorting categories of behavior based on description seldom opens an illuminated path to insight. As Drews (1993: 297) wrote, "An asymmetry in the outcome of particular interactions is not a sufficient justification to introduce a dominance concept, either as a descriptive tool or an explanatory mechanism." This is perhaps even truer in attempts to describe nonlinear hierarchies in which kinship can force intransitivities and context determines the outcome, as when offspring dominate their mothers in some situations but not others (Tufto *et al.* 1998).

Even if predictive value is high, accuracy and precision are not confirmation of a definition but a description of how the animals behaved in those circumstances; that is, a definition can have "heuristic value" without explanation (Drews 1993: 299). Science is the business of testing theories. As emphasized, descriptive studies have limited scientific utility unless placed firmly within the context of hypotheses. The ultimate objective should be to address how and why animals behave as they do, which renders behavioral description as half-completed and executed in reverse; that is, data are collected before testable hypotheses have been devised. The large number of structural definitions of dominance relative to functional ones, combined with a history of inconsistent results, is evidence of this deficiency.

1.3 Dominance status and dominance hierarchies

According to Drews (1993: 283), "*dominance status*" [italics added] refers to dyads while *dominance rank* [italics added], high or low, refers to position in a hierarchy and, thus, depends on group composition." From this structural perspective, learning through past encounters, individual recognition, and other important but confounding variables become irrelevant. What the subordinate recognizes is some feature of the dominant individual perhaps associated with a prior attribute. Here clarification is warranted. Note that Drew's structural definition identifies dominance as *an attribute of the pattern of interactions,* and although the submissive animal recognizes some feature of dominance in its opponent's signals it is the exchange of signals and responses – characteristics of *pattern* – that determine respective status within the dyad, not a prior attribute of either individual.

A dominance hierarchy that places individuals of a group into ranks of descending dominance can exist between two individuals or among several, but such relationships are always sums of composite interactions occurring *between* two individuals, not *among* three or more. Even in tight settings the process is sequential, although often appearing to be simultaneous. In other words, a cat confronted suddenly by two antagonists must instantaneously assess first one then the other. However, because dominance-submission prompts interaction, results of isolated dyadic measurements are unlikely to be realistic descriptions. In any case, assessments of dominance hierarchies in captive domestic cats have limited application to knowledge of relationships in free-ranging cats because the focus is limited to aggressive encounters. As Kerby and Macdonald (1988: 72) pointed out, "None of these studies shed light on the workings of a hierarchy in the cat's natural history, and none has reported the subtle behavioural cues one might expect to signal the *status quo*. ... "

The truth of this last statement casts an antinomic shadow. We need to know what mechanisms make dyadic interactions in isolation different from those in groups and elucidate why an animal that seems dominant in one situation is not in another. Perhaps the answer is simple, an inability to evaluate the status of more than one conspecific simultaneously relative to your own, as mentioned in the paragraph above. Consider humans at a cocktail party. What looks superficially like multi-person interactions are actually shifting dyads of focus. One individual speaking while the rest listen is a monologue. Humans behaving socially communicate as dyads using dialogue. The word "trialogue" is a comparatively minor entry in the English lexicon. Cats are no different. After watching kittens, West (1974: 433) wrote, "In play involving three or more individuals the nature of the play patterns allows for only peripheral interaction by the 'third' member."

Experimentally, dominance status of cats is assessed through "tournaments," the objective being to seek the underlying dominant-subordinate structure within a group. This is evaluated by placing pairs of cats from the same group in an "arena" together for predetermined periods until every individual has been tested against each of the others in round-robin competition. Resultant scores are expected to reveal a pattern. One troubling aspect is that the results of such tests are artifactual by taking place outside the only context that really matters, which is the group itself. Although individuals presumably interact in dyads, any outside influence has been walled off. Another is the problem of apparent linear rankings sometimes being indistinguishable from randomness (Appleby 1983, but see Jameson et al. 1999).

Hierarchies are of two basic kinds, neither especially relevant to the lives of free-ranging cats. A *transitive hierarchy* describes a linear (i.e. straight-line) scale of dominance, or "peck-order," in which animal A is dominant to animal B, and B is dominant to animal C. Consequently, A is dominant to C. This can be expressed symbolically as $A > B > C$. An *intransitive hierarchy* is similar to the first: $A > B > C$ except that $C > A$, implying a nonlinear looping back of the dominance order. The further an intransitive hierarchy deviates from linearity, the more intransitive it becomes (i.e. the greater the possibility that intransitive loops will increase with the number of criteria). Intransitive relationships are common, surfacing during dyadic encounters when the outcome is determined by two or more factors (Petraitis 1981). Perfectly linear hierarchies occur most often in small groups; that is, groups having <10 members (Chase et al. 2002, Drews 1993). The larger the group the more its pattern slides toward intransitivity (Jameson et al. 1999).

Rank based on prior attributes (Section 1.2) is thought to influence rank-order within a linear hierarchy, but these factors alone are not its building blocks (Chase et al. 2002). Thus the linear hierarchy of a society can form, disintegrate, form again, and remain consistently linear even if half the members change rank with each iteration (Chase et al. 2002). In other words, linearity must be driven by factors and forces other than those easily measured and observed in pairwise contests. This shortcoming, combined with confounding by winner and loser effects (reciprocal reinforcement), bystander effects, the stringent mathematical conditions required to produce linearity if based on prior attributes, and doubtless other factors, call into question the relevance of testing dominance-submission using pairwise interactions and extrapolating the results to the group.

As hinted above, to account for what makes one animal dominant and another subordinate in social species ultimately requires evaluation at the group level. Two hypotheses can be considered. The first is deterministic by stating that an individual's position in a hierarchy is more or less decided in advance by features that enhance its capacity to dominate. This is the *prior attributes hypothesis*, elements of which were described previously, and although it forecasts linear social structures this is not always the result (Chase et al. 2002). As noted before (Section 1.2), prior attributes can include behavioral or physical characteristics (e.g. aggressiveness, age, body mass, sex) or a mix of these. If its pertinent elements can be identified and limited, then the individual with the highest prior attributes score presumably emerges dominant over the others. The animal scoring second-highest ranks second, and so forth. Often an animal predicted to be dominant based on a prior attribute (e.g. body mass) turns out to be submissive (Winslow 1938).

The *social dynamics hypothesis* is more stochastic and predicts nonlinear social structures. It states that social interaction among members of the *whole group* and not its paired components drives the formation of hierarchies, and that hierarchical structures emerge from causative factors other than prior attributes (Boyd and Silk 1983, Chase et al. 2002). Specific social interactions culminating in intransitive hierarchies possibly include (1) winner and loser effects in which winners or losers of earlier contests assume a pattern of winning or losing later ones (Chase *et al.* 1994, Hsu and Wolf 1999) and (2) bystander effects during which conspecific bystanders observing individuals interact with others adjust their own behavior (Johnsson and Åkerman 1998, Silk 1999). Seen from this perspective, hierarchies become self-organizing based on group dynamics *within* the social system and not derived entirely from any prior attributes of its individual members (Theraulaz *et al.* 1995).

Each hypotheses has elements of validity, and the two might exert complementary effects (Chase *et al.* 2002). Until recently, support for the social dynamics hypothesis came mostly from models showing that restrictive mathematical conditions are necessary to produce linear hierarchies based solely on prior attributes (Landau 1951). Chase *et al.* (2002: 5748) concluded: "Linear structures should not be assumed to result simply from variation among individuals or from cumulative conflicts among pairs of individuals." They advocated instead that investigators "look at patterns of interaction across whole groups and understand how these patterns produce hierarchy ladders."

Stated differently, inherent properties as observed in individuals or dyads are not indicators of social structure and therefore unable to represent it. Going further, they might not even indicate dominance, at least not the transitive kind. Statistical analyses of dyadic interaction are based on paired comparisons. In tests like Appleby's (1983) the null hypothesis states that from among a group of paired comparisons the chances of any individual winning is random. If the test statistic is then sufficiently large to reject the null the alternative hypothesis simply presumes a transitive underlying structure. Any interpretation that the dominance structure is *actually* transitive falls outside the capability of the analysis and must be incorrect. As Tufto *et al.* (1998: 1489) explained, "rejection of the null hypothesis of randomness implies only that the alternative hypothesis is a better description of the dominance structure among the individuals being studied."

Winner and loser effects could also be termed *reciprocal reinforcement* because each individual of a pair potentially "trains" the other to perform as dominant or subordinate (Flannelly and Blanchard 1981, Spotte 2012: 54). This situation arises, for example, during paired competition for food under arena conditions. The first to reach the reward and eat it (the "winner") is scored as dominant, the "loser" as subordinate. Repeated trials usually yield consistent results once the participants become acquainted, and the same is true between evenly matched strangers (Chase *et al.* 1994; Hsu and Wolf 1999; Winslow 1944a, 1944b). Such findings could be artifactual, the animals having learned to solve the problem efficiently (i.e. without strife); that is, outside the assumption of a prior attribute and therefore independently of the experimental design. Instead of revealing true social relationships each animal "trained" the other to retain its respective status, which was then reinforced in subsequent trials. The result is less a hierarchy than the illusion of one. Many times during dyadic interactions competing cats end up sharing the reward more or less equitably or after some harmless nudging and pushing (Winslow 1944b).

The issue is further confused by striking individual differences in any group of cats. Some strive consistently to be more competitive whether they win or lose dyadic contests. Others seem to give up, and still others vary their effort depending on intensity of the competition. Winslow (1944a: 311) wrote, "In general … the form of social interaction elicited in cats … depended upon the nature of the social relationship that had existed between the competitors prior to the tests." This, and the fact that a cat's performance changes when the competitor is removed (Winslow 1944a, 1944b), indicates to me that dominant-subordinate relationships can be created artificially simply by placing cats in abnormal situations; that is, conditions of forced interaction. Free-ranging cats ordinarily avoid each other, thereby circumventing conflict and leaving no outward evidence of either dominance or submission.

It seems unlikely that either transitive or intransitive relationships exist as permanent fixtures among free-ranging cats. Laboratory observations to the contrary appear to result from confinement and crowding. The original descriptions of a transitive dominance (i.e. the original "peck-order") hierarchy were by Schjelderup-Ebbe (1922) in domestic fowl, which unlike cats are unquestionably social. The transitive relationship is therefore one of descending dominance-submission and truly hierarchical. Sustaining such a system requires every individual to recognize, accept, and remember its status. Once in place a transitive relationship is mainly peaceful, the dominant animal seldom fighting to retain its rank. This asymmetric pattern is consistent and unidirectional, and contrary to territoriality (Chapter 2) the direction of agonistic behavior is independent of location (Kaufman 1983).

Even confining a group of cats in a small space does not predictably induce agonism. Among the most common interactive behaviors is none at all; that is, mutual indifference, each individual spending most of its time alone (Hart 1978, Podberscek *et al.* 1991). Eight adult males (seven castrated) were maintained under laboratory conditions in a cage 2.2 m high × 1.62 m wide × 3.9 m long. Three walls had shelves; there were litter boxes on the floor. Of behaviors recorded, agonistic interactions accounted for just 1% and consisted of hissing. When cats played it was usually alone; when grooming, 90% was self-grooming (i.e. licking itself).

Liberg (1980: 347), remarking on Leyhausen's (1965) comments about social organization of solitary mammals, suggested that territoriality could develop at low population densities, "but at higher densities this might change into a dominance order system." Say and Pontier (2004) seemed to agree when speculating that urban strays are not territorial. This is certainly true of gray wolves, which form hierarchies in captivity where space is limited but are territorial in the wild (Spotte 2012: 90–107, 221–227). I doubt the existence of a similar pattern in the domestic cats. Location reliably predicts most dominant-subordinate relationships in territorial species, but evidence of territoriality in domestic cats is unconvincing (Chapter 2).

Some investigators have postulated that the population density of free-ranging cats – and consequently social organization – is influenced strongly by the abundance and distribution of food (Macdonald 1983, Liberg and Sandell 1988). I disagree. Evidence seems to show that only the first is true and that cats gathered around waste-disposal sites and feeding stations are aggregations of asocial individuals, not a society as the term is usually understood (Chapter 3). Dominance hierarchies supposedly form in groups of stray cats gathered near clumped food (Dards 1983, Liberg *et al.* 2000, Natoli and De Vito 1991, Say *et al.* 2001). However, clustered resources (e.g. feeding sites, garbage bins, waste-disposal dumps) do not always forecast the competitive interactions that typify social species. Male strays at crowded urban locations, for example, are often more tolerant of each other than feral males at rural areas of low population density (Liberg 1980).

Izawa *et al.* (1982: 377) reported neither avoidance nor agonism among ~200 strays around fish-waste dumps at 125-ha Ainoshima Island, Fukuoka Prefecture, Japan, where individuals fed together, and "several cats [as many as 19] often ate the dump of fishery wastes or large fish nose to nose. … " Page *et al.* (1992) saw no evidence of food competition among strays living at Avonmouth Docks, Bristol, United Kingdom. Garbage was plentiful, supplemented with food set out regularly by employees. Cats did not crowd the feeding areas, some rarely visiting them at all. Others checked the feeding sites consistently, including some that occupied small home ranges (Chapter 2) centered around them. Denny *et al.* (2002) did not see evidence of male dominance at a waste-disposal site on the central highlands of New South Wales, Australia. Yamane *et al.* (1997) attempted to show that age determined feeding "rank" among male strays, but the data, so far as I could tell, revealed merely a vague outline of pattern, not actual evidence of dominance traceable to underlying cause and effect. Interaction among cats of all ages was peaceful.

Observations of farm cats chasing off the competition might be interpreted as follows, although my comments are speculative. Strange cats are attacked and chased by both resident male and female farm cats (Macdonald and Apps 1978). Females are more tolerant of females of their own group than of females from neighboring farms, with males seemingly tolerant of males from both their farms and others (Turner and Mertens 1986). Farm cats are not always fed consistently, nor does the amount provided always meet their nutritional requirements. The cats might be defending their food supply, which in its meager outlay has become more valuable and easier to protect than the endless pickings at a waste-dump site.

Cats are not overtly competitive. Weaned kittens and adolescents of 4–6 months are generally tolerated by adults and often allowed to feed first (Bonanni *et al.* 2007, Yamane *et al.* 1997). Urban strays apparently did not compete for food with red foxes or stray dogs at Avonmouth Docks (Page *et al.* 1992), and Beck (1971) observed stray cats and dogs, along with rats, foraging peacefully on garbage in a Baltimore alley within centimeters of each other.

If this situation seems confusing, laboratory tests and field observations offer little in the way of clarity. Baron *et al.* (1957: 59) reported "fairly linear" hierarchies emerging from food competition tests among confined cats that knew each other but found no particular association between dominance and aggression. The rankings were roughly consistent whether the number of cats was two, three, or four, but among both captive and free-ranging cats transitivity is generally weak (van den Bos and de Cock Buning 1994, Laundré 1977, Liberg 1980, Natoli *et al.* 2001). Leyhausen (1965) reported a dominance hierarchy at feeding. According to Cole and Shafer (1966) the most food-motivated and aggressive individuals in food competition tests were not necessarily the most dominant during other interactions. Masserman and Siever (1944) concluded that aggression during food competition tests resulted more from frustration at failing to obtain food than a means of obtaining it. Laundré (1977) reported a similar situation and the formation of a female dominance hierarchy among farm cats fed intermittently. Baron *et al.* (1957) did not see an association between dominance and aggression using food competition tests. When the dominant animals of three groups competed among themselves, Baron *et al.* (1957: 64) wrote, "there was no apparent relationship between the status of the leaders as measured by their food-getting success and the aggressivity that they demonstrated while competing together." Accordingly, these findings "generally confirm the naturalistic impression of the cat as asocial and individualistic in its interactions with other cats."

1.4 Dominance–submissive behavior

Nearly all the 90 or so visual and tactile behaviors documented for wolves also occur in domestic dogs (Scott 1950: 1013–1015, Table 1, 1967; Scott and Fuller 1965). The cat's seem impoverished in comparison, including several inconsistently interpreted elements of dominance-submissive behavior. Slaps and hisses have been classified as low-intensity aggression (Dards 1983) and as submission (de Boer 1977b). Wails, yowls, and piloerection are seen as submissive responses (Dards 1983). Caterwauling has been linked loosely with aggression, although some subordinate cats caterwaul too, and growling can signal either dominance or submissive behavior (de Boer 1977b). Shimizu (2001: 88) described "threatening vocalizations" as "lasting sounds, like a dog howling." These were emitted by free-ranging, same-sex adults while staring at each other, and were identical to the yowl. However, Shimizu (2001) considered the yowl of adult males to be different from the sound made in the breeding season (Chapter 4). During laboratory tests of dominance an especially aggressive male might briefly mount a less aggressive male competitor (although without intromission), inducing it to become passive (Winslow 1944a). Males mounting each other has also been reported in strays (Yamane 1999).

As defined by Fox (1975: 413), "A *display* [italics added] is a composite of different units or actions (e.g. tail and ear positions and movement, angle of body, crouch, forward lean, back arch, etc.)." Visual signaling between individuals is apparently less important to cats than to social monkeys. Young stump-tail macaques (*Macaca speciosa*) blinded shortly after birth were harassed by sighted conspecifics. The cause was "lack of comprehension of visual signals" (Hyvärinen *et al.* 1981: 4), requiring them to be separated from the sighted group. In contrast, Crémieux *et al.* (1986: 231) reported that "The social behaviour of the blind cats with the other blind and control [sighted] cats was almost normal." Although the experimental cats were tested as adults they had been purposely blinded as kittens and were oblivious to visual signals. The importance of visual signaling needs to be assessed. Whether the absence of signals like tail-up (see later) go unnoticed when tail-less cats fail to provide them is unknown. The only conclusively demonstrated use a cat has for its tail is maintenance of balance (Walker *et al.* 1998). *In the following descriptions, hyphenated terms in italics refer to distinctive signals in a kind of shorthand used elsewhere in the text.* On meeting, two males might sniff noses (*nose-sniff*) before displaying signs of agonism, but this behavior declines with increased familiarity (de Boer 1977b). Aggression among cats that know each other is minimal: 14 instances in ~1200 interactions seen in four barn cats comprising an adult male and three adult females (Macdonald and Apps 1978). According to Dards (1983), greeting behavior between cats of either sex consists of any or all of three components: (1) raising the tail vertically (*tail-up*), (2) nose-sniff, and (3) rubbing heads (*head-rub*). In tail-up the tail, normally carried at an angle of ~45 degrees below horizontal, is lifted to horizontal or higher (sometimes to 90 degrees with a slight curl at the tip) when encountering another cat. Before or after nose-sniff another part of the body, particularly the perianal region might be sniffed (*perianal-sniff*). A head-rub (Fig. 1.1) might extend from the other cat's head laterally along its body. Females are about twice as likely to greet males than to be greeted by them in return, and greetings initiated by females are often more intense (Dards 1983). The male's response is likely to be less intense, if he responds at all.

Male–male interaction is generally splenetic, ranging along an ascending continuum from mutual avoidance to tolerance to aggression (Dards 1983). Rarely is it openly friendly. Even if neither male runs away, agonistic encounters are usually restricted to transient displays of agonism, vocalizing, or mutual disregard. What the vocalizing component signifies is uncertain. Bonanni *et al.* (2007: 1371) included in dominance

Fig. 1.1 Head-rub. Source: © Ruzanna Arutyunyan | Dreamstime.com.

behavior "ritualized vocal duels and real duels" without defining either term. Several observers have described the cat's agonistic behaviors (Bonanni *et al.* 2007, Dards 1983, de Boer 1977b). In assuming a posture of dominance a male stands straight (*stand-straight*), supposedly for the function of appearing taller. The pupils of his eyes constrict (*pupils-small*). He moves with exaggerated slowness, *stiff-legged*, hindquarters seemingly higher than the shoulders. Cole and Shafer (1966: 49) called this "strutting behavior." The tail is arched near its base (*arch-tail*), body hair in some degree of piloerection, tail hair usually moreso. He emits low wails escalating into yowls (*yowls*), lashes his tail from side to side (*tail-lash*), holds his head high (*head-high*) with chin pointed down (*chin-down*) and jaw chomping rhythmically (*chomp*). Alternatively, he makes licking motions (*lick-lips*) or smacks his lips (*smack-lips*). He might also strike at his opponent with a paw (*paw-strike*). Sometimes cats sit and assume agonistic poses while displaying some of these signs (Fig. 1.2).

A male displaying submissive behavior half-sits (*half-sit*) or crouches (*crouch*), sometimes with chest and abdomen pressed to the ground (*ventrum-pressed*). He might lick-lips or smack-lips, lie down, even roll onto his back, occasionally paw-striking (*paw-strike*) or hissing (*hiss*) at the other cat, although dominant cats also hiss during agonistic encounters. The head is pulled back and kept low (*head-low*), the ears flat to the sides (*ears-flat*) or laid back or folded (*ears-folded*) against the head (Fig. 1.3). From any of these postures he might yowl or spit (*spit*) at his aggressor.

If a fight breaks out it happens suddenly (de Boer 1977b). The two antagonists grip each other face to face and scratch with all four legs. Usually the dominant then leaves while the subordinate remains in a defensive posture (e.g. crouch). Afterward, and sometimes when fighting has not occurred, both cats sit facing each other, alternately opening and closing their eyes slowly (*blink*).

According to de Boer (1977b), behaviors that seem associated with either aspect of agonism – that is, dominance or submission – can include exploration (*explore*) and sniffing of the area (*sniff*), spraying urine (*spray*), rubbing against objects (*object-rub*, Fig. 1.4), grooming themselves (*self-lick*), sitting (*sit*), assuming a Sphinx-like posture

Fig. 1.2 Agonistic encounter between two strays on an urban street. Source: © Andris Daugovich | Dreamstime.com.

Fig. 1.3 A cat displaying simultaneous agonistic expressions, hiss and ears-flat. Source: © Georgiy Pashin | Dreamstime.com.

(*sphinx*) while blinking, and sitting opposite the opponent while looking askance (*look-away*).

Sometimes elements of dominance and submission are mixed (Dards 1983). A submissive cat might stand up instead of crouching, although with its back arched (*arch-back*). It might also display piloerection and lift a forepaw partly off the ground as if preparing to slap. The combination of arch-back, piloerection, *arch-tail*, and the neck flexed (*neck-flex*) is the embodiment of the cartoon Halloween cat, which I shall call *halloween* (Fig. 1.5). It can be expressed with or without a forepaw raised (*forepaw-raised*) or stepping sideways (*side-step*). Males sometimes stand before each other with heads averted (*head-avert*), or one of the two might head-avert with chin-down and chomp. Encounters ordinarily end not in a fight but with one or the

Fig. 1.4 Cats rub against objects for unknown reasons, but deposition of scent (i.e. "marking") is probably not among them (Chapter 2). The so-called cheek glands of cats have never been described. Perioral glands on the lips could potentially deposit scents, but their function in this regard is unconfirmed. Source: © Astrid228 | Dreamstime.com.

Fig. 1.5 Kitten displaying arch-back, neck-flex, piloerection, and side-step, in combination called halloween. Source: © Tatyana Chernyak | Dreamstime.com.

other retreating, often walking away slowly and stiff-legged. The departing individual might be the aggressor (Konecny 1987a), in which case evaluating which of them "won" the putative contest is unclear.

1.5 Dominance in free-ranging cats

Probably all dyadic interactions are potentially contentious, but as mentioned at the start, dominance in asocial species is likely to be situation-specific. Here are seven examples. (1) Bonanni *et al.* (2007) reported directional dominance relations based on correlation of aggressive and submissive behavior from observations of 13 cats (males and females); both sexes displayed aggressive behavior more or less equally. (2) Adult female cats are sometimes aggressive toward strange females and young males (Macdonald *et al.* 1987). (3) Females with kittens often attack males (Natoli 1985a). (4) Castration can cause diminished dominance behavior and increased submissive behavior (de Boer 1977b). (5) In a group of Roman strays the females were consistently more aggressive than males when near food, often dominating them, although the males seemed otherwise dominant (Bonanni *et al.* 2007). (6) Among female farm cats, most agonism was seen at feeding sites, not over food specifically but as a result of crowding the location in expectation of being fed (Panaman 1981). (7) Baron *et al.* (1957) found that feeding a "dominant" cat prior to a food-competition test sometimes caused it to lose status.

I question whether dominance assessment in free-ranging cats is worthwhile. Experiments using dyadic outcomes to establish rank-order (e.g. Bonanni *et al.* 2007, Natoli 1985b) incorporate an unknown factor, the presumption of evolutionary relevance. Unless hypothesis-based, such endeavors yield results amenable only to conjecture, leaving even the matter of sociality unaddressed. For example, if devising rank-orders is intended to reflect kinship bias in females, does high rank offer privileged access to resources like food and shelter? If so, does rank-order demonstrably affect fitness? The fact that female cats sometimes live in "kinship groups" does not necessarily ameliorate conflict. Inherited relationships failed to bestow detectable privileges on subordinate females among rural Swedish house cats, some of which succumbed to conspecific pressure (including aggression) and dispersed to nearby houses that had no cats (Liberg 1980). Neither was kinship bias evident during competitive feeding experiments in laboratory cats (Masserman and Siever 1944). Before kinship is presumed to affect dominant-subordinate relationships or relationships of any sort in free-ranging cats, its possible effects must be separated from familiarity and the two variables tested and evaluated independently.

Stating that directionality observed between familiar individuals reflects dominance status and then claiming that mutual history is the proximate cause of their respective ranks is tautologous, and tautological statements are not empirical (Popper 1968: 85). Actually, learning by both dyadic members is the proximate cause of such dominance relationships, its existence shown by consistent directionality of the asymmetry (Bernstein 1981).

Prior attributes do not reliably determine dominance even in barnyard fowl that form clearly transitive hierarchies (Schjelderup-Ebbe 1935), and the same is true in the less structured hierarchies of both captive and free-ranging cats (Baron *et al.* 1957, Panaman 1981, Winslow 1938). Cole and Shafer (1966: 48) thought that "it is recognition of the overt behaviors which serves as the important cue for the development of dominance-subordinance relationships." Agonistic behaviors (e.g. aggression, submission) were on display more often during arena situations than when cats were tested in pairs, suggesting social stimulation. Cole and Shafer (1966) mentioned how cats in dyadic food competition trials displayed declining interest once the dominance relationship had been established.

Many investigators seem to view dominance and dominance hierarchies as ends in themselves or treat them as intervening variables mediating other behaviors (Hinde 1978, Hinde and Datta 1981). To Seyfarth (1981: 448) they are "simply shorthand, descriptive terms used by observers to describe what they have seen. ... " Dominance hierarchies are sometimes taken for granted and presumed to exist even when no evidence has been presented (e.g. Macdonald and Apps 1978). To accept the existence of dominance relationships does not require acceptance of dominance hierarchies too, especially transitive ones; their causes might be completely different (Bernstein 1981). Deviation from linearity is common in most species (Hinde 1978).

Are hierarchies even relevant in behavioral assessments? Certainly not without evidence obtained using streamlined definitions and testable hypotheses. Baron *et al.* (1957: 65) wrote, "Systematic manipulation of experimental variables such as motivation and social learning of individual subjects holds greater promise for comparative studies of behavior. ... " And Bernstein (1981: 428) warned, "Shotgun correlational techniques and closely reasoned logical arguments of what *should be* [italics added] the case ... will not prove that agonistic dominance ranks are a factor in social organization." Moreover, in abbreviated dominance hierarchies (e.g. only two ranks, high and low), identifying rank-order adds nothing beyond what we already know. This is apparently the situation in cats when females and males, and just females, are kept together in small groups: one or more individuals appear dominant and the rest seem to rank lower with nothing separating them in terms of status (Baron *et al.* 1957, Laundré 1977, Panaman 1981, Rosenblatt and Schneirla 1962, Winslow 1938).

Rosenblatt and Schneirla (1962: 453) speculated that "Dominance relations in the cat appear to be more a matter of indifference of one animal towards another, in which the more active animal appears to be the more dominant, than the end-result of a series of encounters in which mutual relationships are worked out among the individual animals." Such casual disregard of conspecifics is inconsistent with the behavior of truly social species even in confinement. Spotted hyenas (*Crocuta crocuta*) captured as infants were taken to California and reared in separate peer groups. Captivity denied them sustained maternal influence, the opportunity to hunt, and presented other unnatural conditions (e.g. no chance to acquire skills by learning from older, experienced conspecifics). Nonetheless, they arranged themselves into a natural hyenid system characterized by powerful social facilitation (e.g. group eating and drinking, group defecation and scent-marking, group greetings), a complex array of social signals, female dominance, matrilineal organization, inherited social status, and dominance hierarchies in which males were subordinate (Glickman *et al.* 1997). Such complex and coordinated interactions are not required of asocial species like the domestic cat (Chapter 3).

It seems reasonable to ask whether dominance is best measured in terms of arena tests involving competition if such tests are not controlled for the effects of facilitation. Winslow (1944a: 297) defined *competition* in laboratory settings to be situations "in which winners and losers are selected on the basis of their speed or strength in executing the experimental task, with the consequence that winners receive the reward and losers are denied it." He defined *social facilitation* as "situations in which an increment in the activity results simply from the presence of other individuals. ... " According to Ward (2012: 223), facilitation occurs "where individuals are more likely to express a given behaviour, or express it a greater rate, in the presence of conspecifics." Therefore, in dyadic contests is the first animal to the food truly dominant or does the other animal's mere presence facilitate a faster response by the so-called winner? In sorting this out we might conclude that "dominant" and "subordinate" have little meaning when cats are tested in arenas, and that "social facilitation" implies some standard of interaction rarely encountered in a species that behaves as tolerantly asocial most of the time.

Leyhausen (1965, 1973) doubted that cats form rigid dominance hierarchies, yet believed that dominance status becomes established among free-ranging cats. He noted how familiar males often entered each other's spaces peacefully. This could be expected only if the occupied areas represent home ranges but not territories, which by definition are *defended* spaces (Chapter 2). As to interactions between strange males, Leyhausen (1973: 127) wrote, "Adult tomcats meeting for the first time are liable to engage in fierce fighting regardless of the season." In his opinion, fights between males are never territorial disputes because they take place on neutral ground. This seems to me an impossibility because any location occupied by a cat obviously comprises part of its home range even if only temporarily, and although a home range is not defended it is anything but neutral. The presumed hierarchies formed as a result of these dyadic encounters are, in Leyhausen's opinion, "absolute" and represent fixed dominance status. Afterward, certain "rules of the road" dampen future aggression. As an example, Leyhausen (1973: 125–126) wrote, "If the inferior [subordinate] cat has already entered a commonly used passage before the superior cat arrives on the scene, the latter will sit down and wait until the road is clear; if it does not, its superiority may be challenged successfully." No empirical evidence was presented to bolster any of these claims.

Relationships in presumed cat hierarchies can fail to show transitivity in other ways. In a hierarchy comprising eight laboratory cats, two individuals accounted for 59 aggressive events, 51 attributable to one cat, eight to the other (Cole and Shafer 1966). The incidents seemed unrelated to dominance rank (the investigators ranked these animals third and fourth).

The putative transitive hierarchies in stray males reported by Natoli (1985b), having been based on paired comparisons, were probably invalid. According to Richards (1974), if a species is thought to be truly social, constructing transitive hierarchies based on dyadic interactions is inappropriate. She listed three criteria for evaluating dominance–submission. First, because accurate assessment of social ranks requires a high frequency of social interactions the species studied must be clearly social. This makes the domestic cat a questionable candidate (Chapter 3). Second, in laboratory settings captivity precludes escape, and observations made of captive groups can yield an artificially inflated number of "social" encounters, particularly of the agonistic kind. Third, the group studied must be stable, its membership unchanged over a long period.

Testing an asocial species as if it were social can yield puzzling and inconsistent results. Cole and Shafer (1966) assessed dominance in eight laboratory cats (males and females) using food competition. The subjects were tested in all dyadic combinations and then as a group in a room familiar to them. Hierarchies from the two experimental configurations were different. In fact, the cat that emerged as dominant from extensive round-robin dyadic competitions ranked lowest when all cats were in the room together. This individual, which had been eager and motivated in paired comparisons, was "aloof and relatively placid in his behavior" in a group setting (Cole and Shafer 1966: 47). In the end the study failed to show "whether a cat is dominant because of his ability to make the required response in a competitive situation or whether the efficiency of making the response is dependent on his position in the dominance hierarchy."

Dominance hierarchies for both male and female cats have been reported in urban strays (Bonanni et al. 2007, Devillard et al. 2003, Natoli and De Vito 1991, Say et al. 2001), and male strays supposedly live in established linear dominance hierarchies (Dards 1983, Natoli and De Vito 1991). These conclusions are doubtful. Such rankings depend on the number of individuals that can be defeated by the cat designated as "dominant" in dyadic contests, which does not correlate exactly with the number of times the dominant actually achieves a victory or with the number of wins minus the number of losses (Bernstein 1981). In addition, expected success becomes increasingly inconsistent with declining rank-order even if transitive hierarchies are considered to exist. In a group of free-ranging cats, some might not interact with others or be absent during observation periods. As a result, frequency tables in which encounters are recorded then contain missing values, making statistical assessments of transitivity difficult or impossible (Bonanni et al. 2007, Jameson et al. 1999).

The importance of familiarity is hard to understate, and perhaps we should be as wary of ranking cats as trying to herd them. Mutual recognition might be the foundation of their associations. Prior interaction in some fishes can predict the results of dominance encounters (Chase et al. 1994). A fish that wins is likely to win again if a second opponent is presented shortly thereafter, and this "winner effect" diminishes over time, noticeably so after 1 h. How long the memory of an encounter lasts in cats is unknown, but evidently quite a while. Of a group of four laboratory cats, Masserman and Siever (1944: 9) wrote: "Once established, the dominance hierarchy in feeding responses was found to persist in all possible combinations … even after weeks of rest from experimentation."

2 Space

2.1 Introduction

To repeat and expand definitions used in the Preface for categorizing free-ranging cats, I follow previous conventions for dogs (Spotte 2012: 107–108; also see Dickman 1996a). *Feral cats* are wild animals. They exist independently of humans, avoid human contact, and survive by hunting and scavenging. *Strays* are ownerless cats – varying in fear response from wild to tame – that live in urban, suburban, and rural areas and depend indirectly or directly on humans for food and shelter. I classify "farm cats" as strays if fed and feral if not. Untamed farm cats that are not fed are therefore considered feral even if they use farm buildings intermittently to rest or give birth. Some strays might be born ownerless; others are abandoned, leave home voluntarily, or are forced out of their homes by intraspecific competition.

Free-ranging cats take shelter in farm buildings, vacant buildings, abandoned motor vehicles, under porches and steps, or in nearby woods and fields, living mainly on garbage. They might hunt occasionally or be fed regularly or intermittently by cat aficionados. The third category comprises *house cats* let outside by their owners. They depend entirely on humans for food and shelter, and if they scavenge, hunt or find temporary shelter away from the home it is seldom from need. Any such categories are artificial, especially to the cats. By these definitions, free-ranging cats at Santa Catalina Island, California, shifted back and forth from feral to stray, the latter fed by cat lovers along the coast, the former living in rugged inland areas and existing on small animals (Guttilla and Stapp 2010).

Free-ranging cats are not merely generalist predators (Chapter 9). Their generalist tendencies extend to habitat occupancy and use of space. Those living on their own can be found in swamps and marshes, in deserts, along the windswept cliffs of high-latitude islands, and in humid tropical forests. On islands they occupy most habitats (Medina et al. 2006). A cat can live wherever prey is of suitable size and abundance and shelter sites are adequate. How free-ranging cats make use of space can be distilled to a single question: do they occupy home ranges or are they territorial? Draped across the answer is the nature and effect of dominance and how – or whether – it modulates interactions between individuals (Chapter 1). These issues might have come into focus sooner were it not for the use of vague terminology and inductive methods of experimentation.

As emphasized in the last chapter, a useful concept must be predictable in ways that observations can be explained, not merely described. The scientific method does not start with observation and experiment and then proceed to general, or universal, statements; it begins with universal statements formulated as theories, which are then distilled to hypotheses. A *hypothesis* is a simple testable statement derived from a theory and amenable to falsification by empiricism. A theory strengthens each time it resists falsification through hypothesis testing (Popper 1968: 141). The notion that knowledge *as* theory emerges from patterns in nature is backward, as seen from the misplaced reliance on observation and data analysis to formulate universal statements like those claiming to have decided whether cats have dominance hierarchies, are social creatures, and so forth. Transposed to the present context, we have adequate descriptions of *what* free-ranging cats do, but mainly speculation and conjecture about *why* they do it. Much of the problem arises because the associated definitions are too vague or complex to be falsifiable. Popper (1968: 142) wrote: "Simple statements, if knowledge is our object, are to be prized more highly than less simple ones *because they tell us more; because their empirical content is greater; and because they are better testable.*" The utility of this should soon be apparent.

2.2 Space defined

Philopatry is attachment to place, and most animals are philopatric. A *home range* is the space an animal uses in its daily activities to find food and shelter and to reproduce. A *territory* is the part of a home range that is defended, ordinarily against conspecifics but sometimes against other species too. Unwanted encroachment is deterred actively by *competitive exclusion* and passively by *noncompetitive exclusion* (i.e. when leaving a mark or just showing up is sufficient to discourage encroachment). A bird's territory often comprises only the nest and immediate surrounding area, although the home range might be much larger. For stationary packs of gray wolves, home range and territory are synonyms because the home range *is* the territory, and the entire area is defended against conspecifics (Spotte 2012: 90). Leyhausen (1965) considered free-ranging cats to occupy home ranges of which smaller parts comprised territories. As I discuss elsewhere in this chapter, Leyhausen's assessment is doubtfully valid.

The definition of territory has a long and tortuous history (Kaufmann 1983, Maher and Lott 1995). Expansion of its meaning to fit specific requirements (e.g. Kaufman 1983) is always accompanied by dilution and loss of both accuracy and precision, resulting in greater confusion instead of increased clarity. In hypothesis testing, less is more. Hypotheses stated simply, directly, and devoid of contingencies are the most vulnerable to falsification, which also makes them exceedingly powerful when not refuted by experimentation. Noble's original definition of a territory as "any defended area" (Noble 1939: 267) allows hypotheses about the theory of territoriality to be devised and tested: for any given species and under the conditions observed a space is defended or it is not. Nothing else need be asked. Also with emphasis on conciseness, Burt (1943: 351) defined a home range as "the area, usually around a home site, over which the animal normally travels in search of food." Prior to Leyhausen (1965), he then identified a territory as the protected part of a home range. One stipulation of Brown and Orians (1970: 241) is that "exclusive occupancy of an area and territoriality are not synonymous." Much has been written about cat home ranges and whether the absence of overlapping boundaries in some instances is evidence of territoriality (see later). The relevance of this point is dubious. As Brown and Orians emphasized, "Nonoverlap of home ranges might also be caused by mutual avoidance, by preference for an unexploited food supply, by physical barriers, or by differing habitat preferences." Newer techniques and equipment such as proximity loggers should help clarify the association between overlap and actual interaction (e.g. Robert *et al.* 2012).

To many investigators the concepts of home range and territory are frequently confused, misused, not distinguished, or applied interchangeably (e.g. Barratt 1997a, Bradshaw and Brown 1992, Dards 1978, Genovesi *et al.* 1995, Harper 2004, Jones 1989, Langham 1992, Remfry 1981, Zaunbrecher and Smith 1993). Natoli (1985b: 302), for example, stated after watching urban strays: "I never observed territorial disputes among adult females. ... " Later in the article (p. 303), after remarking how home range overlaps were tolerated by all stray cats of this group, she noted that adult females cooperated "in defending the territory against stranger cats. ... " By implication the sum of home ranges across age and sex composed a territory that adult females defended against strange cats. This last idea has not been tested empirically, but its usefulness seems questionable considering the impunity with which adult males roam between and among cats aggregated around waste-disposal sites. After monitoring ~200 strays feeding at fish-waste dumps at Ainoshima Island, Japan, Izawa *et al.* (1982) reported that no individual occupied an exclusive resting site, nor was there any evidence of spaces being defended.

To Jones (1989: 7), feral cats are territorial. He wrote: "Although there can be an overlap of home ranges between individuals, strong territorial behaviour is suggested by urine spraying, pole-clawing, cheek-rubbing and sometimes direct conflict." The problem is, these activities have yet to be linked ineluctably with territoriality. Jongman (2007: 194) considered free-ranging cats of both sexes to be territorial, "with most territories centered around a food source." Some evidence exists that strays occasionally defend their feeding sites (Macdonald and Apps 1978, Yamane *et al.* 1994), but whether such behavior constitutes actual defense of spatial boundaries is unconfirmed. Izawa *et al.* (1982), who studied the Ainoshima Island cats prior to Yamane and colleagues, emphasized the absence of agonism among them. Laundré (1977) did not see evidence of territorial behavior in a group of Wisconsin, United States, farm cats. The cats occasionally *allogroomed* (i.e. licked each other), rubbed against one other, or nose-sniffed. Most aggressive encounters (97.5%) were recorded during milking sessions while the cats waited to be fed from a single dish. Aggression in these situations could center on food, not space (Spotte 2012: 111), or individual space around the food bowl. In another instance of casual usage, Corbett (1978: 269) labeled feral cats in the Monach Islands, Scotland, as territorial, their territories "based on rabbit stronghold areas" but offered neither a logical basis for this decision nor supporting data.

Langham and Porter (1991: 756) gave an arbitrary definition of territory after night-tracking feral males, writing of one cat that its "movements for two consecutive nights showed that the boundary areas were patrolled, representing a territory." Employing similarly loose terminology, de Boer (1977b: 229) remarked that during arena tests in a laboratory setting his cats considered the space a "joint territory because marking often occurred, the animals rarely slept there, and were never offered any kind of food." Hall *et al.* (2000: 19) stated: "Home ranges showed little overlap, suggesting a territorial social system." Macdonald (1983: 380) referred specifically to "exclusive female territories" of free-ranging estrous females, not to their home ranges. None of these authors mentioned defense of space, a territory's single defining property.

Ignoring the core of a term's definition in favor of superfluous contingencies can fray meaning beyond recognition, like the idea that a home range can become a territory simply by the occupant shifting its areas of usage periodically (Edwards *et al.* 2001, Konecny 1987a). Powell and Mitchell (2012) and Powell (2012: 887) proposed viewing home ranges as "the cognitive maps that animals maintain and update." This makes a lively metaphor, but the home range as mobile GPS with the output data subjected to our understanding of animal cognition is impossible to falsify and still shackled to measurable space. We might, for example, monitor a mouse using functional magnetic resonance imaging while it runs a maze, but the data would appear as a map of the brain with certain areas brighter than others, and the interpretation would still be based on spatial pattern and not cognition. Edwards *et al.* (2001: 99) wrote, "The periodic shifting of 24 h home ranges within the bounds of the larger long-term home ranges by male feral cats in central Australia may be a manifestation of territoriality, such movements providing the mechanism by which the large territory can be covered on a regular basis." The time spent near a home range's boundaries does not make a territory of the area enclosed, nor does consistent use of the space or failure of boundaries to always overlap (see later). Adding contingencies allowing for expanded definitions (e.g. two animals can use the same defended space by occupying it at different times) dilutes the concept with another layer of complexity and conjecture, in effect creating an auxiliary hypothesis and presenting still more impediments to falsification. Auxiliary hypotheses are acceptable only if they keep intact the original level of falsifiability and are not devised merely to help preserve an established system of thought (Popper 1968: 83).

My evaluation of the literature is based on the definitions of home range and territory by Brown and Orians (1970), Burt (1943), and Noble (1939) discussed earlier. By using the terms carelessly investigators both past and present have labeled free-ranging cats as territorial without regard for the implications (e.g. Biró *et al.* 2004; Churcher and Lawton 1987; Corbett 1978; de Boer 1977a, 1977b; Devillard *et al.* 2011; Edwards *et al.* 2001; Espinosa-Gayosso and Álvarez-Castañeda 2006; Fox 1975; Green *et al.* 1957; Hart 1978: 35–36; Hart and Hart 1985: 127; Hendriks *et al.* 1995a; Jongman 2007; Kitchener 1999; Leyhausen 1973, 1979: 217–226; Li *et al.* 2010; Liberg, 1980, 1984a; Lorenz and Leyhausen 1973: 123–130; Mendes-de-Almeida *et al.* 2004, 2011; Miyazaki *et al.* 2006a, 2008; Natoli *et al.* 2001; Neville and Remfry 1984; Pontier and Natoli 1996, 1999; Pontier *et al.* 2005; Rosenblatt and Schneirla 1962; Rutherfurd *et al.* 2007; Say *et al.* 2002a; Verberne and de Boer 1976; Warner 1985; Zaunbrecher and Smith 1993). In attempting to render this perspective irreversible, Pontier and Natoli (1996: 85) stated: "Among behavioural ecologists that study the domestic cat … there is a complete agreement on the fact that it belongs to a territorial species. … " Consensus is not a substitute for empiricism, and the data fail to support this high level of confidence.

Liberg (1980) suggested that free-ranging cats might be either territorial or hierarchical depending on population density. Territoriality would tend to occur more often at low densities. More likely, free-ranging cats living together as house cats or strays, or as captive groups in laboratories, *might* form rudimentary hierarchies if crowded (Chapter 1), establishing home ranges as space permits.

Contrary to the conclusions of some (e.g. de Boer 1977b, Leyhausen 1973), rank-ordering in groups of cats occupying a restricted space is not evidence of territoriality, nor do dominance hierarchies resulting from food-competition tests necessarily reflect real-world situations (Chapter 1). Outright fighting among free-ranging male cats might be most common when a young male leaves his natal group and encounters an older one (Dards 1983, Leyhausen 1965). Leyhausen's descriptions (Leyhausen 1973) of a sodality (*Brunderschaft*) among free-ranging male cats that involve initiations of younger members and the establishment of stable hierarchies functioning to limit aggression seem oddly like college fraternity rites, although females sometimes participated too. Leyhausen (1973: 126) wrote, "Males and females come to a meeting place adjacent to or situated within the fringe of their territories and just sit around." He continued, "They sit, not far apart – 2 to 5 yards or even less – some individuals even in actual contact, sometimes licking and grooming each other." Behavior was generally friendly. Sometimes these gathering lasted all night, but they usually broke up around midnight.

Most species of mammals are not territorial, but instead establish overlapping home ranges (Leyhausen 1965). Free-ranging cats apparently belong in this category. In territorial mammals the exploited resources are usually abundant, predictable, and often clumped (Brown and Orians 1970), habitats and situations descriptive of some group-living domestic stray cats except for the notable absence of territorial behavior.

Opposing those who declare free-ranging cats to be territorial are others who described males, females, or both as mutually tolerant, occupying home ranges that overlap, and making no attempt to defend the space around them (Dards 1978; Feldman 1994; Harper 2004; Langham 1992; Laundré 1977; Liberg 1980, 1984a; Mirmovitch 1995; Panaman 1981; Say and Pontier 2004; Tennent and Downs 2008; Turner and Mertens 1986). Leyhausen (1973) wrote that males are more tolerant than females when their space is invaded by a conspecific. He believed cats to occupy territories within home ranges, as mentioned, but to also form dominance hierarchies. Liberg (1980: 347) thought feral males might be territorial, although he saw no direct evidence of it, noting just that their home ranges were "more or less exclusive." However, exclusive use of space is less indicative of territoriality than of noncompetitive exclusion, in which cats occupying adjacent home ranges practice mutual avoidance. Noncompetitive exclusion is perhaps more common among adult males than adult females (Liberg 1984a, Turner and Mertens 1986).

2.3 Diel activity

The inclination to travel is affected by weather, and cats take shelter and rest for longer periods in wet conditions (Derenne 1976, Harper 2004). Once in motion, a cat travels as required to meet its physiological needs. A feral male monitored on Stewart Island/Rakiura, New Zealand, traversed the length of his home range (7 km) in a straight line over 23 h; another moved the length of his home range (5 km), also in a straight line, in 20 h, for an average speed of 250 m/h (Harper 2004). Average daily distance that five feral cats traveled in the Victorian Mallee, southeastern Australia, ranged from 0.29–1.67 km with no significant differences between males and females (Jones and Coman 1982b). Cats are vagile, athletic, and rate of travel is affected little by terrain (Harper 2004).

Home range boundaries might not be encountered daily. On single nights at Avonmouth Docks a female with a home range of ~18 ha visited 32% of her space, and a male occupying ~38 ha visited 35% of his (Page *et al.* 1992). Both males and females typically follow familiar paths through their home ranges (Anderson and Condy 1974, Izawa *et al.* 1982, Macdonald and Apps 1978, Macdonald *et al.* 1987: 15), indicating that portions of the area are used irregularly. In contrast, Kauhala *et al.* (2006) monitored the nocturnal movements of feral, stray, and house cats in southeastern Finland and recorded 41% of locations near home range boundaries, a finding apparently unrelated to reproduction because the incidence was greatest in summer and autumn (44–48%) and lowest in spring (25%). Individuals of all groups used 93% of their total home ranges throughout the year. At some places in the Simpson Desert, southwestern Queensland, Australia, cats traveled along dune crests and roads preferentially (Mahon et al. 1998). Cattle also preferred traveling on the roads, and their carcasses made convenient scavenging, although the dunes were still first choice.

Use of space varies over the diel cycle and in some locations by season. Feral cats, defined previously as those that survive entirely by hunting and scavenging, often travel farther at night than during the day (Langham 1992). Expanded portions of the home range were traversed at night by suburban house cats in Canberra, southeastern Australia (Barratt 1997a). Avoiding disturbances (e.g. people, dogs) and heat of the day perhaps account for the nocturnal activity of stray and feral cats living in farm buildings and near farms (Langham 1992, Langham and Porter 1991). This was also true of house and farm cats in central Illinois, United States, where humans, dogs, and machinery accounted for 18% of mortality (Warner 1985). Cats in this region were most active in the cool, quiet time from 1800–0800 h. Air temperature is an important seasonal factor, and the activity of many free-ranging cats picks up during the day in winter (George 1974, Jones and Coman 1982b, Liberg 1984a). Feral cats in another Illinois farming region were more nocturnal than house cats living in the same area and more active generally throughout the diel cycle (Horn et al. 2011). Feral cats in the braided river valleys (upper Waitaki Basin) of New Zealand's South Island were most active between 1500 and 0300 h and tended to select mature riverbed habitats (Recio et al. 2010).

Panaman (1981) followed the diel activities of English female farm cats. The animals were fed irregularly. Their budgeted percentage time: sleeping (39.7), resting (22.2), hunting (14.8), grooming (14.5), traveling (2.7), feeding (2.3), and other activities (1.4). Most activity (62.6%) occurred between dawn and dusk, and most sleeping (82.1%) was between dusk and dawn. However, the cats were fed only during daylight hours, and this might have influenced their pattern. A sleeping bout varied widely ($\bar{x} = 114$ min $\pm$ 88 SD) and often ended without apparent cause. The majority of time (81.4%) was spent in a shared core area. When traveling away from this location, all cats moved preferentially through long grass and thickets and along the walls of buildings, avoiding open spaces. Farm cats observed by Macdonald et al. (1987: 19) showed similar patterns when inside the barn, spending most of the time sleeping, followed by self-grooming and just sitting. When away from the barn they hunted in nearby hedgerows.

According to Brothers et al. (1985), a feral male monitored at subantarctic Macquarie Island over 38 h spent its percentage time walking or running (47.2), sitting (29.5), out of sight (6.2), sleeping (3.0), hunting (2.8), feeding (2.2), playing (0.9), interacting with other cats (0.6), self-grooming (0.4), and inspecting rabbit and petrel burrows (0.09). Hunting was difficult to distinguish from walking and running. Feral cats in the Galápagos Islands spent the majority of time (206.5 h of observation) hunting (Konecny 1987b). Feral cats at an Illinois farming region were more active than the house cats (Horn et al. 2011). House cats spent 80% of their time sleeping or "denning" compared with 62% for feral cats. For house cats, 17% of time was expended during low-activity behaviors and 3% at those requiring high activity. In feral cats these numbers were 23% and 14%.

Cats at Avonmouth Docks were active at all times (Fig. 2.1), but mainly at *twilight* (dawn and dusk) and after dark (Page et al. 1992). No tests were made to decipher this pattern, but human disturbance, midday heat, and reduction in the activity or

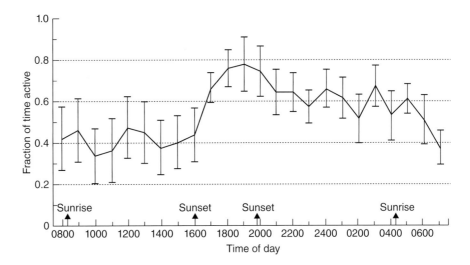

Fig. 2.1 Patterns of mean daily activity by feral cats at Avonmouth Docks, Bristol, UK ($n = 8$). Earliest and latest times of sunrise and sunset are indicated by arrows. Source: Page et al. 1992. Reproduced with permission of CSIRO.

availability of prey in daylight are possibilities. Strays at the Royal Navy Dockyard, Portsmouth, United Kingdom, were mostly crepuscular (Dards 1983). Those at Ainoshima Island, Japan, were crepuscular and nocturnal (Izawa et al. 1982). Daytime resting sites were selected to avoid humans, and times of activity when they temporarily abandoned these sites probably were too. According to Jones and Coman (1982b), feral cats in southeastern Australia were also most active at night and at twilight.

Increased human activity during the day was proposed as the reason urban strays in Brooklyn, New York, United States, were nocturnal (Calhoon and Haspel 1989, Haspel and Calhoon 1989). Feral cats in New Zealand's Orongorongo Valley were active both day and night (Fitzgerald and Karl 1986), as were feral cats at Tasman Island (Brothers 1982). The feral cats of semi-arid Hattah-Kulkyne National Park, Victoria, southeastern Australia, were most active at night (Jones and Coman 1982b).

Under certain conditions human disturbance actually enhances and reinforces higher activity levels in free-ranging cats. Urban strays in Jerusalem, Israel, became most active from 0600 to 0900 h when people were leaving home for work and the hours after dinner (1800–2100 h). During both intervals, food was emptied in greatest amounts into the garbage bins, and the high activity levels of cats and people coincided (Mirmovitch 1995): the cats anticipated food and stayed near the bins.

Feral cats on subantarctic Marion Island and Macquarie Island were mainly nocturnal (Bloomer and Bester 1992, Brothers et al. 1985, van Aarde 1979), as were those occupying farmlands in central Poland (Krauze-Gryz et al. 2012). Feral and house cats on agricultural land near the Warsaw suburbs were also most active at night, especially on fallow land nearest to built-up sites (Romanowski 1988). These cats preferred hunting more frequently in newly harvested fields during July and August, switching to meadows in autumn. Feral females in an agricultural area of New Zealand's North Island occupied either barns and pastures or swamps and willows (Langham and Porter 1991). The former were nocturnal and traveled long distances to hunt. The latter were active both day and night and traveled shorter nocturnal distances. Feral cats in a winter agricultural landscape in central-western New South Wales were active both day and night (Molsher et al. 2005).

2.4 Dispersal

Dispersion refers to the distribution of individuals or objects in space at a given moment, *dispersal* to movements of animals and not their locations (Brown and Orians 1970). Howard (1960: 152) defined *dispersal* as "the movement the animal makes from its point of origin to the place where it reproduces or would have reproduced if it had survived and found a mate." A *random pattern* of dispersion is described by $\sigma^2 = \bar{x}$ where σ^2 = variance and $\bar{x}$ = mean. It indicates that the sampled area – regardless of its shape or size – is homogeneous for the species observed and that neither the space nor its properties influences behavior. Deviation from randomness infers nothing about mechanisms of dispersion (Brown and Orians 1970) and is merely descriptive. For example, departure from a random pattern might be caused by heterogeneity of the habitat, rather than some inherent social tendency like attraction or avoidance between or among individuals.

Clumping is common in nature (Ludwig and Reynolds 1988: 21). A *clumped pattern* (also called *aggregated pattern*), is defined by $\sigma^2 > \bar{x}$ and indicates heterogeneity, requiring categorization of the sampled space; in other words, behavioral patterns of the species observed have been selected for certain properties of the habitat, either favorable (where individuals occur) or unfavorable (where they are absent). The property of a clumped pattern is purely descriptive, predicting a higher probability of individuals being present at some locations than others. Agreement with the negative binomial can then be tested, and the extent of clumping assessed using various indices of dispersion based on the ratio of variance to mean (coefficient of variation).

In a *uniform pattern* (also called *regular pattern*) $\sigma^2 < \bar{x}$ and indicates that every point in the sample is as far from each of the others as possible considering the density of points. Uniform dispersion in nature is rare. The important idea in all three distributions is that identifying a pattern of dispersion tells us nothing about its underlying cause.

Age and distance of dispersal might differ between free-ranging cats in low- and high-density groups. The latter occur most frequently in urban areas. In a high-density group of strays ($n = 70$) studied at Lyon, France, females dispersed at age 1–2 y, before or after first reproduction, but males were philopatric (Devillard et al. 2003). Disappearing females were in worse physical condition than those remaining, although their reproductive success was the same. The distance some cats disperse can be considerable, to 15 km at one location on New Zealand's South Island (Norbury et al. 1998b).

Free-ranging adult males sometimes disperse more often than females and move farther from the natal site (Liberg 1980). Females might spend their entire lives where they were born, especially house cats and strays. In a low-density rural population in Sweden, 87% of males abandoned the natal home range at 1–3 y, purportedly under pressure from resident males (Liberg 1980). However, 74% of the females remained. Female strays at the Royal Navy Dockyard, Plymouth, United Kingdom, were highly philopatric, one group even staying in and around a building while it was razed and rebuilt (Dards 1978). Males born into this group of strays dispersed at age 1–2 y, and future contact was unusual (Dards 1978, 1983). Dispersing males typically settle in areas where an adult male does not presently reside or within part of an established adult male's home range (Liberg 1980). When a resident feral male dies or otherwise disappears, he is soon replaced by a transient (Liberg 1980). If a feral male dies or emigrates, his home range is usually taken by another (Langham and Porter 1991).

Cats might move short distances or not at all if food is freely available. Some males from a group of strays in central Rome moved only a short distance away and were later seen in neighboring aggregations (Natoli 1985b). Male-biased dispersal was recorded for strays occupying the Lyon park, but permanent immigration from outside the park by either sex was rare (Devillard *et al.* 2004). Cats were associated with either of two groups stationed ~500 m apart, and "dispersing" males simply changed groups. Dispersing females, in contrast, left the park. In the Lyon population, <12% dispersed annually, and only 5 individuals immigrated during 8 y of monitoring (Devillard *et al.* 2003).

2.5 Inbreeding avoidance

Several hypotheses attempt to explain why animals emigrate from natal sites, a prominent one being to lessen the chances of breeding with close relatives. As defined by Keller and Waller (2002: 231), *inbreeding* describes "various related phenomena ... that all refer to situations in which matings occur among relatives and to an increase in homozygosity associated with such matings." In addition to consanguineous matings within populations, some authorities include random genetic drift among populations (Keller and Waller 2002). In either case the result can be *inbreeding depression*, or reduced fitness caused by breeding with relatives. Among its many manifestations are lower birth weight, reduced survival and reproduction, reduced resistance to disease and predation, and environmental stress (Keller and Waller 2002). Any of these can arise when harmful mutations accumulate, a particular problem in small populations where genetic selection is less effective (Lynch *et al.* 1995). It is important to note that inbreeding does not disturb the frequencies of alleles in a population; rather, it changes the frequencies of genotypes, increasing the proportion of homozygotes relative to heterozygotes (Keller and Waller 2002). Of harmful mutations arising continuously in all populations, most are at least partly recessive, and increased homozygosity permits greater expression of the *genetic load* (i.e. the fraction of the genotype in which mean fitness is influenced and altered by such processes as mutation and recombination). The consequence is then inbreeding depression.

Dispersal as a mechanism of inbreeding avoidance is unlikely to affect either the timing or pattern of emigration in free-ranging cats. Considerable inbreeding was demonstrated in a group of French urban strays, including mother–son, father–daughter, and brother–sister matings (unpublished data cited in Devillard *et al.* 2003). Visual evidence of harmful effects was not apparent, although mortalities traceable to genetic causes are rarely detectable if mitigated by increased susceptibility to environmental factors (Keller and Waller 2002). Inherent behavioral or ecological mechanisms to avoid inbreeding have yet to be identified in free-ranging cats, and mating among close relatives elsewhere is hardly a rare event (e.g. Devillard *et al.* 2011, Dreux 1970, Kerby and Macdonald 1988, Macdonald and Apps 1978, Pontier *et al.* 2005, van Aarde 1979), suggesting that selection has not favored dispersal over inbreeding (Moore and Ali 1984).

I can think of a reason why this might be so, but my comments are purely speculative. They have to do with sociality – or rather, its absence – as a factor in the lives of free-ranging cats. Were we to constrain all notion of dominance (Chapter 1) within the boundaries of sociality, where I believe it belongs, animal societies are either high-skew or low-skew (Reeve *et al.* 1998). Breeding in the first category is limited to one or a few dominant breeders; the second allows more egalitarianism, and a greater proportion of adults participate in reproduction. Social animals live together and interact frequently. Not all encounters necessitate displays and subsequent acknowledging of dominant or subordinate status. To keep the group together, dominant members sometimes make concessions. One of these is yielding sufficient reproductive opportunities to subordinates so they stay satisfied instead of emigrating or challenging for a higher position in the social order. Optimal-skew models are designed to reveal how much leeway this requires, its reinforcement modulated by aggression. Theory tells us that (1) skew increases with increasing relatedness between dominants and subordinates, (2) successful reproduction outside the social group decreases, (3) a subordinate's contribution to productivity of the group increases (requiring less compensation from the dominant members), and that (4) the relative fighting ability of a subordinate decreases (references in Reeve et al. 1998).

If cats are asocial then dominance has little meaning in their lives, as argued in Chapter 1. An animal that spends most of its time alone as an adult is neither dominant nor subordinate. As such, (1), (2), and (4) above become irrelevant in the case of free-ranging cats, and (3) is demonstrably untrue.

Part of what defines dominance is resource-holding power, or authority over resources like mates, food, shelter, water, and space. Cats might growl at each other around the food bowl or hiss if a shelter space is invaded, but they do not "guard" resources, including prospective mates (Chapter 4), nor are they territorial. When the rules of sociality fail to apply then social factors thought to alleviate or mitigate inbreeding are probably irrelevant. Where feral cats live completely apart from humankind, their population densities are often low, home ranges large, and heterozygosity is sustained less by inbreeding avoidance than the difficulty of finding mating partners. We see this situation collapse in suburban and urban locations where cat populations are large and densities high because clumped food artificially constricts home-range size (e.g. Natoli 1985b).

None of these social models – high-skew, low-skew, or optimal-skew – applies easily to free-ranging cats. Reeve et al. (1998: 276) proposed, in addition, an incomplete control model in which dominant and subordinate individuals compete directly and share breeding opportunities because the dominant is unable – or unwilling – to monopolize the group's reproduction. Efficacy depends on equitable mating being a sufficient incentive for a subordinate to tolerate the status quo. The incomplete control model seems a good fit with the asocial life-style of free-ranging cats because of its low responsiveness to ecological constraints and the hypothesis that "reproductive skew should decrease with, or be insensitive to, increasing genetic relatedness among group members. ... " Nor would any sort of hierarchy be required.

The effects of both inbreeding and inbreeding depression are most noticeable in small populations where outsized fractions of the genetic load are potentially fixed by drift. When individuals later disperse the result can be inbreeding among populations. During inbreeding both within and among populations the frequencies of individuals homozygous for identical alleles increase. Without the prospect of immigration into other populations, all insular free-ranging cats eventually become inbred. Nonetheless, populations of feral cats founded by just a few individuals or that have experienced drastic reduction in numbers do not necessarily lose genetic diversity beyond that required for population growth (Devillard et al. 2011, Pontier et al. 2005), in part because harmful mutations are selected against, effectively purging a population of some of its genetic load (Keller and Waller 2002).

Such purging is efficacious, especially against lethal components, and associated inversely with population size (Keller and Waller 2002). If purges fail to occur in a timely manner, harmful alleles are more likely to become fixed, adding to the drift load (i.e. reduced mean fitness resulting from local genetic drift). Descendants of no more than 4 domestic cats released onto 6600-km^2 Grand Terre, Kerguelen Islands, in the 1950s to control mice (Derenne 1976) have multiplied to ~7000 (Devillard et al. 2011). The number fluctuates but has probably approached carrying capacity (Pontier et al. 2005). Short-term diversity (i.e. allelic richness) followed population trends, although the effect was small (Devillard et al. 2011). Variability seemed best explained by the rate at which local populations changed and not the shifting in numbers.

Hypotheses supporting inbreeding avoidance as a functional mechanism driving dispersal are almost always based on the observation or assumption of differential emigration between sexes, or sex-biased dispersal (Moore and Ali 1984). Generally, this means that one sex stays home and the other leaves, provoking natural separation among close relatives and lessening the chances they will mate. Usually the females stay put and males move on (Clutton-Brock and Lukas 2012) with the result that reversal of this pattern (i.e. males are philopatric and females disperse) supports the inbreeding avoidance hypothesis, the idea being that inbreeding depression ceases to be a problem if one sex or the other moves far enough away from the natal area (Moore and Ali 1984). Three practical difficulties arise. First, patterns of single-sex emigration are less common than ordinarily thought. Second, dispersal distances often do not preclude mating. Third, emigration by either sex can be explained just as well using criteria other than inbreeding avoidance (e.g. intrasexual competition for food, mates, and other resources). In many species of birds and mammals, the distance of sex-biased dispersal does not affect the odds of inbreeding, and in feral and stray cats there appears to be no consistent pattern of males dispersing farther than females, the reverse, or even whether one sex consistently leaves or stays.

That inbreeding can cause loss of heterozygosity leading to compromised reproduction and survival is not in question. However, the extent of its potential importance where free-ranging cats are concerned is unknown. Limited observations imply – but do not confirm – that estrous cats tend to avoid mating with close kin, although not with distant relatives (Ishida et al. 2001), but whether this constitutes a pattern is doubtful. As mentioned, others have seen no such inhibitions in free-ranging cats, nor are they apparent among cats kept in confinement (Connelly and Todd 1972). Also as mentioned, many island populations were started by a few founders, in which case all descendants are related. For example, the estimated population on Marion Island at one time exceeded 2000 cats, all descended from 5 animals released in 1949 (van Aarde 1979), and in a one-day survey of the Courbet Peninsula of Grand Terre (Kerguelen Islands) made by helicopter in austral autumn, Dreux (1970) counted 30 cats, 20 of which were either black or black and white, which he attributed to a few founding individuals.

Dispersal, which is usually necessary for outbreeding when outbreeding is possible, actually appears to be maladaptive, entailing numerous risks associated with emigration, among them additional stress, exposure to predation and disease, reduced foraging efficiency, lower reproductive rates, loss of contact with familiar locations and conspecifics, and higher mortality (Clutton-Brock and Lukas 2012, Moore and Ali 1984, Packer 1986). In contrast, inbreeding is neither inherently dangerous in this broad somatic sense nor maladaptive genetically (Moore and Ali 1984). Thus philopatry and inbreeding perhaps fit evolutionary theory better than dispersal from natal areas to reproduce elsewhere.

Female mammals in particular benefit from dispersal when advantages clearly outweigh the liabilities of staying behind (Clutton-Brock and Lukas 2012), although none has yet been shown to influence how free-ranging cats behave. The first of these possibilities is to escape competition from other females over resources, especially breeding opportunities. This is an unlikely problem considering the promiscuous mating system of cats (Chapter 4) and the absence of clearly functioning dominance hierarchies even where cats aggregate around clumped food. The second is to avoid competing with kin, but evidence of kinship stimulating the *voluntary* dispersal of any species of mammal is slim. As to the third, females sometimes disperse to keep immigrant males from killing their offspring. Male infanticide is usually associated with persistent polygyny and intense male–male competition during mating, in combination with the restricted capability of females to evade males (Clutton-Brock 2009). Neither condition pertains to free-ranging cats even on crowded islands (Apps 1986b). Although documented (Macdonald *et al.* 1987: 24–25), male infanticide is an occasional aberration, rendering it a doubtful factor affecting dispersal. Fourth is the possible benefit of dispersing to gain access to unrelated males (i.e. for outbreeding). However, inconsistent sex-biased dispersal across populations makes this factor problematical. Furthermore, I found no evidence of cats preferentially choosing unrelated mating partners but substantial proof that inbreeding is common. If either inbreeding avoidance or outbreeding preference among cats has behavioral markers, these have yet to be isolated, identified, and described.

Clutton-Brock and Lukas (2012: 482) also discussed species differences in female philopatry, noting that linear female dominance hierarchies are often influential and simply "reflect the fact that females typically remain in the same group for most or all of their lives." A suggestion that habitual female dispersal occurs most often in species in which the breeding tenure of individual males (or all related males) commonly exceeds the age at which most females breed, causing them to emigrate and avoid breeding with close relatives, is also questionable in the case of free-ranging cats. Females of species that attain maturity while their fathers are still reproductively active are more likely to disperse at adolescence (Clutton-Brock 1989). This does not apply to domestic cats, although in some groups of strays the females emigrate while males remain in the natal area (Devillard *et al.* 2003). Once again, the promiscuous mating system of cats (Chapter 4) works against such a system, making it an unlikely variable. That longevity of adult males can easily exceed the age at which females are first receptive hardly seems relevant, considering the inconsistency of female philopatry and promiscuity of both sexes.

As discussed, inbreeding between close relatives can cause inbreeding depression via homozygosity and increased exposure to harmful recessive alleles. Clutton-Brock and Lukas (2012) mentioned heightened pre- or post-natal mortality, birth defects, and the reduced fecundity and breeding success of offspring all potentially lowering reproductive success and ultimately fitness within a lineage. Conversely, inbreeding can increase the benefits of kin selection, promoting altruism and sociality and having a positive effect. In a strange twist the expected altruism from being surrounded by relatives can itself appear as inbreeding depression, signs of which might be variable forms of competitive incompetence favored by kin selection (Moore and Ali 1984). Furthermore, as noted by Moore and Ali (1984: 96), "When a previously outbred population begins to inbreed … increased homozygosity will expose deleterious recessive alleles accumulated during outbreeding and inbreeding depression will be relatively severe." Over time these are excised from the lineage through fitness-related factors, and inbreeding depression attenuates or disappears. The alternative is population extinction. Inbreeding depression is not evident in populations never exposed to outbreeding because harmful alleles are unable to accumulate. Of course, sociality and cooperation are not evidence of kin selection, and neither factor is a demonstrable species trait of the domestic cat.

As stated, the mating system of free-ranging cats is promiscuous, not monogamous or polygynous (Chapter 4). This alone is enough to discount the probability of an inherent selective mechanism to prevent mating with close relatives. A polygynous male has a smaller investment in his offspring than the females with which he mates (Trivers 1972), and his loss from inbreeding depression also will be less (Smith 1979). As Moore and Ali (1984: 100) wrote, "It is therefore unlikely that aggressive male enforcement of inbreeding avoidance against the 'wishes' of a female would evolve." And naturally, if both sexes are promiscuous their mutual loss is mutually diminished. Were we to consider all aspects of dispersal in terms of evolutionary theory then females and not males ought to consistently vacate the natal area, assuming dispersal's function is the prevention of inbreeding. The behavior of chimpanzees (*Pan troglodytes*) matches this pattern (references in Moore and Ali 1984); that of free-ranging cats does not.

Experiments to assess inbreeding avoidance must be controlled for familiarity between or among test subjects; otherwise, we have no evidence of its validity. Cross-fostering with rodents has shown that unrelated individuals reared together later avoid mating with each other, but relatives reared apart do so readily. Results like these demonstrate that "inbreeding avoidance" could actually depend on familiarity, a behavioral factor, and not genetics (Pusey and Wolf 1996).

2.6 Home-range boundaries

Whereas some free-ranging cats share home ranges and foraging areas, others do not. Of home ranges that overlap, especially of males, some barely do (Smucker *et al.* 2000). Home ranges can therefore be overlapping to varying extents or exclusive. However, exclusive is not to be taken as territorial, as in defense of space. Because few general statements are possible about how cats use their occupied spaces, this short section consists mainly of examples. What they reveal most saliently is the absence of any consistent pattern, with a notable exception: home ranges are smaller and tend to overlap more where food is clustered than when resources are scattered (e.g. Konecny 1987a, 1987b; Langham and Porter 1991; Smucker *et al.* 2000).

A *neighborhood* comprises overlapping home ranges and the paths connecting them. A neighborhood's size is therefore synonymous with its "habitat connectivity" *sensu stricto* (Le Galliard *et al.* 2005: 207). Cats on Marion Island left well-worn paths through their home ranges (Anderson and Condy 1974), and such networks of narrow trails are often clearly visible in areas elsewhere used regularly by free-ranging cats.

In rough terms the extent to which home-range boundaries overlap varies inversely according to population density, with sex of the residents exerting a looser effect. In the vicinity of Canberra, southeastern Australia, home ranges of a group of farm cats overlapped extensively, especially around feeding and sleeping sites (farm buildings), and so did those of the suburban cats sharing the same household, whether related or not (Barratt 1997a). Home ranges of suburban cats from separate households did not overlap, and males practiced active avoidance. Turner and Mertens (1986) reported similar findings in Swiss farm cats.

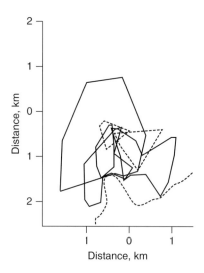

Fig. 2.2 Home ranges of radio-collared feral cats ($n = 14$) at Tagus Cove, Isabela Island, Galápagos Islands, Ecuador. Solid lines, males; dashed lines, females; lowest dashed line, coastline. Distances above and below zero on ordinate indicate north and south, those on the abscissa are not described, but apparently west to the left and east to the right. Zero on the abscissa is presumably the "core" area. Source: Konecny 1987a. Reproduced with permission of John Wiley & Sons.

The home ranges of feral males might overlap (Liberg 1980), and male–male home ranges not only overlap frequently but encompass those of females (Harper 2004). A feral male in northern Italy had a large home range (639 ha) that included the home ranges of three females and overlapped slightly those of two other males (Genovesi et al. 1995). These cats in general had large home ranges with broad male–female overlap. Feral males and females at one location in the Galápagos Islands had home ranges that clearly overlapped (Fig. 2.2). Estimated overlap of home ranges (males and females combined) at Stewart Island/Rakiura, New Zealand, was 36% with little variation between sexes (Harper 2004): male–male home ranges overlapped with those of other males by ~36%, female–female by ~31%. At Catalina Island, California, United States, male–male and male–female overlap was more extensive than female–female overlap, and this included both intact and neutered individuals (Guttilla and Stapp 2010). The opposite was true in the montane wet forest of Mauna Kea, Hawai'i, United States, where sharing of home ranges was more common among females (Smucker et al. 2000).

Molsher et al. (2005) reported considerable overlap of space use between young and old adults. Feral cats monitored at an agricultural area of New Zealand (North Island) hunted mainly at night, when they ranged over a substantially larger area than during the day (Langham and Porter 1991). Home ranges of adult males overlapped little but included the home ranges of females. Home ranges of cats using the barns were larger than those that stayed exclusively outdoors. On small (224-ha) Dassen Island, South Africa, cats of all ages shared home ranges, any of which could be occupied by up to 12 individuals (Apps 1986b).

In group-living, free-ranging cats, the spaces of adult males often overlap those of other group members (Warner 1985). Farm cats on North Uist, Scotland, were said to live communally, occupy overlapping home ranges, and share hunting areas with neighboring cats (Corbett 1978). Home ranges of feral females in agricultural areas often overlap (Genovesi et al. 1995, Langham and Porter 1991). Sometimes the overlap is most pronounced among related females (Langham and Porter 1991), although whether this is a consequence of genetics, familiarity, or happenstance is undetermined. Male–male home ranges overlapped considerably among urban strays in Jerusalem, Israel, and home ranges of females that fed from the same garbage bins overlapped 80% (Mirmovitch 1995). Conversely, 20% overlap was observed in females that fed only part of the time from the same bins using common resources at different times. Nonetheless, overlap among female groups was ~36%. In contrast with the females, males did not form aggregations, but winter overlap of home ranges among 16 individuals was 92.9%. Strays of both sexes that occupied the University of KwaZulu-Natal's Howard College campus (also an urban conservancy) had overlapping home ranges (Tennent and Downs 2008).

As mentioned, the feral cats monitored on Dassen Island had overlapping home ranges (Apps 1986a, 1986b). There was indication of active avoidance, but none of territoriality or agonism. Overlapping home ranges were common in strays at Avonmouth Docks, but actual contact among individuals was made in only 27.5% of locations where overlaps occurred (Page *et al.* 1992).

Home-range overlap in female strays from the same group is common (Izawa *et al.* 1982), as it sometimes is among males (Izawa *et al.* 1982) and among males and females of the same group (Izawa *et al.* 1982, Laundré 1977, Natoli 1985b), but ranges of females usually overlap little, if at all, between groups (Izawa *et al.* 1982, Liberg 1980, Macdonald and Apps 1978, Say and Pontier 2004, Warner 1985, Yamane *et al.* 1994). This situation can also apply to male–male and male–female overlap of individuals from different groups (Izawa *et al.* 1982). Male–male and male–female home ranges of feral cats in New Zealand's Orongorongo Valley overlapped, and the overlap among females was considerable, including those that were pregnant or lactating (Fitzgerald and Karl 1986). Kauhala *et al.* (2006) reported that the nocturnal home ranges of feral, stray, and house cats in southeastern Finland not only overlapped, they overlapped with those of European badgers (*Meles meles*) and raccoon dogs (*Nyctereutes procyonoides*).

Where space is not limiting and food is widely distributed, the home ranges of feral cats often show little overlap (Jones and Coman 1982b), an observation attributed to "active spacing out" or to territoriality (Kerby and Macdonald 1988: 69). The first, which implies home ranges and active avoidance, is more likely. Overlap in home ranges was atypical among feral cats in central California, although it sometimes occurred (Hall *et al.* 2000).

2.7 Determinants of home-range size

Home ranges of free-ranging cats vary considerably in size (Table 2.1), those of adult males ordinarily being largest (Kerby and Macdonald 1988). Without perspective such information is simply a compilation of numbers. A place to start is by assessing the factors likely to influence home range size. I can think of several, including (1) sex, (2) body mass, (3) trophic group to which a species belongs (herbivore, omnivore, carnivore), (4) population density, (5) abundance and distribution of food, (6) latitude, and (7) precipitation.

Mammalian males generally have larger home ranges than females. In only one of 27 studies did the mean home-range areas of mammalian females exceed those of males (Harestad and Bunnell 1979). Domestic cats conform with this pattern, and males – whether feral, strays, or house cats – typically range over larger spaces than females (e.g. Churcher and Lawton 1987, Dards 1978, Fitzgerald and Karl 1986, Guttilla and Stapp 2010, Harper 2004, Izawa *et al.* 1982, Jones and Coman 1982b, Konecny

Table 2.1 Representative mean home-range sizes (HR, ha) and densities (cats/km²) of free-ranging cats from the literature. Various methods were used in the computations, not all of them comparable (I show data for minimum convex polygons, or MCPs, when multiple methods of measurements were used). Variances were omitted because some were missing from the sources; others were inconsistent or unstated (i.e. whether ±SEM or ±SD). Also see Liberg et al. (2000: 122–123, Table 7.1), Nogales et al. (2004: 318–319, Appendix 1).

Location	Habitat	Category	HR (♂), ha	HR (♀), ha	HR (♂ + ♀), ha	Cats/km²	Source
Ainoshima Island, Japan	Urban	Stray	0.8–1.45	0.4–0.6	–	2350	Izawa et al. (1982), Yamane et al. (1994)
Avonmouth Docks, Bristol, UK	Urban	Stray	15	10.3	–	9.5–15	Page et al. (1992)
Baltimore, Maryland, USA	Urban	Stray	–	–	1.51–7.43	–	Oppenheimer (1980)
Bristol, Gloucestershire, UK	Urban	House	–	–	–	229	Baker et al. (2005)
Brooklyn, New York, USA	Urban	Stray	1.97	1.07	–	488	Haspel and Calhoon (1989)
Caldwell, Texas, USA	Suburban	House	–	–	–	1.5	Schmidt et al. (2007)
Canberra, Victoria, Australia	Suburban	Stray	–	–	0.86–23.38	–	Barratt (1997a)
Canberra, Victoria, Australia	Suburban	House	–	–	0.11–43.56	–	Barratt (1997a)
Central California, USA	Rural	Feral	–	–	31.7	–	Hall et al. (2000)
Champaign-Urbana, Illinois, USA	Rural	House	1.8	1.92	–	–	Horn et al. (2011)
Champaign-Urbana, Illinois, USA	Rural	Feral	157.01	56.59	–	–	Horn et al. (2011)
Cousin Island, Seychelles Islands	Rural	Feral	–	–	–	220	Veitch (1985)
Dassen Island, South Africa	Rural	Feral	–	–	11–63	16.5–22.8	Apps (1983, 1986b)
Devonshire, UK	Rural	Stray	56	40	–	6	Macdonald and Apps (1978)
Galápagos Islands (Baltra), Ecuador	Rural	Feral	–	–	0.50	–	Phillips et al. (2005)
Galápagos Islands (Isabela), Ecuador	Rural	Feral	149	35	–	–	Konecny (1987a)
Galápagos Islands (Santa Cruz), Ecuador	Rural	Feral	497	338	–	–	Konecny (1987a)
Gödöllő Hills, Hungary	Rural	Feral	149	41–328	–	–	Biró et al. (2004)
Grand Terre, Kerguelen Islands	Rural	Feral	–	–	–	3.7–6.7	Derenne (1976), Pascal (1980)
Grand Terre, Kerguelen Islands	Rural	Feral	–	–	–	1.5	Pontier et al. (2005), Say et al. (2002a)

Location	Setting	Type				Reference
Herekopare Island, New Zealand	Rural	Feral	–	–	–	118	Fitzgerald and Veitch (1985), Veitch (1985)
Isla Isabela, México	Rural	Feral	–	–	–	113	Rodríguez et al. (2006)
Israel (northern/southern)	Rural/urban	House	72.2	3.25	–	–	Brickner-Braun et al. (2007)
Jerusalem, Israel	Urban	Stray	0.75	0.3	–	–	Mirmovitch (1995)
Kerguelen Islands	Rural	Feral	–	–	–	0.5–6.65	Brothers et al. (1985)
KwaZulu-Natal, South Africa	Urban	Stray	–	–	3.7–10.8	23.4–40	Tennent and Downs (2008)
L'ile aux Cochons, Crozet Islands	Rural	Feral	–	–	–	2.5–9	Derenne and Mougin (1976)
Lyon, France	Urban	Stray	1.51	0.18	–	–	Say and Pontier (2004)
Macquarie Island	Rural	Feral	–	–	–	3.65–7	Brothers et al. (1985), Jones (1977)
Macquarie Island	Rural	Feral	–	–	–	20.4	Copson and Whinam (2001)
Marion Island	Rural	Feral	–	–	–	10.61–13.85	van Aarde (1979, 1980)
Mauna Kea, Hawai'i, USA	Rural	Feral	2.78	1.685	–	–	Smucker et al. (2000)
New Zealand, North Island	Rural	Stray	240	48.239	–	–	Langham and Porter (1991)
New Zealand, North Island	Rural	Feral	140	80	–	–	Fitzgerald and Karl (1986)
New Zealand, South Island	Rural	Feral	–	–	225	–	Norbury et al. (1998b)
New Zealand, South Island	Rural	Feral	–	–	178–2486	–	Recio et al. (2010)
Northern Territory, Australia	Rural	Feral	2210	–	–	0.1	Edwards et al. (2001)
Nouvelle Amsterdam	Rural	Feral	–	–	–	11	Brothers et al. (1985)
Portsmouth, Hampshire, UK	Urban	Stray	8.4	0.8	–	–	2	Dards (1983)
Revinge, Sweden	Rural	House	–	30–40	–	2.5–3.3	Liberg (1980)
Revinge, Sweden	Rural	Feral	990	206	–	–	Liberg (1984a)
Rome, Italy	Urban	Stray	–	–	–	1300	Natoli (1985b)
Sacramento Valley, California, USA	Rural	Feral	–	–	–	12.5	Hubbs (1951)
Santa Catalina Island, California, USA	Urban/Rural	Stray/feral	–	–	150	–	Guttilla and Stapp (2010)
Stewart Island/Rakiura, New Zealand	Rural	Feral	2083	1109	–	–	Harper (2004)
United Kingdom	Urban/rural/suburban	Stray/feral/house	–	–	–	500	Beckerman et al. (2007)
Victorian Mallee, Victoria, Australia	Rural	Feral	620	170	–	–	Jones and Coman (1982b)
Warsaw suburbs, Poland	Suburban/rural	Feral/house	–	–	–	1.5	Romanowski (1988)
Wisconsin, USA	Urban/rural/suburban	Stray/feral/house	–	–	–	10–14	Coleman and Temple (1993)

1987a, Liberg 1980, Macdonald and Apps 1978, Mirmovitch 1995, Molsher et al. 2005, Page et al. 1992, Schmidt et al. 2007, Warner 1985, Yamane et al. 1994). Home ranges of adult males also exceed those of juveniles (Hall et al. 2000) and of subadult and males classified as "subordinate" (Fitzgerald and Karl 1986, Langham 1992).

Exceptions are there. For example, one feral female in the Gödöllő hills region, Hungary, had a much larger home range than a male (Biró et al. 2004), and male and female feral cats in an agricultural area of New Zealand's North Island had diel home ranges statistically similar in size (Langham and Porter 1991). Sex did not influence the nocturnal home-range sizes of feral cats in southeastern Finland (Kauhala et al. 2006). Urban strays in Brooklyn, New York, had home-range sizes that did not vary statistically by sex (Haspel and Calhoon 1989), findings similar to those for suburban cats of Caldwell, Texas, USA (Schmidt et al. 2007); feral cats at three locations on New Zealand's South Island (Norbury et al. 1998b); the semi-arid Victorian Mallee of southeastern Australia (Jones and Coman 1982b); central-western New South Wales (Molsher et al. 2005); suburban and farm areas near Canberra, southeastern Australia (Barratt 1997a); feral and house cats in an Illinois farming region (Horn et al. 2011); and feral cats occupying a riparian preserve in central California (Hall et al. 2000).

As to prior attributes (Chapter 1), neither sterilization nor body mass affected movement, home-range size, or extent of overlap in home ranges of cats that shifted between the stray and feral states on Santa Catalina Island, California (Guttilla and Stapp 2010). Size can be a strong determinant. Molsher et al. (2005) found a positive association between home-range size and body mass, as did Page et al. (1992). However, home-range size does not generally correlate directly with age or group size in either house cats or strays (Liberg 1984a). Dards (1978: 246) wrote: "Group range sizes are probably dependant [sic] on the resources available, and seem to be independent of the number of cats in the group." Feral cats had bigger home ranges than house cats in an Illinois farming region, and home-range sizes within each group were statistically similar between sexes (Horn et al. 2011). In another study, feral females with home ranges that included buildings (e.g. barns) were smaller than the home ranges of cats living completely outdoors (Langham 1992).

Liberg and Sandell (1988) reported a negative association between home-range size and number of cats. If food abundance is indeed a relevant factor influencing population densities of free-ranging cats this has yet to be shown in an ecologically useful context. The real issue is not the relationship of areal space to number of cats, but between abundance of food and abundance of cats. Instead of mentioning food we could just as easily use the categories of feral, stray, or house because each is defined ultimately by food availability. One exception bears mention: food at Avonmouth Docks was not limiting and yet the number of cats remained low (Page et al. 1992). Home ranges of feral cats can swell or shrink seasonally depending on availability of prey, or even change locations. At Little Barrier Island, New Zealand, feral cats shifted to lower altitudes in autumn and winter to feed on Polynesian rats (*Rattus exulans*), forest birds, and insects, moving to higher altitudes in summer when burrowing petrels came ashore to reproduce (Girardet et al. 2001).

Home-range size in a group of urban strays correlated directly with male reproductive success, measured as number of kittens fathered (Say and Pontier 2004), information revealing nothing about cause. Which factor represented the independent variable was left an open question. Correlation merely describes an association between or among variables. In this case home-range size and reproductive success correlated directly. In the absence of hypothesis testing, the likelihood of reproductive success determining home-range size or of home-range size determining reproductive success are equally likely.

Cats of all species are carnivorous (Chapter 7). For reasons about to be discussed, their home ranges tend to be larger than those of herbivores and omnivores. One hypothesis states that home-range size correlates roughly with food distribution, being largest where food is scattered and diminishing along a continuum of increasing patchiness toward clumping (e.g. Kerby and Macdonald 1988). Logically, abundant food should reduce the home-range sizes of free-ranging cats (Horn *et al.* 2011, Tennent and Downs 2008). All factors considered, only diet exerts a significant effect. In the case of carnivores, home range increases with metabolic requirements regardless of taxonomic affinity, and meat-eaters need larger spaces in which to forage than herbivores and omnivores (Gittleman and Harvey 1982).

Basal metabolic rate (BMR) is the energy generated by a fasting organism at rest. Physiologists describe it in terms of a power law and scaling relationship: BMR increases with body mass raised to some exponent, shown by a power equation:

$$BMR = aM^k \tag{2.1}$$

where a is a scaling constant (i.e. the allometric coefficient, or intercept), M represents body mass, and k the scaling exponent. Depending on the value of k, metabolic rate regressed against body mass is either allometric and curved ($k \neq 1$) or isometric and linear ($k = 1$). In isometric relationships, metabolic rate and body mass scale in direct proportion; that is, in a straight line. Kleiber's law sets k at 0.75 (e.g. West and Brown 2005, White 2010). However, this approach has two serious deficiencies: (1) it uses regression analysis, which adjusts the intercept (a) and slope (k) to minimize deviation of the data about the mean; and (2) it wrongly presumes that the intercept stays constant regardless of species or size, remaining invariable whether the animal in question is a mouse or a cow (Heusner 1982a, 1982b). Others (e.g. Heusner 1982a, 1982b; White and Seymour 2005) argued that k should be 0.67, which takes into consideration the ratio of surface area to body mass and thus the amount of body surface available to exchange heat with the environment. Some have tried to measure a cat's surface area, the effort by Vaughan and Adams (1967) of dividing the body into 12 areas being one of the most thorough. They sheared the hair from 42 freshly killed cats of 0.42–5.85 kg, covered the space with cloth adhesive tape, and took measurements with a compensating planimeter. The relationship was then expressed by a linear regression equation:

$$BSA = 388.4M + 896.5 \tag{2.2}$$

in which BSA = body surface area (cm^2) and M = body mass (kg). The correlation is strong ($R = 0.95$), although a curvilinear fit of the data seems better (Hill and Scott 2004: 690, Fig. 1).

Still other investigators refuted both constants and argued that a single value of k is impossible (Burness 2010, White 2010), which makes sense especially if the availability of local food resources is accounted for (Marquet *et al.* 2005). Confusion over a "correct" value of k arises because the relationship between body mass and metabolic rate is actually nonlinear, a convex curve on a logarithmic scale and therefore not a pure power law (Kolokotrones *et al.* 2010).

Herbivorous animals are primary consumers, subsisting on plants. Carnivorous animals subsist by eating other animals, making them secondary consumers. In terms of scaling, the potential energy available to carnivores is less than for herbivores; in other words, more energy is potentially present per unit area of home range for primary production, and a herbivore can fulfill its energy needs in a smaller space compared with a carnivore of similar size. Omnivores, which obtain energy from both plant and animal tissues, fall somewhere between. Imagine a feral cat in a habitat containing just one species of prey, a herbivore about its own size. A rabbit, for example. Rabbits can feed on any number of plants; the cat can only eat the rabbits. Animal flesh contains more energy than plant tissue on a mass basis. Nonetheless, total potential energy available to the rabbit exceeds that to which the cat has access. On this alone it follows that fewer feral cats than rabbits can occupy the same landscape.

Harestad and Bunnell (1979) analyzed the relationship between home-range area and body mass of the three trophic groups. My comments are based on their assessment with an emphasis on carnivores. As noted earlier, recent publications bring specific aspects of metabolic scaling into finer resolution, although not from the standpoint of home-range size.

Herbivores, omnivores, and carnivores have different energy requirements, and each group varies from a universal value of $k = 0.75$, the exponent expected if size of the home range depends directly on BMR. From a literature search on large mammals, Harestad and Bunnell (1979) derived the following associations using Burt's (1943) definition of home range (Section 2.2), where H represents spatial area in hectares: $H = 0.002M^{1.02}$ (herbivores), $H = 0.59M^{0.92}$ (omnivores), and $H = 0.11M^{1.36}$ (carnivores). Carnivores thus require larger home ranges than herbivores and omnivores of similar body size. Home-range size increases in near-linear fashion with body mass in herbivores and omnivores ($k = 1.02$ and 0.92), but the slope of the regression for carnivores rises abruptly with body mass ($k = 1.36$) and differs significantly from the regression slopes of the other two groups. If the prey's home-range size increases linearly then that of the predator must exceed linearity, unless the prey becomes more productive or its home range overlaps with another prey species to compensate. Therefore, a unit rise in body mass results in a greater increase in home range for carnivores compared with herbivores and omnivores.

If an animal uses the minimum space to sustain its daily energy needs, $R(kJ/d)$, and if the space provides usable energy at a rate $P(kJ/d/\text{unit area})$, home range can be expressed by

$$H = R/P \qquad (2.3)$$

It then follows that R is proportional to BMR, or

$$R = aM^k \qquad (2.4)$$

The expected value of P should decrease with an animal's increasing energy requirement, R, which increases in turn with body mass; that is, big animals obviously need more energy than little ones. This situation relates to how food resources are distributed (i.e. their degree of "patchiness") and places carnivores at a disadvantage by forcing them to move around in search of food instead of staying stationary and grazing or browsing. Also assume that k remains constant by trophic group and only a, the proportionality factor, changes.

To bring theoretical aspects of Equation 2.3 into line with empirical findings from the literature necessitates letting P become a decreasing power function of M (Harestad and Bunnell 1979). In herbivores the relationship can be expressed as

$$H = a\left(M^{0.25} M^{0.75}\right) \tag{2.5}$$

Combining Equations 2.3 and 2.5 yields

$$H = M^{0.75}/a'/M^{0.25} \tag{2.6}$$

where a' is kJ/d/unit area of home range instead of ha/M (Equation 2.5). The new equation reveals a habitat's diminishing capacity to meet the energy needs of larger animals. For herbivores and omnivores the value of P in Equation 2.3 is proportional to $M^{0.25}$; for carnivores it is proportional to $\sim M^{0.5}$.

The domestic cat, a medium-sized carnivore, demonstrates modest sexual dichotomy of body mass at adulthood. Females, being the smaller sex, have greater mass-specific metabolic demands than males but lower daily requirements on mass-based total food intake. For discussion purposes, consider females to be two-thirds the size of males (i.e. ~ 3.0 kg vs. ~ 4.5 kg). Even at this slightly excessive disparity female domestic cats are proportionately bigger than most large carnivorans, in which females are only ~ 0.33–0.50 the body mass of conspecific males (Harestad and Bunnell 1979). In these other species the home ranges of females are expected to be ~ 0.24–0.39 that of males based on $M^{1.36}$. The actual proportion, however, is ~ 0.52, indicating that differences in sex-related sizes of measured home ranges are less than predicted from differences in female–male body mass.

I borrowed a few data ($n = 16$) for free-ranging cats from the literature (paired female vs. male home ranges) and compared the results with those of Harestad and Bunnell (1979). The value obtained was nearly identical ($\bar{x} = 0.53 \pm 0.345$ SD, range 0.003–1.317), although with a larger standard deviation and range but nonetheless conforming with predicted relative size for large carnivores (Harestad and Bunnell 1979: 393, Table 2). I emphasize that domestic cats are not directly analogous in this respect to the cougar (*Felis concolor*) and other big carnivorans. The relationship between metabolic rate and body mass is obviously nonlinear because values between small and large animals vary sharply (Kolokotrones et al. 2010). Without knowing where the domestic cat fits on a sliding scale of size, the close association with their value could easily be coincidental, but if real it demonstrates that decreasing variation in sexual dichotomy tightens the relationship between body mass and proportional differences in home-range size. Either way, the inherently appealing notion that simple proportionate scaling of home-range size for domestic cats or any carnivoran based on relative body mass yields an unreliable measure.

Pockets of data from the literature tentatively confirm that "weight alone may account for a large portion of the differences between male and female ... home ranges" (Harestad and Bunnell 1979: 399). Haspel and Calhoon (1989) reported home-range sizes of urban strays in Brooklyn, New York, to be directly proportional with body mass, those of males ~ 1.5 times those of females. Liberg and Sandell (1988: 85, 88) arrived at a different conclusion, deducing that home ranges of males, which from their literature search averaged 3.5 times those of females, "corresponds to a body weight 5.3 times that of females." They concluded, "we interpret this as a clear indication that food is not determining range size for males, at least not directly." This is correct in the sense that neither does *BMR*, but wrong if the implied expectation was a linear relationship.

After pointing out how substituting *H* for *BMR* in Equation 2.1 produces a significantly different value of *k*, Harestad and Bunnell (1979: 398) stated: "Indeed, there is no compelling reason to believe that metabolic rate should govern size of home range ... independent of the distribution of the food resource." Both factors decline with increasing body mass, being steepest for carnivores. Increases in metabolism with activity (mainly the search for food) would appear in the proportionality factor, *a*, and not the exponent, *k*. Exponents >0.75 indicate that with increasing body mass a free-ranging mammal expands the size of its home range beyond what might be expected from Equation 2.3.

Further discussion of this subject is warranted. Home-range size in mammals increases with body mass, as shown by Equation 2.3, the value of *k* varying significantly from 0.75 by trophic group with the greatest differences between herbivores and carnivores. In their study, Harestad and Bunnell (1979) reported 95% confidence limits for carnivores extending from 1.04–1.68. Also according to Harestad and Bunnell (1979: 397), Equation 2.3 identifies animals occupying low-productivity habitats as having larger home ranges than "predicted by the generalized relationship between home range and body weight and vice versa." We can probably accept that home-range size decreases and boundary overlap increases as food becomes increasingly clumped and abundant (e.g. Apps 1986b, Kerby and Macdonald 1988, Natoli 1985b). At a population level, the density of cats, which is influenced directly by food abundance and dispersion, can also correlate negatively with size of the home range (Edwards *et al.* 2001). This being the case, it is not surprising that the size of a cat's home range also varies inversely with the extent of its dependence on humans (Horn *et al.* 2011, Schmidt *et al.* 2007), which itself merely camouflages the fact that by feeding house cats and strays, humans supply part or all of their daily nutritional requirements. The home ranges of house cats are usually smallest, followed by those of strays given supplementary feedings or living where ample garbage is available. Feral cats typically have the largest home ranges because they need to hunt.

A general pattern is for home-range size to increase with latitude regardless of trophic group (Fig. 2.3), the main driving force being seasonal changes in primary productivity. Home-range sizes of feral cats in an Illinois farming region changed seasonally, but not those of nearby house cats (Horn *et al.* 2011). In contrast, Kauhala *et al.* (2006) failed to detect any seasonal variation in home-range size among feral, stray, and house cats in southeastern Finland.

Because primary production depends on precipitation, the expected trend is for home-range size (again, for any trophic group) and rainfall to correlate inversely (Harestad and Bunnell 1979), and animals living in arid regions tend toward larger home ranges than those in wetter locations. These statements are broadly consistent with Equation 2.3. In terms of factors defined previously, seasonal changes in home-range size are therefore influenced by the value of *P*, considering that *R* remains relatively constant. As *P* decreases with declining temperature, the size of the home range should increase, with the effect on smaller carnivores being greater than on larger ones.

Fig. 2.3 Index of area used per g of body mass (M) normalized for trophic status and total mass as a function of latitude. Derived from reconfiguring $H = aM^k$ to give $I = (H/a)^{1/k}/M$ where H = home range (ha), a = a scaling constant (i.e. the intercept), M represents body mass, and k the scaling exponent (see text). On the graph A = expected size of the home range, B = observed limit of home-range expansion. Source: Harestad and Bunnell 1979. Reproduced with permission of the Ecological Society of America.

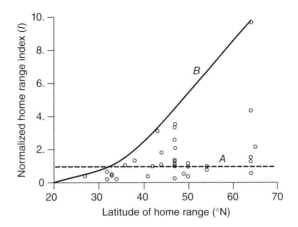

Other factors might be relevant too, although the appropriate tests of assessment have not been made. Home ranges of female cats supplying food to growing kittens can increase even if supplementary food is provided (Turner and Mertens 1986). Home ranges are likely to be smaller for castrated males and all cats disinclined to hunt (Turner and Mertens 1986), but this is not always true (Barratt 1997a). According to Langham and Porter (1991), feral cats sheltering in barns in a New Zealand agricultural region had larger home ranges than cats sheltering outdoors. Reasons for the difference were not determined.

That home-range size and population density correlate inversely (Edwards *et al.* 2001, Turner and Mertens 1986) is an artifact of human interference with regard to strays and house cats. For example, the relationship failed to hold among farm cats given supplementary food three times daily (Turner and Mertens 1986: 43), indicating that home-range sizes of females might correlate more closely to "household density than to cat density *per se*, and that this has little to do with supplemental feeding."

A surfeit of food did not alter the home-range sizes of urban strays in Brooklyn, New York (Haspel and Calhoon 1989), probably because the main food sources (uncovered garbage bins) were scattered through the area instead of clumped, and also because available food more than met daily dietary requirements. This suggests that although food abundance influences the size of the home range, its effect is ultimately limiting (Haspel and Calhoon 1989), making it the same as any resource.

In one respect an urban stray or house cat mimics the humans who made its life-style possible. No specific factor forces it to stay, and yet it does. It stays because of the resources, trading hunting's uncertainty for easily acquired food. By attaching itself to a garbage bin or food bowl, all advantages of having a large home range are relinquished, assuming there actually would be any under these circumstances. Anyway, mobility is costly, exerting a negative effect on birth rate (Le Galliard *et al.* 2005).

2.8 Habitat selection

Selection of habitat is influenced by available shelter spaces for resting, especially in unfavorable climates (Brothers *et al.* 1985, Derenne 1976, Harper 2004), and by cover through which to travel (Dards 1981, Medina and Nogales 2007). The location and availability of shelter and cover affect how habitats are used (Alterio *et al.* 1998,

Brickner-Braun et al. 2007, Calhoon and Haspel 1989, Dards 1981). Stewart Island/Rakiura is New Zealand's most southern island (47°S, 167°50′E). Its climate is cold and wet. Of the areas used by 10 monitored feral cats, 62.9% were podocarp-broadleaf forests, 21.1% subalpine shrublands, 10.5% *Leptospermum* shrublands, and 5.4% riparian shrublands (Harper 2004). Preference for forested areas was attributed to the availability of abundant sites offering shelter from the rain and low temperature. Subalpine shrublands offered less shelter and were used most often in dry weather. When feral cats on subantarctic Campbell Island died out, the cause was attributed to inadequate shelter (Harper 2004). Feral cats on subantarctic Marion Island and Grand Terre, Kerguelen Islands, took shelter in burrows of their prey – seabirds and rabbits (Brothers et al. 1985, Derenne 1976, Devillard et al. 2011, Pontier et al. 2002, van Aarde 1979). Cats at Jarvis Island in the mid-Pacific Ocean used burrows of wedge-tailed shearwaters (*Puffinus pacificus*) for shelter (Kirkpatrick and Rauzon 1986). Those at Macquarie Island and Marion Island sheltered in rabbit burrows, sometimes returning to the same ones inside their home ranges (Brothers et al. 1985, Derenne 1976). In contrast, feral cats on New Zealand's South Island used an average of 11.5 different dens (range 6–17) over 287 d of monitoring with no differences in occupancy between sexes (Norbury et al. 1998b). Similarly, feral cats monitored in central-western New South Wales did not use permanent dens (Molsher et al. 2005), and feral cats at Mauna Kea, Hawai'i, did not make use of permanent resting sites (Smucker et al. 2000).

Urban strays use abandoned buildings (Calhoon and Haspel 1989). Cats elsewhere take refuge in hollow trees or logs, rabbit warrens, under bushes, and in dense thickets (Jones and Coman 1982b, Molsher et al. 2005). Free-ranging cats in suburban and rural areas occupy hollow logs, drainage ditches, offal pits, and buildings (Fitzgerald and Karl 1986, Langham 1990). Feral cats in central California preferred riparian areas and buildings and avoided cultivated fields (Hall et al. 2000). Feral cats in an agricultural area of northern Italy hunted in meadows and also avoided areas of cultivation (Genovesi et al. 1995), as did feral cats in Illinois (Warner 1985) and farm cats in central Poland (Krauze-Gryz et al. 2012). The Italian cats preferred sites with cover: tree lines, thickets of reeds, and patches of vegetation along drainage channels, and the Polish cats preferred woods to farmlands. Cover is also important during times of activity. When English female farm cats traveled away from their core locations, they moved preferentially through long grass and thickets and along the walls of buildings, avoiding open spaces (Panaman 1981). Doing so helped conceal them from predators and prey alike. Israeli house cats and strays living in desert settlements had small home ranges and traversed them using the cover of trees and bushes (Brickner-Braun et al. 2007).

Sandell (1989: 175) suggested that "maintenance of exclusive ranges should be the best tactic when females are dense and evenly distributed." Territoriality in these circumstances conforms with theory (Port et al. 2011, Spotte 2012: 90–106), except that cats are not territorial. I fail to see how home ranges can remain exclusive unless defended. For example, the individual aggregations of stray females at Ainoshima Island, Japan, observed by Izawa et al. (1982), occupied home ranges centered around piles of garbage, and stray females observed by Dards (1983) shared a common core area with home ranges that overlapped almost completely. The cats were fed by dockyard workers, and Dards did not assess how the food was distributed. It might have been less clumped than at locations monitored at Ainoshima Island, but the point is perhaps irrelevant considering the crowded conditions at both locations.

As mentioned, home-range dimensions are largely reflections of habitat resources, probably explaining why house cats, which have no need to forage, range over smaller areas than feral cats (Horn *et al.* 2011). The effort expended tabulating space use by free-ranging cats (tables in Liberg and Sandell 1988) has produced considerable descriptive information, all of it devoid of explanatory power. Such will be the case until home-range data are applied as dependent variables for testing hypotheses relevant to limiting ecological factors (e.g. food distribution and abundance, reproductive capacity, survival to maturity, longevity).

The descriptive information so far points to expected patterns where food is concerned: female home-range size and female density correlate negatively, provided ample food is available and food is clumped (Liberg and Sandell 1988). Except on islands where ground-nesting birds are abundant (Chapter 9), feral cats must rely on scattered resources, and their home ranges are necessarily larger (Liberg and Sandell 1988).

In theory, home ranges are expected to be discrete where food is constant and distributed evenly, but overlapping when availability and dispersion vary (Sandell 1989). On the basis of food alone, discrete home ranges are predictably smaller than overlapping ones (Sandell 1989). In free-ranging cats it seems that home-range size is influenced by the abundance and distribution of food and the presence of estrous females (Sandell 1989). Food is necessary all year, mating partners only during the breeding season. If so, and home-range sizes based on food alone are already no larger than necessary, then a male might need to expand his range when starting the search for estrous females. Alternatively, he can maintain a permanent range that overlaps with those of prospective mates. In either situation a male's home range is probably larger than required to supply sufficient food (Sandell 1989).

Statements indicating that defense can extend from space to individuals (i.e. that males "defend" breeding females) appear to be a misuse of terminology in the case of cats. "Defend" them from what, other males? This hardly seems practical in a promiscuous mating system (Chapter 4). Evidence presented so far indicates that cats do not defend spaces either cooperatively or independently, nor do they defend other individuals, the exception being a mother cat protecting her kittens. Defense of space or individuals is simply not a consistently identifiable component of the domestic cat's behavioral repertoire. Overlapping home ranges and indifference toward reinforcing boundaries refute the notion of territoriality as the term is defined by any reasonable and falsifiable definition.

Among mammals generally, male territoriality, at least during the breeding season, might be favored if females breed synchronously (Clutton-Brock 1989); conversely, expanding the home range to encounter females – as male cats supposedly do – is advantageous if females come into estrus asynchronously (Clutton-Brock 1989). However, this is not always the case. Haspel and Calhoon (1989) found that male and female strays in Brooklyn, New York, had home ranges statistically similar in size. Females seldom ventured far from core areas. Males spent more time at the peripheries of their home ranges.

2.9 Scent-marking

Cats express behaviors consistent with scent-marking familiar spaces, but exactly what this means has not been demonstrated conclusively. Some compound (or compounds) in conspecific urine are recognized early in life. Kittens do not show interest in the urine of other cats until ~5 weeks, at which time some display a flehmen response (Kolb and Nonneman 1975), and most display it by 6–7 weeks. At 5 weeks or younger they often react to strange urine by emitting distress calls or trying to escape to a safe location.

During the *flehmen response* a cat raises the head, pulls back the lips into a "grin," opens the mouth halfway, and breathes slowly while assuming a stare. This behavior is ordinarily linked with sexual behavior and is typically shown by adult males (e.g. males might display a flehmen while examining a pool of urine left by an estrous female), although females display on occasion too (Hart and Leedy 1987). Spayed females that have been "masculinized" by injection with testosterone propionate show increased flehmen displays around intact females treated with estrogen (Hart and Leedy 1987). A strange cat is not necessary to elicit the response, and cats occasionally display the flehmen while examining their own urine, especially if the sample is older than 1 d (de Boer 1977a). An older scent-mark is usually wetted first to fluidize it, either by licking or touching the nose to it (Verberne and de Boer 1976), and a flehmen response is thought to involve the transport of fluids (Hart and Leedy 1987). If substances are functioning as putative pheromones (Chapter 9), they are more likely wafted into the vomeronasal organ (VNO) as fluids, not aerosols (Spotte 2012: 59).

Felinine (2-amino-7-hydroxy-5, 5-dimethyl-4-thiaheptanoic acid), a branched-chain sulfur amino acid, appears in the urine of domestic cats starting about age 6 months (Tarttelin et al. 1998). The substance is excreted in large amounts in the urine of intact male domestic cats, and felinine or one or more of its degradation products is thought to be a putative pheromone (Chapter 4) supposedly serving as a territorial marker and sex attractant (Hendriks et al. 1995b, Li et al. 2010, Tarttelin et al. 1998), although neither function has been confirmed. Both felinine and its precursors are absent from the urine of Pantherinae, the big cats (Hendriks et al. 1995b: 582, Table 1; McLean et al. 2007). Felinine's direct function (if any) is therefore unknown. Although degradation is slow (Hendriks et al. 1995a), some breakdown products might be responsible for the characteristic "catty" odor of cat urine (Hendriks et al. 1995b, Miyazaki et al. 2006a, Rutherfurd et al. 2007), especially the urine of adult males. Felinine concentration correlates positively with the circulating testosterone level of intact adult males (Hendriks et al. 1995a, 1995b; McLean et al. 2007; Tarttelin et al. 1998), which also excrete substantially more than adult females and castrated males (Hendriks et al. 1995a, Rutherfurd et al. 2007, Tarttelin et al. 1998). Intact males produce an average of 122 mmol felinine/kg M/d ($\pm$23.6 SEM), according to (Hendriks et al. 1995a), respective values for neutered males, intact females, and neutered females were 41 ($\pm$8.4 SEM), 36 ($\pm$7.3 SEM), and 20 mmol/kg M/d ($\pm$3.8 SEM). On a mass per volume basis, intact males excreted an average of 2.0 g felinine/L of urine ($\pm$0.55 SEM). Overall urine volumes per 24 h were not significantly different among subjects.

Cats are *proteinuric* (Chapter 7), excreting large quantities of proteins along with their urine (0.5–1.0 mg/mL) of which ~90% consists of *cauxin*, a carboxylesterase-like protein (Miyazaki et al. 2003, 2006a, 2006b, 2008). Two other forms of nitrogen in domestic cat urine are urea-nitrogen and ammonia-nitrogen (Cottam et al. 2002). Intact males excrete higher concentrations of cauxin than females or castrated males (Miyazaki et al. 2003, 2006a, 2008). An intact adult male of 4.2 kg excreted cauxin at a rate of 27.8 mg/kg M/d, but this value dropped 90% 4 weeks after castration (Miyazaki et al. 2006b). Cauxin's function (if any) has not yet been discovered, but its presence

Fig. 2.4 Cats commonly "sharpen" their claws on trees and posts. Its putative function of influencing the behavior of other cats by leaving olfactory or visual signs is doubtful. Source: © Momentsintime | Dreamstime.com.

is also associated with the strong scent of cat urine, and its role in scent-marking has been postulated but not demonstrated (Miyazaki *et al.* 2006a).

Several authors have mentioned the "interdigital glands" of cats. These structures supposedly transfer scent from the forepaws to trees and posts during scratching episodes (Fig. 2.4), simultaneously leaving visual marks of presumed relevance to other cats passing by (e.g. Bradshaw and Cameron-Beaumont 2000, Feldman 1994), either warnings or simply proof of presence. These make interesting conjectures; unfortunately, no anatomist has evidently seen, much less described, "interdigital glands" in cats or any other carnivoran (Spotte 2012: 259–260). Consequently, their proposed function can only be speculative.

As described by Schummer *et al.* (1981: 465), the carpal organ of the cat is "about 2.5 cm above the carpal pad and consists of 3–6 non-pigmented sinus hairs. … " These authors also provided an illustration (Schummer *et al.* 1981: 478, Fig. 362). Schummer *et al.* (1981: 490) later referred to the hairs as vibrissae, attributing a sensory function to them, which is probably accurate (Chapter 9). The cryptic description (p. 465) states only that "the carpal organ of the cat is thus a touch organ combined with scent glands." The carpal organ and associated carpal pad does not serve any known function except, as mentioned elsewhere (Schummer *et al.* 1981: 465, 496), the claw is a possible climbing aid. Whether the carpal organ could serve to transfer scent – or whether it even produces a functional scent – is unknown.

Clawing trees and posts is also supposed to "sharpen" a cat's claws (e.g. Feldman 1994), and Kitchener (1999: 250) even wrote, "Scratching is thought to combine the functions of a manicure with scent marking. … " How tree bark could accomplish this by acting as the abrasive required for a "manicure" has yet to be explained and demonstrated. This is especially problematical if, as Feldman (1994) claimed, soft, nonabrasive bark makes the preferred scratching surface. The sharp points of claws of the forepaws are maintained by periodically shedding worn horn caps from tips of the claw sheaths. Scratching objects might assist this process (Mellen 1993), which is well described and perhaps even necessary to maintain the claws in proper condition for hunting and climbing (Homberger et al. 2009), although any relationship to signaling is doubtful.

Natoli et al. (2001) pointed out that references to nonexistent glands of cats are prevalent in the literature. Putative glands in the cheeks and elsewhere on the skull (e.g. between the ears) have never been found and probably are not there. Cats, alone among domestic animals, possess perioral glands, which are hypertrophied sebaceous glands in the skin around the mouth, the lower lip in particular (Schummer et al. 1981: 463). Their function is unclear. According to Natoli et al. (2001: 142), "Sweat glands are well developed on the toe and sole pads of cats, but it is well known that they are not utilized to mark in other species of mammals (Schummer et al. 1981)." Nowhere did Schummer and colleagues claim this. Nor are they responsible for another comment attributed to them by Natoli et al. (2001: 142): "Schummer et al. (1981) state clearly that, among the general skin modifications, the cutaneous scent glands that can be utilized by the domestic cat to mark the territory and/or conspecifics, are the perioral … tail and preputial glands, as well as the carpal organ and the anal sacs." In fact, Schummer and colleagues rarely mentioned "marking" by domestic cats, evidently content to describe anatomical features and leave speculation about function to others.

Verberne and de Boer (1976) wrote that among the skin's sebaceous glands are those in the cheeks. Cheek-gland "secretions" were thought to be rubbed onto protruding objects in the environment where they release putative pheromones, functioning similarly to compounds in urine-spray and saliva. Such conclusions, typically presented as fact, are conjectural, predicated on questionable observational data such as the amount of time males spend sniffing the urine of estrous vs. nonestrous females or the frequency of their flehmen responses. An attractive component in female urine does not make it a *de facto* pheromone, especially when its chemical composition and neuronal receptors have yet to be identified, nor is observation of behavior alone adequate to determine function even if they have. To my knowledge, cats do not possess cheek glands. Nonetheless, Mellen (1993: 162), citing Verberne and de Boer (1976) as her source, claimed that "male domestic cats can differentiate phases of the estrous cycle of females from cheek gland secretions." Cats are said to transfer saliva to "mark" objects during cheek rubbing (references in Mellen 1993), behavior for which I found no evidence. Nor has the groups of sinus hairs (Schummer et al. 1981: 490) been shown to include glandular functions (Fig. 2.5). These clusters comprise 2 cheek bundles, 8–12 presumed tactile hairs above the eyes, and the vibrissae (i.e. "whiskers") of the upper lip (4 rows on each side of the face for a total of ~30 hairs). Any potential function is probably sensory (Chapter 9).

Mature cats have a gland located on the back at the base of the tail. As Schummer et al. (1981: 466) described it, the tail gland "is formed by an accumulation of large, very lobulated, hair follicle-sebaceous glands." These include apocrine scent glands that are more developed in males than females. It seems to me that transferring scent from this location would be difficult unless a cat rolls on its back, which females do after copulating (Chapter 4). Whether rolling behavior transfers volatile compounds to the surfaces touched by the tail gland or has any other function is unknown.

Perhaps the most commonly mentioned glands are the anal sacs, which Schummer et al. (1981: 467) described as "spherical or ovoid structures of 6–8 mm diameter."

Fig. 2.5 Vibrissae (sinus hairs) of a cat. Source: © Yuri Tuchkov | Dreamstime.com.

Their lining, composed of stratified squamous epithelium, is perforated by efferent ducts of the sebaceous gland complexes and tubular apocrine glands (Greer and Calhoun 1966, Schummer *et al.* 1981: 467). The surrounding smooth and striated muscles empty the glands on contraction. The liquid they contain is a secretion composed partly of fatty and serous matter and cellular debris (Greer and Calhoun 1966) and marked by a pungent, unpleasant odor. In the description of Schummer *et al.* (1981: 468), "The scent substances adhere to the voided faeces and are utilized for territory marking and individual recognition." Their conclusion has yet to be verified, nor is it certain that the anal-sac contents are always excreted with feces. Greer and Calhoun (1966: 773), on whose description of the anal sacs Schummer and colleagues based theirs, pointedly ended the abstract of their report with this sentence: "The function of the anal sacs and their glands is unknown."

The preputial glands comprise hair-follicle glands, free sebaceous glands, aveolar, and tubular scent glands (Schummer *et al.* 1981: 468). Sebaceous glands predominate in cats. As the name suggests, the preputial glands occur in the prepuce, or skin surrounding the clitoris.

Cat urine presumably transmits to other cats knowledge of the donor's sex, age, health, and reproductive status (de Boer 1977a, Li *et al.* 2010, Molteno *et al.* 1998), but whether it actually does any of these things is undetermined. Observational evidence is inconclusive without physiological confirmation and a subsequent explanation of mechanism. Sorting this out seems especially pertinent if the examining cat and donor are the same individual and its behavior (e.g. sniffing, flehmen response) is no different than when the donor is another cat. Such situations are not uncommon (de Boer 1977b, Verberne and de Boer 1976), raising the point of whether recording superficial behaviors like sniffing urine (e.g. Natoli 1985a) produces only observational data devoid of explanatory power. Attributing function to spraying when none has been identified is not worthwhile. Thinking of spraying as serving a "purpose" is even less productive, leading into a teleological labyrinth. Kitchener (1999: 250), for example, wrote that "tigers selected particular tree species for spraying urine onto and they mostly selected the underside of trees which leaned over by 8–24° from the vertical, so that the marks were protected from the rain."

3 Interaction

3.1 Introduction

Present evidence indicates that free-ranging cats should be considered solitary – not social – until it can be demonstrated that solitary living is inferior to group living based on relative fitness of the two life-styles. In their study of the evolution of social monogamy using 2288 species of mammals, Lukas and Clutton-Brock (2013: 526) concluded: "Phylogenetic reconstruction shows that, in the common ancestor of all mammalian species, females were solitary and males occupied ranges or territories overlapping several females." This situation is common among free-ranging cats.

3.2 The asocial domestic cat

Solitary living characterizes 68% of the breeding females of living mammalian species (Lukas and Clutton-Brock 2013). In many instances, differences of opinion about the cat's social organization (or lack of one) can be traced to observations yielding descriptive data instead of experiments grounded in theory. Descriptive studies are sometimes useful to identify elements that can later be tested empirically, but when assessed in the absence of hypotheses they explain nothing. The result has been a literature in which descriptive results are difficult to compare because they lack the power to explain. In the absence of unambiguous definitions and rigorous hypothesis testing, it seems hardly surprising that cats have been described as territorial or not (Chapter 2), and as solitary, social, both, or maybe.

Domestic cats have been classified traditionally as asocial (e.g. Apps 1986b, Baron *et al.* 1957, Corbett 1978, Fox 1975, Mellen 1993, Mendoza and Ramirez 1987, Moelk 1979, Rosenblatt and Schneirla 1962, van Aarde 1978, Verberne and de Boer 1976), which hardly makes them indifferent to conspecifics. As Mellen (1993: 151) pointed out, "such solitary existence does not preclude the 'asocial' members of the family Felidae from possessing a rich repertoire of communicative signals. ... " A scientist's task is to decipher these using rigorous experimental methods, making certain that any conclusions are sufficiently narrow to not overstep the data. Many investigators, having ignored this basic mandate, promote the idea of domestic cats interacting within a complex social system (e.g. Bonanni *et al.* 2007; Deag *et al.* 1988; Leyhausen 1965, 1973, 1979; Liberg 1980; Macdonald 1981; Macdonald *et al.* 1987; Robinson 1992), a premature conclusion. As I intend to demonstrate, supporting evidence is not there, but this has not quelled the flow of general statements. Bradshaw and Brown (1992: 54S) wrote that "over the past ten years several studies, in various countries, have established that, under appropriate conditions, cats can form complex, stable societies. ... " Jones (1989: 9), for example, considered feral cats to be solitary, although "under conditions

---

Free-ranging Cats: Behavior, Ecology, Management, First Edition. Stephen Spotte.
© 2014 John Wiley & Sons, Ltd. Published 2014 by John Wiley & Sons, Ltd.
Companion Website: www.wiley.com/go/spotte/cats

of increased urbanisation Cats are capable of more complex social interactions, enabling them to live at much higher population densities." To date, no one has yet defined such societies (their nature and limits), including how they differ from opportunistic aggregations of conspecifics gathered around clumped food (e.g. Laundré 1977) and grown tolerant through familiarity and regular meals. Cafazzo and Natoli (2009) claimed that free-ranging cats live not just in social groups but occupy hierarchical positions within these groups. Their data were unconvincing. Packer (1986) did not mention domestic cats when surveying the social ecology of felids, but of the rest he considered only lions to form societies. Mellen (1993) basically agreed and included the cheetah. Kitchener (1999: 236) named the lion, cheetah, and domestic cat "as showing a high degree of sociability. ... " Advocates of domestic cats as social animals have tended to place them on a continuum bounded by "solitary" at one end and "social" at the other without separating either concept into defined, testable components.

No one who promotes cat sociality has tested any presumed component for demonstrable evolutionary benefits. Macdonald (1981, 1983: 24–25) and Macdonald et al. (1987: 21–22), for example, listed as evidence of sociality the direction and frequency of friendly and agonistic interactions and frequency of cats sleeping together, none of which is direct proof of a complex social system or offers a test of fitness, nor were such associations examined for correlation with extraneous factors that might reveal tentative evidence of function. Sleeping together, for example, could be a thigmotactic response to ambient air temperature and relative humidity. Leyhausen (1979: 131) pointed vaguely to rank-order around the food bowl, shifting of hierarchies with increased crowding, and alternating use of common spaces. Rank-orders in cats might not form except in confinement, and even there we have good reasons to doubt their existence (Chapter 1). Assuming they did exist in free-ranging cats, situations such as Leyhausen mentioned are still not evidence of complexity or even of social order, and the alternating use of common spaces actually shows aversion, not affiliative behavior.

Increased tolerance of conspecifics is not necessarily a sign of willing social interaction. Even among lions, supposedly the most social felids, the trend toward sociality is less well developed than in wolves, the signaling system that might reduce aggression comparatively unstable and less predictable. A higher level of sociality has developed in lions living on open grasslands that support intense concentrations of large prey – in other words, a concentrated source of food – although as Kleiman and Eisenberg (1973: 647) wrote, "lion social groups show neither a high degree of integration of the two sexes nor a clear division of labour within the pride." In the more solitary cats – and here I include domestic cats – affiliative behavior (e.g. allogrooming, head-rub, nose-sniff) would occur mainly in the rearing phase, such contact being less common otherwise (Kleiman and Eisenberg 1973). Affiliative interactions (Fig. 1.1) among domestic cats, when they happen at all, appear most often under crowded conditions (e.g. laboratory groups, strays around clumped food, several cats occupying a single location like a barn or temple). Even then the overall impression is of mutual disinterest. Feral cats are less easily observed, and from what we know their preference is to avoid close encounters except during courtship and mating. Pair-bonding, the foundation of sociality in monogamous species, is absent from the repertoires of all the cats. The only persistent interaction occurs over a few weeks between mother and offspring (Chapter 5) and of littermates with each other. Hunting is the métier of all cats. In the small species, including free-ranging feral cats, cryptic hunting with emphasis on scattered prey and a purely carnivorous diet encourage isolation by dispersion and mitigate against sociality (Kleiman and Eisenberg 1973).

Even among social animals, the interests of individuals are always out of synchrony with those of the group (Alexander 1974). If cats are indeed complex social creatures, then confirmation should be apparent through experiments arranged to reveal some fitness advantage of sociality as opposed to a life of solitude. What we have instead are unconfirmed statements like that of Fitzgerald and Karl (1986: 79): "Sociality seems to be strongest where man provides food and shelter and weakest among feral cats having dispersed prey and plentiful shelter." Cats, in Fitzgerald and Karl's interpretation, are social to a varying extent despite an equally plausible interpretation that they could be solitary along a scale of increasing or diminishing tolerance.

Contrary to prevailing opinion (e.g. Pontier and Natoli 1996), a sighting of stray cats around clumped food or a convenient shelter site is not a sign of sociality without loosening its vernacular usage to include opportunism combined with a capacity to form aggregations. In such instances, more than one species can participate. Multi-species aggregations are common where food is both predictable and abundant. Examples include multiple species of birds at bird feeders; gray wolves (*Canis lupus*), coyotes (*Canis latrans*), and ravens (*Corvus corax*) at large ungulate kills (Atwood 2006, Ballard *et al.* 1997, Harrington 1978, Hayes *et al.* 2000); stray dogs, stray cats, and rats in urban alleys where garbage is dumped (Beck 1971, 1973); and gray wolves, striped hyenas (*Hyaena hyaena*), golden jackals (*Canis aureus*), and griffon vultures (*Gyps fulvus*) at Israeli carcass dumps (Gittleman 1989, Mendelssohn 1982). I would hesitate to call these "social gatherings" and hold the same opinion of an aggregation of cats foraging at a waste-disposal site. Of the interactions between jackals and hyenas, Gittleman (1989: 196) wrote that where large quantities of food are conveniently available, "individuals congregate while eating and remain together afterward even though they inhabit extensive areas." He continued, "In east Africa, where these species forage independently of humans, both are primarily solitary (or occasionally seen in pairs). ... "

The principal factor separating feral cats from strays is the difference in dispersion of their food. This dichotomy is artificial, of course, but not more so than the presumption of an aggregation of cats attracted to a feeding site constituting a social group. All cats lived independently before humans started feeding them, either directly by hand or indirectly by making garbage accessible. When we say that free-ranging cats feed opportunistically, it includes food supplied by humans, access to which is circumstantial. If food is clumped, they aggregate near it; if not, they search elsewhere. Circumstances drive these patterns, not an inherent mandate to be part of a social order.

Foraging aggregations could also demonstrate that an animal like the cat – solitary by choice in the feral state – is versatile enough to exploit more abundant resources as a stray or pet without actually sacrificing its predilection for solitude. The dilemma is then determining whether cats seen together constitute a true society or simply individuals gathered to feed or mate, each following its personal agenda. How cats behave during intraspecific encounters is of interest, but more important is whether the nature and persistence of their interactions qualify them as solitary or social in the context of evolutionary theory. That a cat inhabits the same location in the company of others, for example, does constitute sociality regardless of whether food is abundant, nor is it proof of social ties or even permanent philopatry. Free-ranging cats arrive and depart on their own schedules. Well-fed house cats sometimes leave home if the cost of competition (e.g. stress, risk of injury) with house mates exceeds the advantages of ample food and shelter. In such cases the uncertain benefits of living elsewhere outweigh the certain disadvantages of staying.

That the domestic cat is social at all, much less socially complex, is doubtful. Aggregating around clumped food, interacting in groups with minimal agonism, retaining more or less stable numbers at a given location, occupying overlapping home ranges (Chapter 2), and associating with relatives do not make a society. As Alexander (1974: 328) observed, "Group living … is like extended juvenile life … and the attribute evolves only because benefits specific to the organism and the situation outweigh what appear as automatic detriments." In social species these detriments can include lower reproductive rate caused by extended juvenile development, functional sterility from relegation to subordinate status, increased transmission of infectious diseases and parasitic infestations, cuckolding of dominant males by subordinates, increased likelihood of detection by predators, food competition, reduced individual food intake, and greater risk of injury by conspecifics (Alexander 1974, Gittleman 1989).

Increased fitness must be shown before any evolutionary advantage of group living is accepted, and the literature reveals no such benefit for domestic cats. Relevant assessments might be whether living in the company of conspecifics measurably enhances longevity, reduces the incidence of disease, or increases some predefined measure of reproductive success. Experiments to test such factors in truly social species are easily confounded by the same variable under investigation (i.e. sociality), simply because separating an individual from the group can reduce its *relative* fitness. I doubt this would be the case for cats. As Brown and Orians (1970: 240) wrote, "In species in which the individual is the primary social unit, the fitness of isolated individuals is normal for the species."

3.3 Solitary or social?

As mentioned, the domestic cat has been classified traditionally as solitary and largely asocial (Section 3.2). Recent investigators have sought to change this picture through detailed observations of how cats interact, pointing to subtle behaviors and interpreting them as glimpses of social complexity. The truth in either case depends on how several crucial terms are defined and applied in context. Feral adult female cats are still thought of as solitary (Jones 1989), especially those in rural areas living at low population densities and hunting scattered prey. In contrast, rural, suburban, and urban strays sometimes live at high population densities when food is abundant and clumped (Liberg *et al.* 2000). The availability of food in stable, clumped patches is therefore believed to be responsible for these gatherings (Izawa *et al.* 1982, Yamane *et al.* 1994). The extent to which free-ranging cats interact seems dependent on habitat quality, which influences food availability and probably determines the size of home ranges (Chapter 2). If garbage is plentiful or cats are fed by humans, the amount of space needed to survive shrinks, allowing larger groups to exist and leading inevitably to more frequent contact.

Before continuing to evaluate any basis for sociality in free-ranging cats, the terms solitary and social must be defined in ways conducive to empiricism. As already discussed here and in Chapter 2, definitions useful to science must be simple, clear, direct, unencumbered by contingencies, and amenable to incorporation into falsifiable hypotheses. Keep in mind that any definition is relative and arbitrary, never absolute, and even the best is serviceable only until a better one can be devised. The objective should be to shrink a term's meaning into a package that is compact and unambiguous enough to test.

We might start by devising a general definition, calling a species solitary if most observations are of solitary individuals. Restrictions can be placed on "most observations" by picking a number (e.g. 80%). Individuals of a social species would then be the reverse, being observed in groups in 80% of observations. Such definitions are testable by collecting the data and evaluating the results statistically, a method itself grounded in arbitrary but consistent and falsifiable inferences. The hypothesis is purely descriptive, offering no measure of fitness. Its only value would be as a starting point for actually testing *why* this species appears to be solitary (or not). In other words, observations alone have little importance; their value is helping point the way toward the truly relevant questions. What do the observations we have in hand tell us?

Interactive behavior consumes only 1–2% of the time budgets of small wildcats in captivity (Mellen 1993). The descriptive literature on free-ranging domestic cats conforms to how we just defined solitary living in simple, arbitrary terms, along with a corollary of minimal interaction. Observations of more than two individuals together typically involve a mother and kittens or transient groups of adolescents (Kerby and Macdonald 1988), perhaps littermates. In southeastern Australia and at subantarctic Marion Island, 90% of adults were alone when sighted (Jones and Coman 1982b, van Aarde 1978), although small adult groups averaging less than three individuals were seen occasionally. Whether these were estrous groups (Chapter 4) or littermates was not stated. Other investigators reported Marion Island adults as usually alone or in pairs, and kittens in groups of littermates (Anderson and Condy 1974). In the Kerguelen Islands, where the population density was low (0.44–2.42/km^2), 95.5% of observations were of solitary cats (Say *et al.* 2002a). Feral cats in an agricultural area of northern Italy were in associations of two or more in just 4.5% of sightings, mainly during the breeding season (Genovesi *et al.* 1995). At Dassen Island, South Africa, 93% of sightings were of lone individuals; cats seen together were mother with kittens or fleeting aggregations exploiting large carcasses (Apps 1986b) or clumped prey like a colony of nesting seabirds (Rauzon *et al.* 2011).

According to Denny *et al.* (2002: 409), rural strays at a waste-disposal site on the central highlands of New South Wales interacted 35 times during 330 observations of feeding, and although "several animals visited the rubbish concurrently, they remained separated, with few instances of cats lying together or feeding as a group." Males sometimes fed or rested near each other, but female–female interactions were never seen. Panaman (1981) observed a group of five female farm cats for 391 h, recording only 17 interactions in 4338 individual records.

Feral cats at Scotland's Monach Islands seldom interacted, and Corbett (1978) considered them solitary but did not provide data. Although food was abundant at Avonmouth Docks, strays exploiting it were rarely in contact (Page *et al.* 1992). Of 13 monitored males (including a castrated animal), all except one intact individual occasionally interacted with the other males, and 10 were sometimes in contact with females. The females never interacted. In Hattah-Kulkyne National Park, Victoria, southeastern Australia, 71% of all cats sighted and 90% of adults were alone (Jones and Coman 1982b). Just 18% of cats recorded were in groups of three or more, probably mothers with litters (no group included more than one adult). Nearly all sightings of feral cats in an agricultural area of central-western New South Wales were of solitary individuals, the exceptions being courting adults and females with kittens (Molsher *et al.* 2005). Adult feral cats at Macquarie Island were mostly nocturnal and solitary (Brothers *et al.* 1985). A feral male singled out for extended monitoring led a solitary life. During 38 h of observation he interacted with conspecifics four times, although other cats were seen occasionally inside his 41-ha home range (Brothers *et al.* 1985). Cats at Marion Island were sometimes seen in aggregations of two to five, which van Aarde (1978: 299) attributed to mothers with kittens: "The age composition of these groups showed a definite seasonal pattern as a result of seasonal breeding."

The pattern (if it can be called that) seems to be a preference for solitude at low population densities and scattered food, and of increasing tolerance – not necessarily sociality – as space becomes restricted. Degree of interaction varies by individual among group-living strays, some cats preferring the proximity of conspecifics, others keeping a greater distance (Natoli *et al.* 2001). Females might show preferences for certain males by spending more time near them (Kerby and Macdonald 1988, Natoli *et al.* 2001). After watching a group of farm cats, Laundré (1977) remarked that when friendly behavior occurred it was generally between a female and a male. Mirmovitch (1995) reported urban females from the same feeding group (i.e. that fed regularly from the same garbage bin) to allogroom, travel and sleep together, and display mutual tolerance when feeding.

Most free-ranging male cats are solitary whether feral, stray, or pets. Dards (1983: 150) considered the mature males she watched at Plymouth Dockyard crowded with strays to be "notable for their lack of amicable behaviour." She wrote, "This is perhaps surprising, in view of the cooperative pairs or groups of males which are found in lions." However, it is not surprising if domestic cats are actually asocial and a large aggregation of strays is a human-induced artifact taking advantage of clumped and abundant food and shelter.

Adult feral females are usually alone when living in low-density populations. As strays occupying high-density areas, they sometimes form female groups, interacting with minimal agonism and rearing litters in the same location. How far propinquity goes toward including cooperation is debatable. The idea that female strays defend their collective young against invading cats (e.g. Macdonald *et al.* 1987, Natoli 1985b) and calling it evidence of sociality is misplaced even if a function (e.g. the possibility of male infanticide) could be demonstrated. Packer (1986: 439) made this clear: "Although communal defense against infanticidal males may be an important advantage of group living in female lions, by itself it cannot explain the distribution of sociality across species: infanticide by males also occurs in tigers and cougars … but females in these species are nevertheless solitary."

On occasion the attribution of cooperative behavior becomes wildly speculative. Izawa and Ono (1986) went so far as to propose that "territories" of female strays are passed down in matrilineal fashion to female offspring and thus inherited, a prospect endorsed by Kitchener (1999: 251), who then claimed, "Both mother and daughter may benefit by combined territorial defence from other females by having an ally close at hand, thereby reducing the area each animal defends directly."

The female's reproductive investment, unlike the male's, is substantial (Chapter 4). She rears the young entirely without his help. Besides the energy required for personal survival is the added burden of producing milk and later foraging to feed her kittens (Chapters 5 and 8). Success is best achieved at locations where food is adequate, especially during gestation and rearing (Sandell 1989). Female distribution – whether in the feral or stray state – appears to center on food resources regardless of reproductive condition. Female house cats and strays become *potentially* more tolerant of each other at high densities when food is not limiting. I qualify this last statement because females in a household crowded with female cats sometimes emigrate and take up residence elsewhere even if fed regularly (Liberg 1980).

Food, its availability and dispersion, is therefore an influential factor – probably the most influential – affecting the distributions of free-ranging cats. Cats are more likely to be seen alone when food is scarce or dispersed (Genovesi *et al.* 1995, Liberg and Sandell 1988). Even urban strays living where food more than meets daily dietary requirements (Calhoon and Haspel 1989, Haspel and Calhoon 1989), but is scattered instead of clumped, were solitary except for females with offspring (Calhoon and Haspel 1989). Stray cats in aggregations waiting to be fed often space themselves in a way that indicates a combination of tolerance and mutual avoidance (Fig. 3.1), and in feral cats avoidance can be maintained even if home ranges overlap (Biró *et al.* 2004).

The manner in which individuals of a species interact during breeding and rearing of young is the principal criterion for how animal societies are classified. On this basis, cooperative societies are rare among mammals (Clutton-Brock 2009). As emphasized, a grouping of animals does not necessarily indicate a society, although it does make the individuals gregarious – that is, tending to gather together – in the sense of a school of fish. In female domestic cats, gregariousness apparently increases with increasing population size, which is typical of many mammalian species (Packer 1986). This accounts for female urban strays interacting more frequently than their feral counterparts in less populated areas, and in the greater overlap in home ranges where population densities are high. The difficulty is trying to understand whether, if a species is indeed solitary, it

Fig. 3.1 Urban stray cats evenly dispersed around a fountain at Córdoba, Spain. Source: © Paul Neyman | Dreamstime.com.

can assemble voluntarily into a seemingly paradoxical unit, an aggregation of solitary individuals. So how is "solitary" defined? Perhaps it really does refer to animals living alone and seldom or never interacting except at certain times, such as the breeding season. Even then, reproductive activity is "social" only because extended solitude is interrupted briefly for procreation. Thus every mammal is social to some extent when it interacts with conspecifics, however fleetingly (Sandell 1989). Or maybe, as Leyhausen (1965: 249) wrote, solitary in the case of all mammals means "they can only be shot one at a time." The fact that any species is solitary nonetheless leaves it still aware of nearby conspecifics, if simply to avoid them. To quote Leyhausen (1965: 257) again, "the only mammal one could conceivably speak of as being socially indifferent is a dead one."

The term "partially solitary," as used by some to describe the cat (e.g. de Boer 1977b), is oxymoronic if taken literally. Sandell (1989) preferred narrowing the meaning of solitary behavior by contrasting it with cooperativeness. He defined "solitary" as the absence of cooperation with a conspecific except when mating. Sandell's concise definition is easily incorporated into testable hypotheses, provided usage is restricted to this single context and a suitable definition of "cooperation" is devised that includes more than function-specific interaction. However, it leaves out any means of testing evolutionary factors.

I define a *solitary animal* as one that spends its adult life alone except when mating and rearing young, does not cooperate with other adult conspecifics, and gains no measurable fitness advantage from group living. The last qualification allows individuals of solitary species to aggregate around clumped resources for purely selfish and function-specific reasons without forming societies.

Most groups of cats consist of those attracted to feeding sites, males attracted by estrous females, or kittens staying near their mothers. For any such aggregations to be considered social they must be shown to compose higher-level evolutionary units (i.e. those leading to developments like altruism).

Social behavior, or sociality, encompasses staying and helping (Le Galliard *et al.* 2005). In most cases, being social imposes twice the cost of living alone. The first imposition is the sacrifice of mobility when individuals are forced to stay home; the second requires some form of altruistic behavior in which a degree of personal fitness is relinquished in favor of enhancing group fitness. A free-ranging cat clearly ignores both costs, often living by itself, shifting its home range at will or after an unpleasant experience like being live-trapped (Jones and Coman 1982b), and evincing no indisputable sign of cooperative behavior; that is, any sign that qualifies as a reliable species character. By these restrictions the current belief that cats are highly evolved social creatures carries little merit, having been based on observations of occasional friendly interactions, formation of presumed dominance hierarchies in confinement, occasional communal nursing, dyadic interactions reflecting putative dominance, and other descriptive findings of doubtful evolutionary relevance.

To Alexander (1974: 326), "Sociality means group-living." Citing Alexander's publication, van Veelen *et al.* (2010: 1240) considered the two essential ingredients of sociality to be "The formation of groups and the evolution of cooperation within them." These factors can also define the term: *sociality* is living and cooperating in groups. An adequate test of its existence must then show increased fitness as a consequence of both factors. The probability of equal fitness bestowed by solitary living – along with not cooperating – represents the null hypothesis, requiring, according to Alexander (1974:

326), a search for "the selective forces causing and maintaining group-living." But suppose the animal being observed is actually solitary, its own clumped pattern an artifact of clumped resources in the habitat? Then any so-called selective force is merely a display of plasticity and not direct evidence of sociality. The power of freely available food to modify behavior has not been given proper weight, but can easily be tested. In the case of urban stray cats, take away the shelter sites, the garbage bins, control for ancillary sources of food like rodents and pigeons (*Columba livia*), and then assess the comparative extent of any apparent sociality, assuming the cats stay instead of vanishing once freeloading becomes unprofitable. This is the case with insular feral cats. Unless fed regularly or living where food is easily accessible, aggregations of foraging feral cats are short-lived (Apps 1986b), an indication that they come together to take advantage of clumped food (i.e. a carcass, a colony of breeding seabirds), not to interact socially.

Living in groups is uncommon among the carnivorans. As Gittleman (1989: 183) wrote, "only about 10–15% of all species aggregate at some period outside of the breeding season. ... " Most carnivorans are solitary, nocturnal, shun open habitats, and when they come together in groups it is almost always for defense against predators or to exploit sources of food (Gittleman 1989). As emphasized, actual group living – and not just gathering around free handouts – must offer some fitness gain over solitude. In addition, it must extend to perceived advantages for staying in *this* group as opposed to *that* one, which could include balancing the risk factors aligned with dispersal to another group, like interacting with conspecifics that are strangers initially and afterward trying to establish social relationships with them; in other words, all the uncertainties faced by a social animal after emigration.

Many birds and mammals disperse from the natal area on becoming independent or sexually mature. Dispersal is obviously risky, although any heightened danger an emigrant faces is offset by breeding opportunities and a chance to pass along its genes directly. Dispersing cats pursue a selfish life-style. Routine social interaction and assistance of relatives are not required by solitary species, and this includes helping rear another's young. Just as important, being alone while feeding and rearing offspring is apparently the default situation, neither a liability nor an advantage.

Degree of tolerance among free-ranging cats seems to correlate negatively with population density; in other words, cats in captive groups *might* form rudimentary dominance hierarchies (Chapter 1), which could reduce aggression; strays simply adjust to living in less space because they can always emigrate. When in the vicinity of ample food, this capacity to stay and adapt is evidently stronger than the urge to leave. The mechanism for sharing space without conflict is perhaps active avoidance without implications of territoriality (Chapter 2). The feral cats of Dassen Island make a good example. The island is only 224 ha, and home ranges overlap considerably. Nonetheless, the cats were still able to avoid each other.

Group-living rural cats were often seen together in parts of their overlapping home ranges but alone in the nearby fields (Kerby and Macdonald 1988). Macdonald *et al.* (1987: 13), in observing a group of 4–9 farm cats, found that attendance of each at the barn was independent of the others, behavior suggesting that "members of the colony were neither seeking nor avoiding meetings. ... " They pointed out that even if such individuals lacked organization they would still come together on occasion and interact. "Therefore, observation of interactions is evidence neither of social structure nor of active sociality." Kerby and Macdonald (1988) proposed that "rudimentary proof" of sociality might be departure from randomness in spacing with respect to resources, but positive evidence of sociality reported subsequently ignored the obvious expectancy of food. Most farm cats observed have stayed near barns and spent comparatively less time in the surrounding fields and woods, even if the farmers seldom supplied their total daily food requirements (e.g. Laundré 1977, Macdonald *et al.* 1987). In other words, food as an intervening variable must always be considered.

Cats by nature are inactive much of the time. Philopatry in farm cats is mainly reinforced by intermittent feeding, which usually occurs near the barn. Cats loafing at the site where someone feeds them can be expected to deviate from random spacing even if space is not limiting. That some cats displayed haptic interaction, sat closer together than random spacing would predict, or seemed to prefer the company of some individuals to others, is not collective evidence of sociality. Based on what they observed, Kerby and Macdonald (1988: 75) overstated the case when concluding, "From these analyses it is clear that farm cat colonies are highly structured in terms of the nature of individual relationships."

3.4 Cooperative or not?

As mentioned earlier, one proposed way of distinguishing solitary from social animals is to substitute "cooperative" for "social." Then, as noted by Sandell (1989: 164), "A carnivore is solitary if it never, except when mating, cooperates with conspecifics; that is, if two or more animals of any given species cooperate to rear young, forage, achieve matings, or defend against predators, the species is classified as cooperative (which, so defined, resembles group living. ...)." Sandell (1989: 165) did not consider group defense of a territory a criterion of cooperative behavior because solitary animals of some species can also be territorial, and "it is very difficult to separate group defense of an area from individual defense of overlapping ranges where the residents tolerate each other." I disagree. Home range and territory have once again been confused. Solitary species can certainly be territorial, but no species defends any part of a home range. Second, group defense of a territory is clearly cooperative in truly social carnivorans such as gray wolves (Spotte 2012: 96–99). If in other species interlopers are tolerated, this is because the spaces they occupy are functioning as home ranges, not territories, in which case the behavior of two solitary individuals with overlapping borders is likely driven by active avoidance instead of aggression. This seems true of free-ranging male cats.

Domestic cats, similar to cheetahs (*Acinonyx jubatus*), sometimes straddle the line between solitude and interaction. Male cheetahs form companionships but females live alone (Caro and Collins 1987, Schaller 1972: 302). It seems clear from the literature that feral male cats are solitary, interacting with other males and with females only as required while foraging and in estrous groups. At no time and under no reasonable condition do they behave cooperatively. A male cat – whether feral, stray, or house – supplies sperm and otherwise makes no parental investment in his offspring (Chapters 4 and 5). He does not provision the female while she lactates, nor does he bring food to her or the young. He forages alone; when not breeding, the size of his home range is evidently dictated by food availability, but during the breeding season his focus also extends to finding estrous females. To locate them, his home range might expand to overlap with their home ranges (Izawa *et al.* 1982). At such times a male's home range is more likely to encroach on those of other males too (Mirmovitch 1995).

According to Avilés (2002: 14 269), *cooperation* comprises "any behaviors (joint resource acquisition, information exchange, communal brood care, predator defense, etc.) that despite any individual costs have a net beneficial effect on group members. ... " Is either group nursing and group "defense" of young, as occasionally reported for domestic cats, a valid example of sociality as identified by cooperation, or is it simply an incidental occurrence without an effect on group fitness? Conversely, could the increased risk of infectious disease (e.g. feline panleucopenia virus) transmission select against communal nursing of litters by reducing fitness (Macdonald *et al.* 1987: 53, Page *et al.* 1992)? Neither question has been put to the test.

In cooperative interactions a task ordinarily accomplished alone becomes more efficient if performed by the group. The effect overall is increased group fitness accompanied by reduced relative fitness of the cooperators (van Veelen *et al.* 2010). Powers *et al.* (2011: 1529) pointed out that viewed superficially, cooperation is still the preferred situation because by definition it raises absolute individual fitness. Any such advantage, however, can be misleading. As they further noted, "individual selection responds to relative and not absolute fitness, and selfish individuals by definition have a greater relative fitness than those cooperators that they exploit."

The point at which trends of cooperation and freeloading (see later) intersect depends on group size. Cooperation is beneficial in small groups, less so in large ones. According to van Veelen *et al.* (2010), how "large" is defined depends on factors like parameter values and the current extent of cooperativeness. Just as there exists an optimal group size for performing any given task, the addition of equally cooperative individuals has diminishing returns by reducing the size of each one's contribution to group success (van Veelen *et al.* 2010).

Animals cooperate by helping each other, behavior most often evident while rearing young. *Cooperative breeders* are those in which group members other than, or in addition to, the parents participate in care of the offspring (Clutton-Brock 2006, Port *et al.* 2011). Among vertebrates, most cooperative breeders are birds (Port *et al.* 2011). *Noncooperative breeders* differ by not providing *alloparental care* (i.e. *helping behavior*). Gregarious mammals are usually *group breeders*: parturition and rearing occur in close proximity with conspecifics, and alloparental care is not provided. In this sense they are noncooperating. As Clutton-Brock (2006: 173) wrote: "Group members may cooperate to defend resources against neighboring groups or to detect or deter predators, but direct alloparental care is limited or uncommon." Adult female cats will sometimes attack other cats that come too near their kittens and are reportedly aggressive toward strangers, mainly other females and young males (Kerby and Macdonald 1988). Adult males are generally intolerant of other males, usually avoiding each other and sometimes chasing younger males (Kerby and Macdonald 1988). They are also less likely to associate with a particular group.

Whether alone or associating in aggregations, cats fit the category of noncooperative breeders because usually only the mother provides parental care. Exceptions occur. Natoli (1985b) reported stray females cooperating to rear young, but did not define "cooperation" or provide data. Evidence of cooperative rearing in Japanese cats was also reported, in particular an instance of kittens being nursed by a cat that was not their mother. The exact nature of the cooperation was vague. Izawa and Ono (1986: 31) wrote, "Although the relationship between the two females was unknown, they cared for their kittens cooperatively and offered mutual help." The authors reported 19 other instances of cooperation during their observations, implying that most participating females were either mothers and daughters or littermates. In some cases the relationships were unknown. I saw this too in my own barn cats when two female littermates gave birth almost simultaneously in the same pile of hay. No sooner were the kittens mobile than they began cohabiting, and the mothers treated them all the same. Does this observation hold any relevance? I have no idea, although a literature search then and more recently failed to uncover even one report describing the effect of "cooperative rearing" on fitness in domestic cats. For the time being this behavior should be called "communal rearing," the term "cooperative" having a specific connotation in biology.

Cooperating conspecifics ordinarily live together during the time cooperation is taking place. Male cats, regardless of category or habitat, seem uncooperative in all respects. Male domestic dogs, which unlike male cats actively seek social contact with other dogs, are indifferent fathers too, and yet most investigators would agree that dogs, moreso than cats, are social. Aspects of the cat's life-style that preclude cooperation have already been mentioned; others will be noted in future chapters.

Conspicuously absent in cats is a large and finely tuned repertoire of signals expected of a social mammal (de Boer 1977b), nor does the cat's general behavior indicate sociality. Members of both sexes hunt alone, not in pairs or groups (Turner and Mertens 1986), and certainly not cooperatively. Adults rarely occupy a common shelter site (Brothers et al. 1985, Norbury et al. 1998b), and their spacing even at high densities, whether grouped around a garbage bin or the coffee table, usually seems arranged to minimize contact. Wolves and domestic dogs, in contrast, display an impressive array of nuanced postures, vocalizations, and facial expressions used to convey affectation. Putative evidence of territoriality by either sex is unconvincing (Chapter 2), and although free-ranging cats share space, they make an effort to avoid encounters.

Sociality has its evolutionary basis at the dynamic, synchronous threshold where group formation and cooperation come together. As Avilés (2002: 14 269) stated, "Once within groups, individuals help one another as a function of their cooperative tendencies." In the case of cats we must ask, what cooperative tendencies? Under certain circumstances (e.g. abundant and clumped food) solitary animals can form loose gatherings having few or none of sociality's trappings. If feral cats wandering across a landscape, strays lurking around a garbage bin, or several cats kept together in a laboratory cage indeed compose a social group, there ought to be evidence of them cooperating in ways that enhance fitness.

Some individuals behave selfishly even among group-forming cooperators. *Cheaters* tend to cooperate less than the average individual while still benefiting from group living (Dugatkin et al. 2003); *freeloaders* have lower cooperative tendencies but are more social; that is, they are inclined to join groups more than average. Their presence lowers potential group fitness. Freeloading has a fitness advantage when cooperation is costly (e.g. risk of injury), but this is countered by a drop in productivity by the supporting group. In gravitating toward groups, freeloaders can cause any group they join to be larger than optimal size, which lowers group fitness too. Freeloaders are more likely to be associated with larger groups, but cheaters tend to occupy groups of all sizes in equal numbers. They lower group productivity by reducing the per capita growth rate, not by making it oversized. Being either a cheater or freeloader requires membership in a group that meets the requirements of cooperation. Typically, cats in groups are either laboratory cats, pets, or strays. In all cases adequate food is usually available, making cheating and freeloading superfluous (i.e. neither beneficial to the individual nor harmful to the group). A feral cat fends for itself, not relying on humans to supply its food and shelter. So far as we know its behavior is entirely selfish. In all these situations cooperation would not be beneficial.

3.5 The kinship dilemma

Theory requires some elements of fitness to increase with group formation, but what would these look like in cats, and how would we recognize them? Kinship has been proposed as an influential component of sociality in domestic cats (e.g. Fagen 1978, Genovesi *et al.* 1995, Langham and Porter 1991, Liberg 1984a). As mentioned, females occasionally nurse one another's offspring (Deag *et al.* 1988; Ewer 1959, 1961; Macdonald 1981; Macdonald *et al.* 1987: 52–53) and supposedly engage in the mutual "defense" of kittens against infanticidal males (Macdonald *et al.* 1987: 53–54, 56). Usually kinship is invoked when explaining such behavior, minus the requisite hypothesis testing necessary to render it believable, such as controlling for familiarity and cross-fostering. For example, no evidence exists that birds can identify kinship among chicks in a brood using genetic cues (Cornwallis *et al.* 2010). That they demonstrate the capacity to distinguish kinship between broods indirectly using prior attributes (e.g. vocalizations) is not exactly kinship discrimination; rather, it is recognition of differences within categories of prior attributes (Chapter 1). Allogrooming among adults is cited as another example of cooperative behavior in cats and, by extension, sociality based on kinship. Neither group nursing nor group "defense" is consistently identifiable – that is, diagnostic – of domestic cats and thus predictable as a species trait, nor has either been linked conclusively with kinship while controlling for obvious confounding variables like familiarity, feedback based on ontogenetic state of the young, and hormonal status and consequent motivation of the mothers. Both behaviors could just as easily represent individual interests. The fact that strays at the same farm are often related and have overlapping home ranges (e.g. Laundré 1977, Macdonald and Apps 1978, Liberg 1980, Panaman 1981) is scarcely evidence of a kinship effect.

Matrilineal groups and kinship effects have been used to partly explain aggregations of strays (Kerby and Macdonald 1988, Macdonald *et al.* 1987), minus direct evidence of altruism in the hamiltonian sense. These proposed associations have been based entirely on observations instead of attempts to falsify hypotheses, which require the manipulation of test subjects and test conditions. Until demonstrated within a tight experimental format, observations interpreted as shared defense of resources by related females (e.g. Liberg 1980, Yamane *et al.* 1994) are evidence neither of kin recognition nor of any presumed reproductive advantage. Yamane *et al.* (1994: 18) overstepped their evidence when claiming that "these defended areas pass to their female offspring. ... " and "group females employ strategies to 'defend' their food and breeding resources from non-related females in order to maximize their reproductive success." None of this was actually tested. Whether individuals of a group are related could just as easily be irrelevant. Cooperation can evolve in a purely selfish context, and relatedness among group members is not always a requirement (van Veelen *et al.* 2010). Finally, neither sociality nor cooperation is evidence of kin selection. As emphasized by Moore and Ali (1984: 108), "mutualism or simple individual competition is adequate for explaining the origin and function of many apparently altruistic, cooperative or nepostistic [sic] behaviours. ... "

Kinship has also been invoked to explain the exclusion of competitors from their owners' homes by house cats (Liberg 1984a), but this too disregards familiarity as a test variable. Liberg's (1984a: 284) statement that female cats "of the same kin group shared a communal home range" is simply a statement about relatedness, not a demonstration of any kinship effect. That alien females were avoided at home-range borders and treated aggressively near the household also does not qualify as confirmation.

Despite implied claims to the contrary (e.g. Denny *et al.* 2002, Liberg and Sandell 1988), the mere fact of relatedness and low female immigration at places of high population densities like waste-disposal sites are not proof of an organized social structure based on kinship. More parsimonious hypotheses might be natal philopatry or reluctance to abandon a reliable source of food. In wild chimpanzees, which are social and live in kinship groups, males are philopatric and females of breeding age emigrate. However, provisioning of one group by investigators disrupted this pattern, and 11 of 16 females stayed instead of abandoning reliable meals (references in Moore and Ali 1984).

Discounting simpler explanations, Liberg and Sandell (1988: 91) wrote that aggregations of stray cats around waste-disposal sites constitute "true social groups." As confirmation, and citing Liberg (1980), Natoli (1985b), and Turner and Mertens (1986), they pointed to reduced immigration of females (compared with males), higher than expected female kinship, and hostility of resident females toward nonresidents. Their assessment remains speculative without a concise, falsifiable definition of "social group," evidence that kinship groups in stray cats enhance fitness, and assessments of how food availability affects philopatry.

Two additional terms need discussing. My premise has been that domestic cats are solitary, not social. Therefore, a group of them is not a society and certainly not a "colony" as the term is commonly applied (e.g. Kerby and Macdonald 1988; Kienzle *et al.* 2006; Mendes-de-Almeida *et al.* 2004, 2011; Natoli 1985b; Natoli and De Vito 1991; Nutter *et al.* 2004; Rees 1981; Robertson 2008; Robinson and Cox 1970; Rosenstein and Berman 1973) but an aggregation, which I have not defined to this point. For purposes here, an *aggregation* is a group of conspecifics gathered to feed or mate and having no stable social order, its individuals not participating in cooperative functions that establish and enhance individual or group fitness. The lone exception is the *family unit* comprising a mother cat and her kittens, which according to Jones (1989: 7) "is the only long-term social group of feral Cats."

Free-ranging domestic cats are most social as kittens and as adult females interacting with their kittens, which is during lactation and shortly after weaning, although in Australian feral cats an occasional family unit (female and kittens) might persist up to 7 months (Jones 1989). Two other "social groups" given by Kitchener (1999: 236) – estrous groups (Chapter 4) and dispersing siblings – are simply opportunistic aggregations too. Keep in mind that an aggregated pattern is the same as a clumped pattern (Chapter 2), defined by $\sigma^2 > \bar{x}$ and indicating heterogeneity. *This is evidence of behavior having been selected for certain properties of the habitat, either favorable (where individuals aggregate) or unfavorable (where they are absent).* The property is purely descriptive and merely predicts the higher probability of individuals occurring at some locations than others. In terms of space, if the dispersion of food at a waste-disposal site is clumped, dispersion of the exploiting cats represents a clumped pattern too. The reason for the dump site is obvious: humans put it there. Why the cats are present should be just as obvious. I say this simply to illustrate the irrelevance of pattern without finer resolution of what occurs inside it.

As mentioned, groups of cats, whether living together by choice (strays) or not (laboratory cats), are often referred to inappropriately as "colonies." Used in biology, *colony* applies to *social groups* of individuals or *social units* like pairs or families. Brown and Orians (1970: 243) specifically excluded from qualifying as colonies any aggregations around clumped resources, retaining "(*a*) group defense against predators, (*b*) group defense of feeding areas, (*c*) ability to exploit a resource not readily captured by solitary individuals, and (*d*) the ability to profit from the foraging success of other individuals by observing where they find food." None of these represents a clear condition for feral, stray, or captive cats except possibly the last, when weaned kittens observe older cats hunting or scavenging. If the definition excludes gatherings around clumped food, then such aggregations are not colonies. Obviously, neither is a group of laboratory cats, which has no control over its situation, making misnomers of such designations as "colony cats," "colonies of cats," and "managed colonies."

Were the domestic cat really a social animal we could test the effect of competition balanced against altruism. Offspring that remain in the natal area instead of dispersing at sexual maturity potentially compete with their parents and erase indirect benefits of altruism (Le Galliard *et al.* 2005, West *et al.* 2002). For example, if urban strays indeed live in matrilineal groups and through kinship acquire some implied but as yet unidentified fitness advantage, then competition for limiting resources could eventually be assessed against the cost of cooperation. The implication (and the test hypothesis) is that matrilineal groups might not be sustainable except where food is present in excess.

What matters ultimately is the net effect (West *et al.* 2002). At some point it becomes advantageous to the parents if their offspring disperse instead of staying home. In practical terms, although dispersal is possible, staying around offers more advantages, at least for the moment. Consider a general model involving three groups of urban strays living near three close but isolated garbage dumps. Assume the females have more restricted ranges than the males and that some of their female offspring remain in the natal area for life, maturing and reproducing. Although food is unlimited, breeding sites are not, eventually causing some females to disperse to the other dumps or perhaps farther. Meanwhile, even if related females behave altruistically, the competition among those left behind eventually cancels any benefits of cooperation (Taylor 1992).

In one sense the tenuous behaviors sometimes interpreted as social seem connected to habitat saturation, which has long been considered a crucial component in the evolution of sociality, influencing such factors as delayed dispersal and skewed reproduction (Kokko and Lundberg 2001, Reeve *et al.* 1998). As explained by Le Galliard *et al.* (2005: 221), "The general view is that habitat saturation drives the joint evolution of philopatry and altruism. ... " Classically, philopatry followed by cooperation evolve sequentially. Territorial species are affected most strongly because dispersal is made difficult by adjacent territory holders. With restricted mobility comes forced crowding. To control agonism, competition between individuals is subsumed by tolerance. In cats, and urban strays especially, dispersal is unobstructed, yet the urge not to abandon the resources at hand is stronger than enforced togetherness, which requires a certain degree of getting along. By *not* competing aggressively for resources (food, shelter, breeding privileges), a free-ranging cat exists inside an isolated bubble, solitary and selfish, surrounded by equally isolated peers behaving similarly.

The mating system of a species (Chapter 4) helps predict whether its members are likely to cooperate in a group setting. Results of a phylogenetic analysis of 267 species of birds by Cornwallis *et al.* (2010: 969) showed promiscuity to be three times greater in noncooperators, and that promiscuity is a "unifying feature across taxa in explaining transitions to and from cooperative societies." For the most part, promiscuity and cooperation have moved in opposite evolutionary directions, which have culminated in promiscuity reducing relatedness through multiple paternity and disrupting selection for cooperation among relatives. In other situations (e.g. when a species is both promiscuous and cooperative) a heavier ecological cost is imposed for kin selection to sustain cooperation, making altruism, as evident by the presence of helpers, less frequent (Cornwallis et al. 2010). We know that domestic cats mate promiscuously (Chapter 4), reducing the likelihood that cooperative behavior, including altruism, is part of their evolutionary history. Evidence is limited, but what there is suggests the absence of altruistic behavior in domestic cats.

In contrast, offspring of altruistic animals that do not emigrate often delay or forego reproduction to help rear the next generation of their siblings. These species are seldom promiscuous. If the mother is monogamous, her older daughter or son shares half its genetic material with younger brothers and sisters, and in outbred populations staying to help raise them instead of dispersing to reproduce is the genetic equivalent of breeding (i.e. its degree of relatedness to its siblings and to its own potential offspring is the same (Fig. 3.2). The cooperative rearing of young is clearly a social endeavor. An important component influencing transition from solitary to social is therefore explainable in terms of how many males a female mates with. Cornwallis et al. (2010) called this the *monogamy hypothesis*. Their findings predicted high promiscuity favoring loss of cooperative breeding and undermining any general belief that group-living cats behave altruistically by relinquishing individual fitness to enhance the fitness of others.

Do domestic cats ever cooperate? Females sometimes give hints of it, but whether their behavior can be called altruistic is doubtful. Macdonald and Apps (1978) observed four group-living farm cats comprising an adult male and female and their

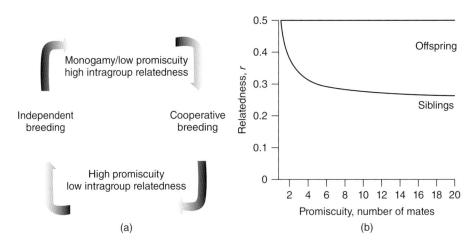

Fig. 3.2 Monogamy hypothesis. (a) Monogamy (including low levels of promiscuity) leads to high relatedness in family groups and tending toward transition to cooperative societies; that is, to an increase in the relatedness term, r, of Hamilton's rule (Hamilton 1964a, 1964b), which posits that cooperation is favored if $rb > c$ where b is the benefit in reproductive success to the relative being helped and c is the cost to the benefactor performing the cooperative behavior. Intense promiscuity leads to low relatedness in family groups, favoring loss of cooperative breeding. (b) Promiscuity and relatedness. Female promiscuity (number of mates) plotted against mean genetic relatedness between potential helpers and either their siblings or offspring. A parent is always related to its offspring at $r = 0.5$. In contrast, as the number of males the mother mates with increases, an offspring's relatedness to siblings decreases from $r = 0.50$ to $r = 0.25$ (from full-siblings to half-siblings). Source: Cornwallis et al. 2010. Reproduced with permission of Nature Publishing Group.

two daughters. The male impregnated all three females. The sisters nursed and groomed the kittens indiscriminately. When all kittens except one of Smudge's (the oldest female) died, one of Smudge's daughters (Pickle) groomed her mother and assisted with grooming her mother's surviving kitten (and Pickle's full-sibling) from the current year's litter. She also took prey to Smudge (a vole, a shrew, and a house mouse), but sometimes kept it for herself, growling if Smudge came near. Macdonald and Apps (1978: 260) wrote, "The willingness of Pickle to share food was apparently signalled by a 'churring' call, similar to that made by Smudge when bringing in food for her kitten." These investigators also mentioned cooperative "guarding" of the kittens but provided no data. Sometimes the adults slept lying against each other.

Because the three females were related and nursed kittens communally, Macdonald (1981) proposed inclusive fitness or reciprocal altruism as possible explanations. However, reciprocal altruism purports to explain helping behavior among unrelated individuals (Trivers 1971), not those related. An example is a report of Leyhausen (1979: 88) of an incident in which unrelated adult females, including a European wildcat, delivered both dead and live prey to the kittens of another captive female. As to inclusive fitness, was Pickle's delivery of prey to Smudge actually helping behavior; that is, was it truly altruistic? Probably not. There was no hint of a dominant–subordinate relationship between mother and daughter, which typifies mammalian species in which helping is common (Clutton-Brock 2009), nor had the mother suppressed her daughter's opportunity to breed, which is also common in social species. Mother cats are most likely to deliver prey when their kittens are 4–6 weeks old (Chapter 5), but rarely to other adults. These events seem to require close timing of the kittens' ontogenetic development with the mother's post-weaning hormonal status (Caro 1980a). Pickle did not nurse her mother's kitten, and grooming it could be attributed to maternal drives not yet extinguished after her own offspring died. In other words, in her present hormonal state she might have licked any kitten, related or not. The maternal urge is strong, and lactating cats can be induced to nurse even pup dogs (Fox 1969). Cooper (1944) mentioned a mother cat that after having lost her neonates was seen pacing, calling, and subsequently trying to "retrieve" adult cage mates.

The "churring" call Pickle made when delivering prey resembles the maternal behavior of mother cats toward their kittens, as mentioned, or in this case displaced maternal behavior directed by Pickle toward her own mother, Smudge. The prey was more likely intended for the kitten and not Smudge, whether or not the kitten was developed sufficiently to accept it (the kitten's age was not given) and might be why Pickle, when carrying prey, sometimes growled at Smudge. Macdonald's comments are thus conjectural, not evidence of any kinship effect. The pattern persisted for ~3 d. Perhaps the best reason to discount altruism on Pickle's part was that she neither delayed nor abandoned her own reproductive opportunities to help rear Smudge's offspring.

Altruism is not a phenomenon we should casually label as consistently advantageous. Overlooked is the fact that most competition is local. To quote Queller (1994: 72), "relatedness is not just a statement about the genetic similarity of two individuals, it is also a statement about who their competitors are." Investigators have seldom considered competition for resources (e.g. food, shelter) when mentioning presumed kinship benefits of group living in free-ranging cats.

In one group of strays, kittens were initiators of social contact to both males and females, females initiated contact with males and other females, and males seldom sought interaction with other cats (Kerby and Macdonald 1988). As mentioned, males of all categories behave selfishly, but female strays sometimes show strong philopatry to natal sites, resulting in mainly local dispersal and putative clustering of matrilineal groups. However, because patterns can be local they do not necessarily represent how domestic cats elsewhere behave, which also means they are not diagnostic of the species. For example, female strays at Lyon dispersed at 1–2 y while males the same age stayed behind (Devillard et al. 2003).

Relationships can be friendly among adult females and their independent daughters (Macdonald and Apps 1978, Natoli 1985a). In these circumstances, negative forces pull against altruistic behavior, making its evolution problematic. Populations showing "viscous," or slow and limited dispersal rates, as female urban strays often do, are confronted by population viscosity's two offsetting effects: increased encounters among relatives, which occur at a cost of heightened competition for resources (Taylor 1992). The first effect (increased encounters) promotes altruism, the second (heightened competition) opposes it. Stasis should result if they balance exactly, imposing local limits on the evolution of altruistic behavior. For kin selection to evolve, relatives must stay together instead of dispersing; alternatively, their dispersal rate must be slow enough to conserve kinship effects.

Any selective advantage of a so-called "altruism gene" (call it p) depends on how often it bestows disproportionate benefits on other copies of itself instead of on copies of non-altruistic alleles. Population viscosity thus promotes clustering of p through relatedness, increasing the likelihood of p's presence in other conspecifics chosen randomly from the neighborhood. Conversely, population viscosity selects against altruism by heightening competition among offspring of parents carrying p. When females cooperate and dispersal of their offspring is both local and slow, p's effective threshold is the point at which any benefit of altruistic behavior equals the cost of expressing it (Taylor 1992). Theory shows this level to be obtainable (Taylor 1992, Wilson et al. 1992) regardless of whether population dispersion is random, clumped, or uniform (Chapter 2). Taylor's model demonstrated that benefits of increased relatedness resulting from limited dispersal through population viscosity are cancelled by the cost of more competition among kin. Dispersal rate therefore has no effect on gene selection for altruism.

Relatedness is a statistical concept (West et al. 2002). Puzzling though it seems, high relatedness is not necessarily characteristic of close relatives (Queller 1994). To use the example of West et al. (2002), it comes down to the scale (base population) on which r (the coefficient of relatedness) is measured. If the genetic markers used are neutral (e.g. microsatellites), the result is essentially a measure of pedigree (Bednekoff 1997), or ancestral lineage. Two full siblings have a relatedness coefficient of 0.5 if determined against a large outbred population. Measured against only other siblings, the frequency will be the same for all individuals: $(p_y - \bar{p})$, and $r = 0$. Assuming the existence of an altruism allele, $r = \Sigma(p_y - \bar{p})/\Sigma(p_x - \bar{p})$ where $\bar{p}$ = the population frequency of the altruism gene, p_x = the frequency of the allele in all benefactors of altruism, and p_y = the allele's frequency in all beneficiaries of altruism.

3.6 What it takes to be social

From descriptions provided by Clutton-Brock and Lukas (2012) the domestic cat fulfills few of sociality's requirements. Although female strays that are philopatric often associate with female relatives, they do not form alliances. Not all aggregations of female strays are stable, consist exclusively of matrilineal relatives, or commonly display evidence of cooperation. Relatedness among adult females during "communal rearing" is not required (Leyhausen 1979: 88), and females of some aggregations of strays fail even to show attachment to place, being more likely to disperse than the males (Devillard *et al.* 2003). Female strays sometimes rear young together in the same location, but to call this "cooperation" (e.g. Izawa and Ono 1986) suggests true sociality, which involves a species-diagnostic system of fitness-enhancing behaviors, a category into which the domestic cat is a poor fit.

The domestic cat is not handicapped by low habitat connectivity even in difficult terrain and unfavorable climates, nor does it pay a high cost in mobility, factors that in combination select for asociality. Limited mobility, dependency on conspecifics after weaning, and altruism are not part of its evolutionary heritage. To survive in depauperate landscapes, the cat expands its home range as necessary or migrates. It seems content living in solitude except for brief interludes to mate or rear offspring.

Kerby and Macdonald (1988: 69) defined a social group as "a number of adults whose interactions are generally tolerant and whose home ranges overlap more than would be expected by chance." This describes in vague terms (e.g. "generally tolerant") something about behavior and use of space but lacks the evolutionary component of a falsifiable fitness benefit. Later the authors seemed to contradict themselves. Noting that if a group of cats comprised only an aggregation, some interaction would still be expected, and rightly concluding (p. 73) that "observation of interactions is evidence neither of social structure nor of active sociality." Two pages later (p. 75), however, is the statement, "From these analyses it is clear that farm cat colonies are highly structured in terms of the nature of individual relations." Putative confirmation was nonrandom spacing with the implication that some individuals preferred and sought the company of others, behavior describing a possible expression of preferences but not evidence of social structure.

It seems that the importance of anticipating food is underestimated during assessments of sociality. Macdonald *et al.* (1987: 50), remarking on the proximity of barn cats to each other being greater than expected by chance, wrote, "The frequency of interaction was greatly in excess of that prompted by clumped resources (feeding in the observation area took only a few minutes each day, whereas the cats interacted throughout the day. ... " This is not in the least unusual and is actually predictable. Habituation to place when reinforced by food – especially if reinforcement is intermittent – can be rapid and persistent in many species. Perhaps the mistake is in thinking our sense of time and theirs is the same. Such gatherings jolt animals from their natural patterns. As more individuals appear and take up room in a finite space, their interactions inevitably increase.

During my early years as a public aquarium curator I sometimes joined the crowds viewing the exhibits. It was a way of assessing their effectiveness from the public's perspective. Among the largest displays was an artificial coral reef containing hundreds of fishes comprising dozens of species. Some species were gregarious and schooled, but others were solitary by nature, even cryptic. One morning I noticed people peering into the three large glass panels hoping to see life where none was apparent. Where had all the fishes gone? The back wall was darkened purposely to focus visitor attention to the front, and every specimen – gregarious and solitary alike – had joined a large group milling around in its shadow as if constrained there. Not one was actually on display. On investigating I was told the aquarists had recently started feeding from the back wall because of convenience. Feedings took place once daily and lasted ~2 min, but there the fishes stayed around the clock in anticipation. They had become conditioned, habituated to that single location. We immediately began feeding from the front of the exhibit, scattering food randomly over the surface so it would enter the water at unpredictable places, but always within the viewing area. Within a day or so all specimens were clearly visible and behaving normally.

To pass the sociality test ought to require more rigorous standards than correlation, including discernible and consistently measurable cooperation and true group formation, not simply local dispersion that differs significantly from a Poisson distribution. I can accept that some behaviors are subtle and infrequent, but a mean rubbing interaction between adult female barn cats of once per 25.3 h of observation (Macdonald *et al.* 1987: 5) seems weak proof of a strong and highly structured social system.

What population structure does the domestic cat possess that promotes and supports cooperative behavior? For an animal to become social, selfish behavior needs to diminish in frequency and be substituted by cooperative behavior (Powers *et al.* 2011). In the case of strays and house cats this must involve more than sharing food that none has expended any effort to obtain. Food scavenged from trash bins or provided by cat lovers falls outside the category of *diffusible goods,* which are items produced or procured by an individual and made available for public consumption (Driscoll and Pepper 2010). These can raise the fitness of all participants but at a cost to the producer. When cats share food provided *for* them, none bears any cost other than the human who tossed his garbage into the bin or bought the bag of cat food. The situation is different when cats hunt, and then food is seldom shared among adults. Fitzgerald and Karl (1986) cited cases in which females scavenged carcasses of rabbits killed by a large male after the male had finished feeding; when the females managed to kill rabbits on their own, the male sometimes confiscated them. In contrast, Macdonald *et al.* (1987: 18) observed cats sharing cached carcasses of large prey (e.g. rabbits, squirrels, pigeons), sometimes feeding simultaneously

Lack of sharing among cats seems to be the rule. As Alexander (1974: 333) wrote: "If food is abundant there is little gain in being able to count on others sharing small finds; if it is scarce there is little gain in sharing small finds with others." Both situations promote individual, not group, behavior. Alexander doubted whether complex social structures could evolve mainly or solely from feeding advantages: "It seems more likely that the feeding behavior, whether competitive or cooperative, is a result of grouping that was originally advantageous for other reasons."

Evidence of cooperation has been proposed in social play (West 1974), resting in contact or close proximity (Izawa and Ono 1986), food sharing (Chesler 1969), distribution of prey to kittens of different mothers (Fox 1975), and, as discussed in Chapter 6, emulative learning (Chesler 1969, John *et al.* 1968). Breeding females "guarding" each other's kittens (e.g. Macdonald and Apps 1978, Macdonald 1983) has been documented only superficially and is not commonly reported. Observations obtained without experimental controls in place cast doubt on presumed fitness components based on unmanipulated interactions between individuals, leaving open the possibility that communal nursing or defense of young might not even be mutualistic, much less altruistic.

The perceived general link between kinship and altruism has tightened considerably in recent years. It once seemed as if all societies could be explained in terms of kin selection, but support eroded with accumulating observations attributed to reciprocal altruism (Clutton-Brock 2006). In addition, it appears as though indirect benefits of kin assistance had been inadvertently inflated. According to Clutton-Brock (2006: 190), these "have sometimes incorporated the effects of helping on direct descendants … as well [as] on collateral kin." The summing of benefits received from kin and those conferred leads to a "double accounting" of advantages derived from kin selection.

Altruism carries a fitness cost; mutualism costs nothing. That "nursing groups" of free-ranging cats are composed solely of relatives is untested. If the individuals are unrelated then helpers are not owed nor eligible for inclusive fitness benefits; in other words, neither mother gains anything from helping the other. Whether related or not, the group exerts no evident pressure on individuals to join, and helpers might or might not be associated through a mechanism other than kinship (e.g. familiarity, similar hormonal status). Cooperation is unenforced in the absence of any real social ties, and even staying is optional: individuals come and go freely, sometimes not returning.

Stephens (1996) listed six conditions for testing reciprocal altruism: (1) the behavior must reduce the donor's fitness relative to a selfish alternative, (2) fitness of the recipient must be raised relative to conspecifics not receiving the benefit, (3) performance of the helping behavior must not depend on receiving an immediate benefit, (4) the first three conditions must apply to both participants of the reciprocal exchange, (5) a mechanism must be in place to detect cheaters, and (6) a large (indefinite) number of opportunities of reciprocity must exist.

The first two conditions make the behavior altruistic, and the third separates reciprocal altruism from *mutualism,* defined in this context as a situation in which the donor is altruistic only if the recipient provides a return benefit without delay (i.e. almost instantaneously). The fourth condition makes the altruistic act reciprocal. The fifth is necessary for altruists to punish conspecifics that fail to cooperate; without it cheaters would always take advantage of altruists, preventing reciprocal altruism from evolving. Consciousness is obviously not a requirement of (5). The sixth condition blocks the "backward induction problem" in game theory (see Stephens 1996: 535, footnote 6).

Detriments of group living, on balance, outweigh any advantages, and sociality is not an unobstructed path to evolutionary righteousness. The benefits it bestows or takes away can be assessed and understood exclusively at the level of the individual, not the group. It is therefore pertinent to ask what advantages might accrue to group-living cats. Alexander (1974) listed three for animals generally: (1) reduced susceptibility to predation through group defense, (2) improved efficacy at finding scattered food, and (3) more efficient use of clumped resources (e.g. food, shelter sites, breeding sites). In the first two, individuals gain from the presence of others. As to (1), cats run when confronted by potential predators such as dogs and humans. Escape is a disorganized scramble, numbers offering no discernible advantage except possibly to confuse the predator. In (2) the assumption is of individuals using others as cover to hide from predators, behavior better exemplified by flocks of birds and schools of fish in which grouping tightens (i.e. individual distance among individuals decreases and those in the middle are presumably safest), not by aggregations of cats, which scatter instead of closing ranks either to hide among each other or to confront the enemy. Of (3) and referring to animals in general, Alexander (1974: 329) wrote, "individuals may aggregate around resources but are otherwise expected to avoid one another or to be aggressive, although they may use the presence of other individuals or aggregations as indicators of resource bonanzas." This certainly applies to stray cats, but such groups display no organizational behavior analogous to a troupe of baboons, a flock of starlings, or a school of herring. Alexander concluded that group living evolves because one of these three general factors, or a combination of them, raises the fitness of those individuals "accepting the automatic detriments of group living above the fitnesses of solitary individuals."

Fagen (1978: 276) called the sociality of cats "facultative," writing that it "may be explained by changes in the spatial distribution, quantity, quality and reliability of limiting resources, food and shelter. ... " He stated that such cats (those I categorize as strays) comprising related "social groups ... may endure for generations in farm buildings and in other human institutions, such as Buddhist temples (wats). ... " Fagen (1978: 278) pushed his case for sociality even further, making the claim that relationships inside these groups border on eusociality, citing "cooperation in caring for young; reproductive division of labor; overlap of at least two generations of life stages capable of contributing cooperatively to the society. ... " These requirements conform with standard definitions of what it means to be eusocial (Crespi and Yanega 1995), but they constitute poor descriptors of cats even if the meaning of eusocial is expanded to include all alloparenting species (see Sherman *et al.* 1995). Division of labor is doubtful among domestic cats, and even group nursing is an occasional event.

Macdonald (1983: 283) wrote: "There are contemporary carnivore societies such as those of ... farm cats in which stable, well defined relationships exist and yet where the benefits of social ties often appear infrequent or small." His statement seems closer to the truth. Macdonald also speculated that the origins of these putative societies might "lie in patterns of prey dispersion which diminish the need to disperse and hence permitted family ties to persist into adulthood ... [thus providing] for subsequent and varied selection directed at sociality *per se*." However, selection could just as easily lean toward the continuation of solitary behavior in a group setting, especially among generally tolerant, nonterritorial individuals like domestic cats. Assuming dispersal is about securing food predictably, then the availability of food in reliable clumps could very well dull any innate or environmental imperative (see Howard 1960) to abandon the natal site and strike out in search of less predictable resources elsewhere.

Moreover, use of the same resource does not require animals to be social. The *Resource Dispersion Hypothesis* (RDH) was conceived originally with territorial species in mind, although it also applies to those occupying home ranges and practicing active avoidance. Using the RDH, Carr and Macdonald (1986) demonstrated that the size of a home range or territory is constrained by dispersion of food; group size is limited independently by food abundance (i.e. richness of the resource). Their model showed groups of solitary animals forming without either benefit of sociality or simultaneous use of the same resource provided participants avoid each other. Common resource use then becomes possible in the absence of any of the disadvantages that sociality poses.

The extent to which free-ranging cats interact depends on habitat quality, which in turn determines food availability and consequently the size of home ranges. However, it should already be obvious that the aggregation itself is not evidence of sociality. When garbage is available or cats are fed by humans, the amount of space needed to survive shrinks, leaving individuals in frequent contact and leading to more encounters. What happens? They adapt. Cost weighed against benefit sometimes determines the extent to which an animal is willing to tolerate conspecifics. This seems obvious in the case of house cats and of strays concentrated around waste-disposal sites. Most smaller carnivorans that display group defense against predators are insectivores or omnivores, and food for these species is relatively abundant (Gittleman 1989). According to Waser

(1981: 234), because insects are found in quickly renewable patches "the cost of social tolerance [by an insectivore] is very small; a mongoose excluding a single competitor from its foraging range will gain only a 1% increase in prey density." To a mongoose the hedge against predation offered by heightened vigilance outweighs any slight benefit gained by defense of resources, and rapid renewal of prey would be a precondition of sociality. In most cases it costs a stray or house cat nothing to share resources; similarly, what a feral cat loses by sharing its home range with other cats is probably minimal except during droughts and other desperate situations when prey is scarce.

4 Reproduction

4.1 **Introduction**

Reproduction in free-ranging cats perhaps coincides with the spring breeding seasons of vertebrate prey with which they evolved, principally rodents, lagomorphs (rabbits and hares), and birds, and in some cases the time of increased activity of reptiles (Apps 1986b, Read and Bowen 2001). Rabbits, for instance, like cats, experience bimodal reproductive peaks during spring and autumn in both hemispheres, usually with a summer hiatus, and reduced reproduction in winter (Poole 1960).

The bimodal reproduction of some free-ranging cat populations might be timed with that of some rodents, although my comment is speculative. If so, it once represented an evolutionary juncture with the domestic cat's recent ancestor that is now conserved. Norway rats (*Rattus norvegicus*) in urban Baltimore reproduced throughout the year, but with maximum peaks in spring (March–June) and autumn (September–October). As shown by Davis (1951), this results in later population spikes of fast-growing juveniles of a size most favored by cats.

Here I outline the physiology and behavior of cat reproduction with emphasis on how environmental factors affect the reproductive cycle.

4.2 **Female reproductive biology**

Female cats mature sexually at 4–12 months, usually between 6 and 8 months (Nutter *et al.* 2004, Rosenblatt and Schneirla 1962), after which reproduction is influenced by body mass (minimum size ~2.3–2.5 kg), nutritional status, breed (long-haired breeds like Persians typically experience longer periods between reproductive cycles), and *photoperiod*, or the number of hours of daylight in every 24-h period (Beaver 1977, Bristol-Gould and Woodruff 2006, Concannon 1991).

Urban cats ordinarily start to breed on attaining physiological maturity, or at 6–9 months for females and 9–12 months for males (Natoli and De Vito 1991; Nutter *et al.* 2004; Say *et al.* 1999, 2001; Yamane 1998). Female feral cats might also breed at 6 months (Derenne 1976), but sometimes delay reproduction (Liberg 1980, Say *et al.* 1999) if food is scarce. Two sibling feral females in New Zealand not given supplementary feeding first reproduced at ages 2 and 3 y (Page *et al.* 1992). Feral females in an agricultural area of New Zealand's North Island matured at ~2.5 kg within 1 y, but generally did not reproduce until their second year (Langham and Porter 1991). In contrast, female strays at the Royal Navy Dockyard, Plymouth, where food was always abundant, reproduced at <1 y (Dards 1978). Occasional female strays in North Carolina that were fed routinely reproduced at 6 months (Nutter *et al.* 2004).

The domestic cat is a long-day, seasonally polyestrous, induced ovulator (Bristol-Gould and Woodruff 2006). In most free-ranging cats the female reproductive

---

Free-ranging Cats: Behavior, Ecology, Management, First Edition. Stephen Spotte.
© 2014 John Wiley & Sons, Ltd. Published 2014 by John Wiley & Sons, Ltd.
Companion Website: www.wiley.com/go/spotte/cats

cycle is initiated during early spring into early summer when hours of daylight begin to equal and then exceed hours of darkness (i.e. *long-day conditions*), although in some groups sexual behavior and reproduction occur throughout the year (Dards 1983, Nutter *et al.* 2004, Shimizu 2001, Wallace and Levy 2006). The quality of oocytes cultured *in vitro* from cats in the northern hemisphere becomes seasonally poor with encroaching short-day conditions (August–October), revealing a depressed incidence of maturation (Spindler and Wildt 1999). Even when the rate of maturation improves, as it sometimes does at other nonseasonal periods (November–January, May–July), the capacity of ova to develop into cleaved embryos is low.

Cats maintained in captive groups under natural light mimic the cycles of free-ranging cats (Scott and Lloyd-Jacob 1955, Tsutsui *et al.* 1990). Adult females of captive groups in the northern hemisphere kept under natural light mated from January–March and again in May or June, producing two litters annually (Scott and Lloyd-Jacob 1955). A group kept under artificial long-day conditions displayed a similar bimodal pattern, reproductive activity peaking in early spring and again in late summer (Robinson and Cox 1970), but in such situations litters can be born during any month (Dawson 1952, Robinson and Cox 1970, Scott and Lloyd-Jacob 1959), as depicted (Fig. 4.1). How long estrus lasts is nonetheless affected by season even if cats are maintained under constant long-day conditions. Females under equal hours of light (L) and dark (D), or 12 h in every diel cycle (12L:12D) had comparable numbers of estrous phases throughout the year, but the mean duration of estrus was shorter in June, September, October, and November than in March, April, May, August, and December, and intervals between estrous phases (6–120 d) varied widely among individual cats (Wildt *et al.* 1978). Furthermore, length of estrus can vary by individual and even in the same individual over sequential reproductive cycles

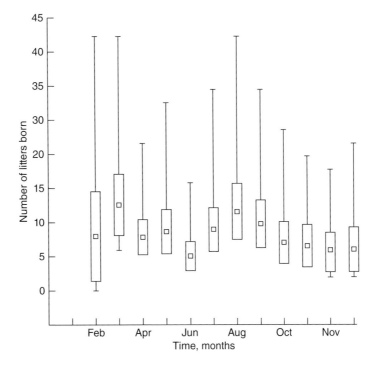

Fig. 4.1 Box and whisker plot of litters born over 10 y in captive cats maintained under continuous long-day conditions (14L:10D). $\bar{x}$ (small squares), $\bar{x} \pm$ SD (large rectangles), $\bar{x} \pm$ 1.96 SD (whiskers) of the minimum and maximum number of litters. Source: Data from Robinson and Cox 1970, table 4, p. 108.

(Dawson 1952). These months are reversed in the southern hemisphere (e.g. Jones and Coman 1982a). Feral cats on subantarctic Marion Island (now free of cats) had litters from September–March with a peak in October (van Aarde 1984); those in central New South Wales also gave birth between September and March (Molsher 2001).

Unlike the dog, which is monestrus, the polyestrous cat proceeds through "waves" of successive estrous phases during the breeding season. Also unlike the dog, which ovulates spontaneously, cats ordinarily require coital stimulation to initiate ovulation (Greulich 1934, Longley 1911, Paape et al. 1975). Required induction by coitus has been interpreted as a conserved evolutionary trait assuring that ovulation occurs only if mating (although not necessarily fertilization) has been successful (Goodrowe et al. 1989).

The female cat's reproductive cycle comprises several phases, which the literature describes using terminology not yet standardized. Concannon (1991: 536), for example, wrote: "The interestrous periods have also been termed metestrus or anestrus in nonovulatory cycles, metestrus or diestrus in nonovulatory cycles or ovulatory cycles, or nonestrus. … " Here I partition the cycle into proestrus (follicular phase), estrus (reproductive phase), interestrus (intraseasonal intervals between estrous phases), diestrus (luteal phase during pregnancy or pseudopregnancy), and anestrus (nonreproductive, or reproductively "quiescent," phase). Wildt et al. (1998) published a brief overview of the domestic cat's reproductive biology and their Table 1, p. 505, summarizes the reproductive characteristics of 40 species of felids. Part of the confusion about use of the terms just mentioned arises from the polyestrous nature of cats: if the female fails to mate she cycles back to proestrus, bypassing later stages and assuring another mating opportunity before the breeding season ends. Polyestrus presents a serious obstacle in the control of free-ranging cats, and a thorough understanding of its biology is beneficial before initiating programs to control populations of feral and stray cats (Chapter 10).

The dog's proestrus is marked by a bloody vaginal discharge and swollen vulva that feels warm to the touch; hence the term "in heat." Visible changes in the external genitalia of adult female cats are much less pronounced. The shifting appearances of vaginal cells are well described (Dawson 1952, Foster and Hisaw 1935, Herron and Sis 1974, Scott and Lloyd-Jacob 1955). In 340 mating tests, Michael (1958) found vaginal smears to correlate well with behavioral patterns of reproduction. Stabenfeldt and Shille (1977), however, considered them less useful indicators because ovulation in the cat is induced instead of spontaneous.

Vaginal smears of anestrous cats yield mainly small nucleated epithelial cells that vary in size and shape. During proestrus the cells increase in number and appear flattened. Few epithelials remain in smears at estrus, and these are large and without nuclei. Cornification proceeds until day 3 or so after coitus, at which time the vaginal epithelium is invaded by leucocytes, signaling impending diestrus. Epithelial cells that are nucleated with large vacuoles are diagnostic of early pregnancy. Smears obtained after implantation start to gradually resemble those seen at proestrus.

Proestrus precedes estrus, and in a normal reproductive cycle lasts 1–3 d (Bristol-Gould and Woodruff 2006, Concannon et al. 1980, Michael 1961, Stabenfeldt and Shille 1977). A proestrous cat attracts the attention of males, which gather around her, following if she walks away, even chasing her if she flees (Dards 1983). Behavioral expressions of proestrus are usually subtle, nothing more apparent than increased affection, rubbing the head, face, and neck against objects, purring, vocalizing, or gaining the attention of adult males but not letting them mount (Concannon 1991, Feldman and Nelson 1996, Michael 1961, Shille et al. 1979). Sometimes elements of estrous behavior (described later) are expressed too (Scott and Lloyd-Jacob 1955).

Proestrus is the time of rapid maturation of the follicles, or *folliculogenesis*, driven by endocrine signaling along the hypothalamic–pituitary–gonadal axis (Bristol and Woodruff 2004). Within this brief 3-d period the pituitary gland releases follicle-stimulating hormone (FSH), which induces folliculogenesis and primes the reproductive system for impending estrus. During folliculogenesis, somatic cells envelope an immature oocyte. An ovarian follicle is composed of several cellular types, but at its center is the oocyte surrounded by granulosa cells. The smallest follicles, consisting mainly of oocytes, are 20–30 µm diameter (Bristol-Gould and Woodruff 2006). Once folliculogenesis is underway, a few larger follicles (ordinarily 3–7) double in size to 2-mm diameter (Foster and Hisaw 1935). Corpora lutea (CL) form eventually from these follicles, reaching diameters of 2.7–3.6 mm (Wildt et al. 1981). With their enlargement, the rest become atretic, degenerating and being resorbed before they can mature (Feldman and Nelson 1996, Liche 1939, Wildt et al. 1981). Domestic cats routinely generate many follicles, and nearly 65% undergo atresia (Wood et al. 1997).

Along with maturation and enlargement of the follicles is a rise in the blood (i.e. circulating) concentration of estradiol (typically determined analytically as estradiol-17β), a major form of estrogen secreted by the ovaries. Circulating estradiol quadruples from 5–20 pg/mL, spiking in the range 40–100 pg/mL over the next few days (Schmidt et al. 1983, Shille et al. 1979, Verhage et al. 1976, Wildt et al. 1981). Keeping the concentration elevated primes the hypothalamic–pituitary–gonadal axis for the subsequent spike in luteinizing hormone (LH) induced by mating (Banks and Stabenfeldt 1982), and correlation between estradiol (its concentration or persistence) and amplitude of the LH response is direct (Wildt et al. 1981). In addition to its other functions (see later), LH induces the follicles to produce testosterone without which they could not generate estrogens (YoungLai et al. 1976). Male hormones have another effect. Administering testosterone to adult females initiates estrous behavior, although subsequent mating fails to stimulate ovulation even when small follicles are present (Green et al. 1957).

The height of follicular activity coincides with size of the surviving follicles (now ~2.5–3.5 mm diameter), the spiking of estradiol in the blood, and cornification of the vaginal cells (Feldman and Nelson 1996, Scott 1955, Shille et al. 1979). If a female does not mate and ovulate, or mates without ovulating, her follicles regress in 6 d or so (Scott and Lloyd-Jacob 1955). Afterward, she promptly returns to polyestrus; that is, cycles back to proestrus and then into estrus once again (Bristol-Gould and Woodruff 2006).

The rise of estradiol during proestrus triggers behavioral estrus (Shille et al. 1979), which lasts 2–10 d, or an average of ~7 d (Concannon 1991). Scott and Lloyd-Jacob (1955) bracketed the estrous phase at 4–6 d using the results of vaginal smears and behavioral changes in combination. Banks and Stabenfeldt (1982) stated 2–8 d ($\bar{x}$ = 5.4 d). Schmidt et al. (1983) gave the mean length as 5.4 d (range 3–8 d). Neither a cat's size nor its age is relevant: duration of estrus does not correlate positively with either factor (Ishida et al. 2001). An estrous cat's behaviors might include affection, crouching, lordosis, rolling, making stepping movements (i.e. "treading"), extending the pelvis, deflecting the tail laterally, and vocalizing (Section 4.5).

The estrous phase can be extended to 6–10 d if the female has no contact with a male (Liche 1939, Scott and Lloyd-Jacob 1955). Female cats are unusual in staying sexually receptive for a time beyond ovulation under conditions of rapidly rising levels of circulating progesterone (Paape et al. 1975) and despite the follicles having developed into functioning CL (Banks and Stabenfeldt 1982). Multiple copulations, which promote ovulation (see later), do not affect the length of sexual receptivity when they occur early in the estrus phase (Wildt et al. 1981). According to Wildt et al. (1981), the average length of estrus in ovulating cats (5.8 d) is similar to that in unmated individuals (6.4 d), and so is the number of follicles (5.0 vs. 5.2). Mating and ovulation early in estrus occasionally shorten this phase by 1–2 d (Goodrowe et al. 1989; Shille et al. 1979, 1983).

Cats ovulate during estrus in response to the pituitary's release of LH into the blood, a detectable rise being evident within 5 min of coitus (Johnson and Gay 1981b). Maximum concentration is attained after ~4 h, culminating in ovulation 25–50 h following coitus (Dawson and Friedgood 1940, Goodrowe et al. 1989, Greulich 1934, Longley 1911, Paape et al. 1975, Stabenfeldt and Shille 1977). Ovulation later than 52 h has also been reported (Wildt et al. 1981). This broad range in timing has been attributed to initiation of sexual receptivity not always being synchronized with pre-ovulatory conditions; that is, mating might occur before the follicular response can be induced by a requisite surge in gonadotropins (Greulich 1934, Paape et al. 1975, Stabenfeldt and Shille 1977).

The concentration of LH is low (<10 ng/mL) until coitus triggers its release (Goodrowe et al. 1989), becoming elevated only in ovulating females (Wildt et al. 1980). An estrous female permitted unlimited access to males copulates several times daily for 3–4 d in succession (Scott and Lloyd-Jacob 1955). Green et al. (1957) reported four intromissions by the same male within 3 min, and 15 in 30 min. Females monitored in a laboratory 2.5–3.0 d into estrus mated 14–20 times during the first 12 h and 30–36 times within the first 36 h (Concannon et al. 1989). Intromission in each instance was brief, lasting 1–27 s. Whalen (1963a) recorded intromissions lasting 2–43 s. These data are consistent with observations of free-ranging urban strays (Ishida et al. 2001). Under laboratory conditions, excited males of an *estrous group* (several males pursuing the same estrous female) sometimes attempt to mount each other or mate with the female simultaneously (Green et al. 1957).

A graphic summary shows how duration of estrus, copulatory stimulus, and changes in associated hormones interact (Fig. 4.2). The number of copulations during estrus and the length of time between them modulates the amplitude and duration of the LH surge (Banks and Stabenfeldt 1982; Concannon et al. 1980; Johnson and Gay 1981b; Wildt et al. 1980, 1981). When a female mates multiple times at intervals of <2 h, LH spikes rapidly and returns to baseline within 12–24 h (Concannon et al. 1980, Glover et al. 1985, Johnson and Gay 1981b). As illustrated (Fig.4.3b), laboratory cats restricted to mating once every 3 h experience a longer LH surge, and LH stays high at 2 h following the first copulation, diminishing to baseline levels by 18 h afterward (Goodrowe et al. 1989; Schmidt et al. 1983; Wildt et al. 1980, 1981). The LH surge is also influenced by how far into estrus the female has advanced when she mates on sequential days (Banks and Stabenfeldt 1982).

Repetition of coital stimulation is often necessary to trigger ovulation (Stabenfeldt and Shille 1977), the probability of ovulation increasing after single matings as estrus

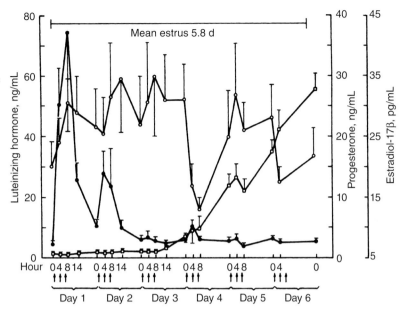

Fig. 4.2 Mean (± SEM) serum estradiol-17β (o), LH (●), and progesterone (□) from 12 estrous periods ($\bar{x}$ = 5.8 d) in mated ovulating females. Each vertical arrow indicates the time of one copulatory stimulus. Source: Wildt *et al.* 1981. Reproduced with permission of the Society for the Study of Reproduction.

Table 4.1 Proportion of females (n = 12) ovulating if mated once or three times on various days of estrus.

Day of estrus	Proportion ovulating	Length of estrus ($\bar{x}$ ± SEM), d	Number of follicles ($\bar{x}$ ± SEM)
		1 mating	
1	1/12	4.9 ± 0.9	4.3 ± 0.6
2	2/12	4.0 ± 0.5	3.7 ± 0.3
3	3/12	5.1 ± 0.2	4.2 ± 0.3
4	4/12	5.9 ± 0.6	3.6 ± 0.4
		3 matings	
1	10/12	4.2 ± 0.5	3.9 ± 0.6
2	10/12	4.2 ± 0.4	5.0 ± 0.5
3	10/12	5.6 ± 0.6	4.1 ± 0.5

Source: Wildt *et al.* 1980. Copyright 1980, The Endocrine Society. Reproduced with permission.

proceeds but rising markedly if multiple matings occur (Concannon *et al.* 1980, Scott and Lloyd-Jacob 1955, Wildt *et al.* 1980). These phenomena are solidly documented (Table 4.1). However, the day of estrus when a female mates, the length of her specific estrous phase, and the number of follicles subsequently formed have no apparent influence on her *capacity* to ovulate (Wildt *et al.* 1980).

Concannon *et al.* (1980) showed that in >50% of cats mating just once, the amounts of LH released (<10 ng/mL in <1 h, declining to baseline levels of 1 ng/mL in <4 h) were insufficient to stimulate ovulation (Fig. 4.4). Cats ovulated successfully when levels continued to rise above 20 ng/mL at 1 h and remained above baseline up to 4 h. Multiple copulations generated higher values of LH that persisted longer. Unrestricted matings over 4 h produced mean LH concentrations 3–6 times higher than single matings, with maximum release at 1–2 h. All these animals produced ova. Mounting without intromission resulted in small amounts of LH released, too small

Fig. 4.3 Mean serum (a) estradiol-17β, (b) LH, and (c) progesterone concentrations during pseudopregnancy and subsequent prolonged luteal phase ($n = 12$). Dark areas represent ± SEM. See source for history of the data. Source: Goodrowe et al. 1989. Reproduced with permission of the Society for the Study of Reproduction.

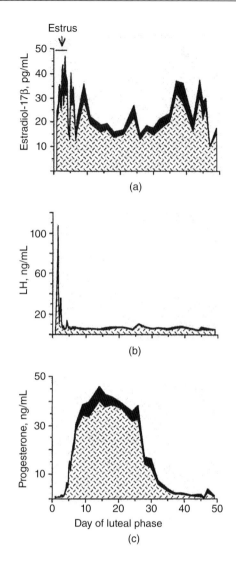

to induce ovulation. Cats mating multiple times therefore release more LH than those mating only once (Concannon et al. 1980, Glover et al. 1985), and the failure of some single-mated females to ovulate can be caused by the release of LH at concentrations too low to be effective or its failure to be released at all (Concannon et al. 1980, Johnson and Gay 1981b, Wildt et al. 1980).

The release of LH has limits. Multiple matings fail to induce additional production after 2–4 h; concentrations decline over the next 2–3 d post-coitus and remain depressed. In one series of experiments the concentrations of serum LH peaked in 2–4 h at 11–280 ng/mL ($\bar{x} = 112 \pm 30$ SEM) 2–4 h after the first mating, diminished steadily thereafter to baseline values (≤ 3 ng/mL) by 20–24 h, and were still low at 36 h (Concannon et al. 1989). The pattern during unrestricted mating indicated that repeated coitus promotes a surge of LH, and continued coitus after 2–4 h resulted in no further release despite adequate amounts retained in the pituitary. A self-priming

Fig. 4.4 Serum LH concentrations in estrous females permitted single or multiple matings. Single matings ($n = 13$) induced LH release adequate to cause ovulation in only half the females. Multiple matings (4 and 8–12 copulations, $n = 23$) released more LH, and all ovulated. Source: Concannon et al. 1980. Reproduced with permission of the Society for the Study of Reproduction.

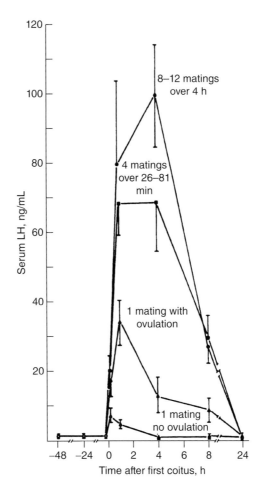

effect of gonadotropin-releasing hormone (GnRH) on LH was proposed (Johnson and Gay 1981b) and later demonstrated (Concannon et al. 1989). Injecting GnRH 36 h after coitus caused an increase in circulating LH.

Release of LH keys the reproductive cycle in all species of induced ovulators (Bristol-Gould and Woodruff 2006). As noted, both the amplitude of its release and its persistence in the blood are influenced strongly by how many times the female copulates (Concannon et al. 1980), but also by the pituitary's exposure to the concentration and persistence of estradiol. In evolutionary terms the functional basis of multiple mating in domestic cats and other felids might be to magnify gonadotropin release from the pituitary, maximizing folliculogenesis and subsequent maturation of the ova (Donohue et al. 1993). Consequently, according to Wildt et al. (1980: 1216), "vesicular follicles rupture in an all or none fashion depending on the adequacy of the mating stimulus in eliciting a LH surge."

Not every follicle eventually ruptures and ovulates to produce a corpus luteum. Wildt et al. (1981) reported that 86.6% of follicles detected on day 1 of estrus did so after repeated matings, resulting in an average of 4.3 CL. In another study (Schmidt et al. 1983), 12 females mated 3 times daily produced a mean of 5.6 preovulatory follicles per cat during estrus, 88.1% of which formed CL ($\bar{x} = 4.9$ CL/cat $\pm$ 0.3 SEM).

Prenatal mortality *in utero* begins early, ~43% of fertilized oocytes failing to survive the first stages of development (Howard *et al.* 1992).

Synchrony of estrus among group-living, free-ranging cats has been reported (Langham 1992, Liberg 1983, Liberg *et al.* 2000; Say *et al.* 2001; Yamane 1998; Yamane *et al.* 1996). This supposedly makes "monopolizing" females by one male easier, a conjecture weakened by commonly high incidences of multiple paternity at many locations, the presumed but undemonstrated fitness advantage of single paternity to a male cat, and lack of documented evidence for the synchronized onset of estrus in females living together. Yamane (1998) claimed estrus was synchronized in strays living in the same feeding group, as it also was among related females in the group. However, the data (Yamane 1998: 244, Fig. 3) could just as easily illustrate that proestrus in a group of cats might not be unusual when breeding is seasonal, in which case the "synchrony" observed is nothing more than a predictable time interval in the normal reproductive cycle.

Overwhelming evidence points to cats being induced ovulators. Synchrony of estrus in free-ranging cats would be possible only if they ovulated spontaneously, experimental documentation for which is unconfirmed. In addition, were females spontaneous ovulators, it would benefit males to be territorial (Clutton-Brock 1989), at least during the breeding season, but we have no evidence they are. However, if males expand their home ranges during the breeding season, as seems to be the case, doing so would be advantageous only if females attained estrus spontaneously in an asynchronous manner (Clutton-Brock 1989) or ovulated only after coitus.

As a polyestrous breeder, the adult female cat experiences recurring estrous phases at intervals through the reproductive season until fertilization is achieved (Scott and Lloyd-Jacob 1955). *Interestrus* describes these repeating "waves" of estrus resulting from the absence of ovulation (Fig. 4.5). The time between any two in succession can be termed an *interestrous interval* (Feldman and Nelson 1996, Wildt *et al.* 1981). Interestrus occurs at the end of follicular function and is marked by circulating estrogen dropping abruptly to <20 pg/mL (Feldman and Nelson 1996). Interestrous intervals can last 7–40 d, although most occur every 8–24 d ($\bar{x} = 17$ d) with >50% of cats cycling again at 14–21 d (Bristol and Woodruff 2004; Concannon 1991;

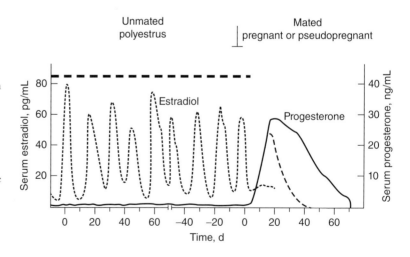

Fig. 4.5 Schematic representation of endocrine changes typically seen during polyestrus in unmated adult female cats and also mated cats in which mating has resulted in either pregnancy or pseudopregnancy. Pregnancy follows a fertile mating; pseudopregnancy results from infertile mating or vaginal stimulation but with induction of ovulation. Solid bars indicate estrus; open areas show potential proestrus. Source: Concannon 1991, figure 5, p. 536. Reproduced with permission of Elsevier.

Concannon *et al.* 1980; Dawson 1952; Goodrowe *et al.* 1989; Paape *et al.* 1975; Scott and Lloyd-Jacob 1955; Verhage *et al.* 1976; Wildt *et al.* 1978, 1981). Polyestrus continues in this fashion through the breeding season or until interrupted by ovulation. During polyestrus (i.e. interestrus) the estradiol concentration rises and falls, spiking at ~59.5 pg/mL and reaching lows of ~8.1 pg/mL (Verhage *et al.* 1976).

Females adapted to a light regimen of 14L:10D and then shifted to continuous light (24L:0D) experienced induced polyestrus after ~45 d, manifested as predictable "waves" of estrus and the expected rise in estradiol. However, interestrous phases were reduced to 1/month compared with 2/month under the original lighting regime (Leyva *et al.* 1989a). Polyestrus stopped abruptly in cats shifted from 14L:10D to 8L:15D, and levels of estradiol fell.

Concentrations of melatonin were lower during interestrus in laboratory cats kept at 24L:0D than in others maintained at 14L:10D, and prolactin was significantly higher in those adapted to 8L:16D than under either long-day regimen. Thus exposure to continuous light stimulated folliculogenesis, but short-day conditions were inhibitory (Leyva *et al.* 1989a). In addition, melanin and prolactin proved significantly lower when cats were in estrus than in interestrus at both long-day conditions, indicating that inhibition of both correlates positively with folliculogenesis (Leyva *et al.* 1989a).

Injection of melatonin stopped polyestrus and folliculogenesis for 2 months (Leyva *et al.* 1989b). At the latitude of Davis, California, where these experiments took place, previous data showed breeding behavior to commence under natural light around 15 January, or ~25 d following the shift to lengthening photoperiod. Leyva *et al.* (1989a: 132) speculated that "a change of photoperiod as occurs in the first 15 d after the winter solstice is sufficient to initiate ovarian follicular activity. … " Melanin implants have been considered as a short-term method of birth control in female cats (Goericke-Pesch 2010).

The results of short-day conditions obtained in a laboratory do not always coincide with what happens in nature. Feral cats that formerly populated a subantarctic island offer a puzzling case. Macquarie Island lies in the Southern Ocean midway between New Zealand and Antarctica. At 54°30'S, 158°57'E it experiences an austral winter of prolonged darkness. Most cats there were born from November–March, but a female with a young kitten was trapped in July 1975, and Jones (1977) speculated that reproduction could occur during other months. Mating in autumn when the days would already be dark followed by birth in mid-winter 2 months later seems unlikely, but is obviously not impossible. As just discussed, laboratory data show that exposing female cats to continuous light stimulates folliculogenesis; follicular activity stops under short-day regimens of 8L:16D (Leyva *et al.* 1989a). The estimated hours of daylight (measured conservatively as nautical twilight) in July on Macquarie Island extends from 0730–1500 h (http://www.antarctica.gov.au/about-antarctica/fact-files/weather/sunlight-hours), or 7.5L:16.5D, conditions under which folliculogenesis and estrus would not be predicted.

Diestrus follows estrus, but only in females induced to ovulate. By encompassing ovulation, this phase requires functional CL. *Diestrus* extends through pregnancy to parturition (Shille and Stabenfeldt 1979, Wildt *et al.* 1981) and, as depicted (Fig. 4.3c), is defined endocrinologically as the time after mating when circulating progesterone is >1 ng/mL. During diestrus, corpora hemorrhagica morph into CL. Ovulation is accompanied by a rise in circulating progesterone, which continues to dominate other hormones until the end of gestation (Feldman and Nelson 1996). Progesterone, the hormone that sustains pregnancy, is produced by the ovaries and later in gestation by the placenta with the aid of prolactin, which in turn is necessary to sustain progesterone production during the last half of gestation (Concannon 1991). Progesterone usually stays <1 ng/mL for the first 24 h after coitus (Glover *et al.* 1985), or until the LH surge when it rises with onset of ovulation at 40–72 h following the spike in LH (Schmidt *et al.* 1983, Wildt *et al.* 1981). A blood concentration >1.0 ng/mL followed by a steady increase is considered indirect evidence of ovulation and luteal function (Paape *et al.* 1975, Shille and Stabenfeldt 1979, Wildt *et al.* 1981). However, because the level can peak anytime at 11–60 d following ovulation and range from 13.5 to 57.0 ng/mL, progesterone makes an unreliable indicator of pregnancy (Schmidt *et al.* 1983). Its concentration declines gradually toward the end of diestrus, falling to ~5 ng/mL near parturition and to baseline concentrations (<1 ng/mL) soon after (Schmidt *et al.* 1983, Verhage *et al.* 1976), but this decline is not necessary to initiate parturition (Schmidt *et al.* 1983).

Nonfertile stimulation of an estrous cat (i.e. ovulation without fertilization) still produces CL, alterations of the uterus mimicking pregnancy and extended diestrus, processes that in combination define *pseudopregnancy,* or "false pregnancy" (Dawson and Friedgood 1940, Foster and Hisaw 1935, Greulich 1934). Pseudopregnancy therefore represents a nonpregnant luteal phase following a sterile mating (Gudermuth *et al.* 1997) or, in the laboratory, either mechanical stimulation of the cervix (Greulich 1934) or diestrus induced artificially by administration of estradiol (Aronson and Cooper 1966, Whalen 1963b). The term is properly restricted to cats demonstrating clinical signs of pregnancy (Stabenfeldt and Shille 1977), and most pseudopregnant cats reveal few of these overtly, such as visible mammary development at ~day 40 (Concannon 1991).

Estradiol fluctuates during proestrus and estrus, but usually stays >20 pg/mL in pseudopregnant cats (Schmidt *et al.* 1983, Wildt *et al.* 1981) as illustrated (Fig. 4.3a). Pseudopregnancy lasts 26–55 d after application of the stimulus, or an average of ~38 d (Wildt *et al.* 1981) during which time polyestrus is suppressed. Paape *et al.* (1975) estimated 30–73 d ($\bar{x}$ = ~41 d), with many individuals displaying sexual receptivity within a week of CL regression, which would be close to day 36 following ovulation. Concannon (1991) gave a mean diestrus of ~45 d. Peak luteal activity occurs around day 16 or 17 (Paape *et al.* 1975). Late into the breeding season, pseudopregnancy can be followed by anestrus (Concannon 1991).

Estradiol, LH, and prolactin stay near baseline concentrations during pseudopregnancy (Banks *et al.* 1983, Paape *et al.* 1975, Verhage *et al.* 1976, Wildt *et al.* 1981). Progesterone, in contrast, rises to 15–90 ng/mL by day 15–25 then declines to baseline levels (<1 ng/mL) by days 30–50 (Paape *et al.* 1975, Verhage *et al.* 1976, Wildt *et al.* 1981).

Estradiol, which spikes at ~60 pg/mL the first day of mating, falls to ~8–12 pg/mL in the initial 5 d afterward (Verhage *et al.* 1976). It stays low during pregnancy, rising slightly just before parturition. By day 39 of diestrus (i.e. 39 d past the spike in LH) estradiol rises, and its concentration fluctuates but stays ~10 pg/mL through the remaining time of the luteal phase (Wildt *et al.* 1981), as shown (Fig. 4.3a).

Progesterone, barely detectable 2–3 d following coitus, peaks at ~24 ng/mL in pseudopregnancy and ~35 ng/mL in pregnancy (Verhage *et al.* 1976). Its concentration remains elevated nearly a month longer in pregnant (Fig. 4.6c) than in pseudopregnant (Fig. 4.3c) cats; that is, throughout gestation, often displaying peaks of 10–60 ng/mL

Fig. 4.6 Mean serum (a) estradiol-17β, (b) LH, and (c) progesterone concentrations during normal pregnancy ($n = 12$). Dark areas represent ± SEM. See source for history of the data. Source: Goodrowe et al. 1989. Reproduced with permission of the Society for the Study of Reproduction.

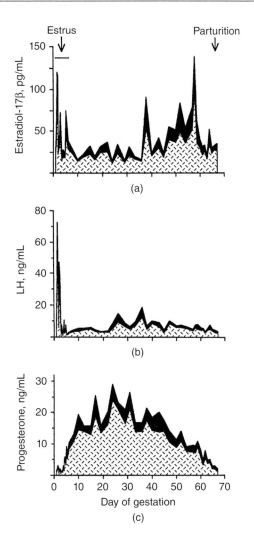

from days 35–45 before diminishing to baseline near or at parturition (Banks et al. 1983, Schmidt et al. 1983, Verhage et al. 1976). The pattern and concentration of estradiol is similar in pregnant and pseudopregnant cats (Schmidt et al. 1983, Wildt et al. 1981). Estradiol rises during the final 9 d of pregnancy, probably a result of fetal secretion as parturition approaches (Goodrowe et al. 1989). In pregnant cats both LH and estradiol-17β demonstrate slight variation within individuals, remaining near baseline concentrations (Goodrowe et al. 1989), as depicted (Fig. 4.6a, Fig. 4.6b).

To reinforce previous comments, females that do not mate, or mate without ovulating, circumvent diestrus and after a week or so cycle once again to proestrus followed quickly by estrus. This is not the case during pseudopregnancy when the CL are functional over 35–44 d (Foster and Hisaw 1935, Paape et al. 1975, Verhage et al. 1976, Wildt et al. 1981). In pregnant cats the CL remain functional through gestation, or ~20 d longer than in pseudopregnancy, regressing as parturition approaches (Concannon 1991, Dawson 1946, Schmidt et al. 1983, Verhage et al. 1976). However, follicles grow and regress continuously, even during diestrus (Schmidt et al. 1983).

Descriptions to this point show clearly that cats are induced, or reflex, ovulators incapable of demonstrating a luteal phase (pregnancy or pseudopregnancy) until after coital stimulation of the vagina and cervix (Greulich 1934, Longley 1911). Such stimulation must be intense to initiate the reflexive release of LH in sufficient concentrations to trigger ovulation (Concannon et al. 1989), and single copulations fail to achieve this about half the time (Concannon et al. 1980, Wildt et al. 1981). Exceptions in confined cats have been reported, but their validity remains unconfirmed. Beaver (1977) mentioned that exposure to other cycling females in laboratory settings possibly stimulates spontaneous ovulation, and tentative evidence was later obtained. Mature laboratory females housed as a group, with or without physical contact, have been shown to ovulate spontaneously (Fig. 4.7); that is, in the absence of prior coitus, sometimes with or without a male but without physical contact (Bristol and Woodruff 2004, Gudermuth et al. 1997). Whether this occurs among free-ranging cats where putative spontaneous ovulation has been linked to "overlapping" reproductive cycles (Yamane 1998, Yamane et al. 1996) awaits validation and seems unlikely (see above).

Anestrus is the nonreproductive phase occurring under short-day conditions when reproduction ceases (Feldman and Nelson 1996) and defines a period of nonestrus, either seasonal or caused by lactation (Concannon 1991). Anestrous females will not mate, evading approaching males or rolling onto their sides and often striking out, snarling, or hissing (Green et al. 1957). Blood concentrations of estradiol and progesterone fall to baseline concentrations (Bristol-Gould and Woodruff 2006). Under natural conditions, anestrus lasts 2–3 months, ordinarily from late autumn through early winter (Concannon et al. 1980). Others reported shorter durations. Longley (1911), for example, stated 3–6 weeks. Lactating females remain anestrous through weaning (Schmidt et al. 1983). Polyestrus can resume 1–4 weeks after the current litter has been weaned (Dawson 1952, Schmidt et al. 1983).

4.3 Male reproductive biology

Age at puberty varies in males. According to Goodrowe et al. (1989: 74) it occurs at 8–12 months, and "Seminiferous tubule mitotic activity is evident when the testes weigh 400–500 mg (at about 5 months of age)." Free-ranging male cats in southeastern Australia show onset of sexual maturity at a mean body mass of 3.2 kg, fully functional males averaging 3.8 kg at 52–56 weeks (Jones and Coman 1982a). Free-ranging urban males in a French survey reproduced at the onset of sexual maturity (~10 months), but their rural counterparts often did not breed until 3 y (Say et al. 1999).

The male's penis has 100–200 cornified papillae on the corpus cavernosum glandis, which appear at age 6–7 months and disappear after castration (Stabenfeldt and Shille 1977). Their function has not been established, and whether they stimulate ovulation is unknown (Fox 1975, Stabenfeldt and Shille 1977). Although male cats are sexually receptive all year, they experience intervals of desensitization of the glans penis that correlate with times when females are not receptive (Aronson and Cooper 1966).

The testes are descended at birth, and combined testicular mass is 20 mg at parturition, 100 mg at weaning, 130 mg at 12 weeks, and 500 mg at 20 weeks when the earliest signs of spermatogenesis appear (Scott and Scott 1957). Spermatids are evident when the testes are ~700 mg; spermatozoa appear when they are >1.0 g. By this time the cat is 30–36 weeks old and 2.5–3.0 kg (Scott and Scott 1957). Subsequently, the testes increase in proportion to body mass, attaining 4.0 g in a 5-kg cat. At 0.08% the gonadosomatic index (testes mass/M) is low compared with most mammals, near the value of human males (França and Godinho 2003).

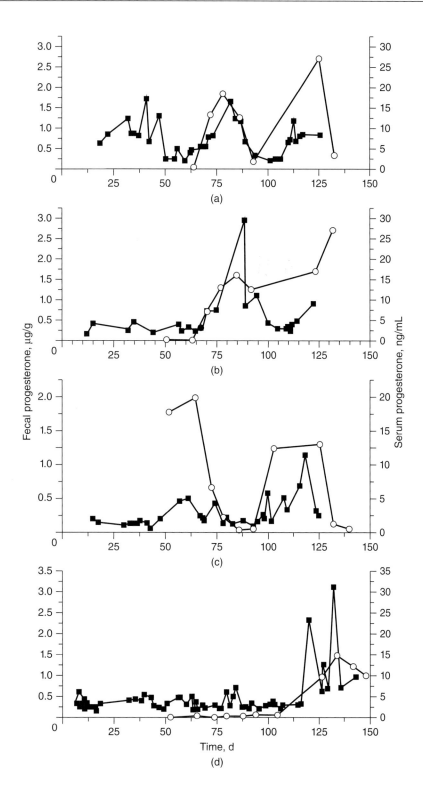

Fig. 4.7 Progesterone concentrations of three females (a, b, c) undergoing spontaneous luteal phases and one female (d) exhibiting coitus-induced pseudopregnancy. Fecal (■) and serum (o) progesterone. Source: Gudermuth *et al.* 1997. Reproduced with permission of the Society for the Study of Reproduction.

Table 4.2 Comparison of testicular and sperm data with significant differences between seasons of high (spring, March–May) and diminished (November–January) gonadal activity. Autumn/winter testosterone concentrations from September–November. Values are $\bar{x} \pm$ SEM, $n = 30$–33, and all seasonal differences were significant at a minimum of $p<0.05$.

Data	Spring	Autumn/winter
Testis mass, g	1.46 (±0.06)	1.29 (±0.05)
Spermatozoa/testis × 10^6	100.7 (±8.3)	74.8 (±7.0)
Motile spermatozoa, %	76.7 (±2.3)	51.6 (±4.3)
Intact spermatozoa, %	49.6 (±4.0)	30.0 (±2.5)
Testosterone per g of testis, µg	0.86 (±0.18)	0.39 (±0.07)
Testosterone/testis, µg	1.43 (±0.32)	0.48 (±0.08)

Source: Blottner and Jewgenow 2007. Reproduced with permission of John Wiley & Sons.

Adult males kept indoors under long-day conditions show no evidence of seasonal reproduction (Beaver 1977) and can fertilize females throughout the year (Goodrowe et al. 1989). Most free-ranging cats under natural photoperiod breed seasonally, in the northern hemisphere between January and July, although actual timing is controlled more by the female's reproductive cycle. In the southern hemisphere (e.g. southeastern Australia) these months are reversed: from October through January females are either pregnant or lactating and breeding takes place in austral spring, most litters being born from September–March (Jones and Coman 1982a). Denny et al. (2002: 408) reported feral cats at a waste-disposal site on the central highlands of New South Wales as breeding "from July through April, and possibly throughout the year." Free-ranging males in both hemispheres produce viable sperm all year (Jones and Coman 1982a, Kirkpatrick 1985, Spindler and Wildt 1999) despite the seasonal shift in photoperiod, and at most are moderate seasonal breeders (Blottner and Jewgenow 2007).

Testis mass and overall sperm production in German free-ranging cats showed slight seasonal variation (Table 4.2). However, more motile sperm were produced during the breeding season, peaking in March. Testosterone concentration varied widely among individual cats, fluctuating over diel cycles and by month, but was elevated in spring, spiking in March and diminishing from autumn into winter (Fig. 4.8 and Table 4.2). Recrudescent steroid production commenced in December preceding meiosis. Canadian free-ranging cats yielded similar results, having greater testicular mass in June than in December and March and demonstrating significant correlation between seasonal testicular mass and circulating testosterone (Kirkpatrick 1985).

Diel fluctuations in testosterone are evident even in laboratory cats kept under constant conditions. In a group sampled over 360 min the range was 0.10–3.25 ng/mL

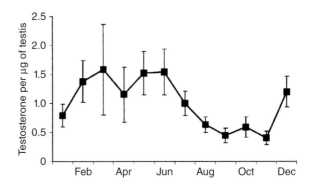

Fig. 4.8 Annual variation in total testosterone concentration ($\bar{x} \pm$ SEM) in the testes of German free-ranging cats. Source: Blottner and Jewgenow 2007. Reproduced with permission of John Wiley & Sons.

within individuals and 0.10–5.90 ng/mL among individuals (Goodrowe et al. 1985). In general, release of androstenedione and testosterone appears to be episodic and without an evident diel pattern (Johnstone et al. 1984, Tsutsui et al. 1990).

As to other hormones, baseline LH concentrations of male cats are similar to those of estrous and anestrous females (Johnson and Gay 1981a), its release controlled by negative feedback from gonadal steroids: LH spikes within 1–5 d of castration (Johnson and Gay 1981a). Administering GnRH initiates a surge of LH and rise in testosterone. Injecting males intravenously with 25 µg GnRH raises the level of circulating LH 20-fold within 10–15 min (Johnson and Gay 1981a); injecting 10 µg intramuscularly elevates it by a factor of 6 within 30 min (Goodrowe et al. 1985). The implantation of slow-release GnRH agonists is a possible method for short-term reproductive control in males (Goericke-Pesch 2010).

The pH range of cat semen is 7.0–8.2 (Sojka et al. 1970). Note, however, that there can be no such parameter as mean, or average, pH of any solution unless the raw data are first altered. Sojka et al. (1970), for example, stated that cat semen has a mean pH of 7.4. However, a series of pH values centered about a mean is impossible; only the range in pH can be a valid expression. Measuring the pH in a series of solutions and dividing by the number of measurements yields a numerical value, but this number is not the mean (Spotte 1992: 602–603). By definition, pH is the logarithm of the reciprocal of H^+ normality. The p stands for power, and H is the symbol for hydrogen. As noted by Kinney (1973), summing a series of pH values multiplies the reciprocals of the normalities. Dividing by the number of samples gives the nth root of the product. The mean can be determined only by changing the original data to hydrogen normalities before dividing by the number of measurements. The logarithm of the reciprocal is then "mean pH."

Semen volume is typically 0.01–0.12 mL with an upper limit of 0. mL, and spermatozoa per ejaculate number ~57×10^6/mL (Sojka et al. 1970, Stabenfeldt and Shille 1977). These counts are considered low. Not surprisingly, motile spermatozoa per ejaculate are lower still at $17–24 \times 10^6$/mL (Howard et al. 1990). Although unlimited opportunities to copulate sometimes reduce sperm concentration and rates of conception, males can ordinarily mate repeatedly over 4–5 d without the numbers of spermatozoa falling (Sojka et al. 1970).

The time limit for fertilization of the ovum is probably 12–24 h (Howard et al. 1992, Stabenfeldt and Shille 1977). Viability of spermatozoa following coitus is 12–48 h (Stabenfeldt and Shille 1977), long enough to span the interval from coitus to ovulation. Cat sperm cells require *capacitation*; that is, they must "incubate" in the female's reproductive tract 2–24 h before becoming capable of fertilization (Hamner et al. 1970). Goodrowe et al. (1989: 77) cited Hamner et al. (1970) but stated the interval as 0.5–24 h. The minimum duration *in vitro* is 180 min (Goodrowe et al. 1989).

Abnormal, or *pleiomorphic*, spermatozoa are usually <30% of the ejaculate in domestic cats kept under controlled conditions (Wildt et al. 1983). This percentage is considered low, and cats of all species demonstrating <40% pleiomorphic sperm are "normal," or *normospermic*; those having >60% pleiomorphic sperm are classified as *teratospermic* (Goodrowe et al. 1989, Howard et al. 1990, Pukazhenthi et al. 2001), a common condition in cats and first reported in South African cheetahs (Wildt et al. 1983). Sperm production (i.e. efficiency) is about 16×10^6/g of testis (França and Godinho 2003). Abnormalities are seen even at the spermatid stage, as are missing generations of germ cells (França and Godinho 2003). At least 70% of all felid species are teratospermic (Pukazhenthi et al. 2001). The reproductive efficiency of male domestic cats decreases with the increased incidence of teratospermia (Long et al. 1996). Although males produce numerous spermatozoa, relatively few (6×10^6/mL) are required to induce conception (Howard et al. 1992). This might be advantageous in populations containing large numbers of teratospermic males.

Motility of the spermatozoa under normospermic and teratospermic conditions is similar (Goodrowe et al. 1989). However, sperm cells from teratospermic cats, even if structurally normal, demonstrate a compromised capacity to bind with the zona pellucida, penetrate it, and fertilize ova *in vitro*, although examination at extremely high magnification (i.e. tunneling electron microscopy) often exposes defective acrosomes in "normal" spermatozoa (Long et al. 1996, Pukazhenthi et al. 2001). Nonetheless, if fertilization can be accomplished *in vitro*, the embryos develop at similar rates and to comparable stages as those from normospermic donors (Pukazhenthi et al. 2001). The low fertility of teratospermic males might be caused by compromised capacitation and a dysfunction preventing the acrosomal reaction (Long et al. 1996). Capacitation precedes the acrosomal reaction. Spermatozoa from normospermic cats required a 2-h incubation *in vitro* to initiate capacitation after removal of seminal plasma, but sperm from teratospermic cats took 2.5 h and afterward completed the acrosome reaction less efficiently (Long et al. 1996).

The cause of teratospermia in domestic cats is unknown, but according to Goodrowe et al. (1989), individuals with this condition have lower circulating testosterone (0.4 ng/mL ± 0.1 SEM) than normospermic individuals (1.2 ng/mL ± 0.1 SEM). Reduced genetic variation has been suggested as a cause of teratospermia in both domestic cats and some populations of wild felids (Pukazhenthi et al. 2001). Sperm quality could also be seasonal. Free-ranging German cats had higher quality spermatozoa in spring than in winter (Blottner and Jewgenow 2007). The most common malformations were in the acrosome (18.4% ± 5.1), defective tails (31.2% ± 6.1), and persistence of plasma droplets (9.5% ± 3.1), but their frequencies did not have a seasonal pattern. Overall, the proportion of normal sperm showed significant seasonal variance, the lowest percentage occurring in December and January. Blottner and Jewgenow (2007: 539) wrote, "This suggests that in unselected cat populations the average proportion of normal spermatozoa seems to be at a level of nearly 50% and shows continuous transitions between normo- and teratozoospermic characteristics." Their statement indicates that for at least part of the year free-ranging cats could straddle the traditional (and arbitrary) demarcation that defines teratospermia as >60% abnormal sperm per ejaculate.

4.4 The cat mating system: promiscuity or polygyny?

The major categories of animal mating systems are monogamy, polygyny, protandry, and promiscuity. Among mammals, >90% are *polygynous*, males mating with the same group of females consistently and keeping some semblance of social bonding afterward (Clutton-Brock 1989, Clutton-Brock and Lukas 2012, Clutton-Brock and McAuliffe 2009). Males and females of *promiscuous* species mate with any available member of the opposite sex, and there is no social bonding after the mating act.

Domestic cats are perhaps least likely to be monogamous. At Grand Terre, Kerguelen Islands, where the population of feral cats is scattered thinly across the landscape, genetic testing showed single paternity of litters to be typical, ostensibly because males are able to "defend" access to females and "monopolize" matings (Say et al. 2002b).

Without behavioral confirmation that monogamy is actually the default mating system at Grand Terre, as the authors proposed, and that males indeed control access to receptive females, a more parsimonious explanation is reduced incidence of encounters between the sexes based on low population density.

Promiscuity is the most typical mating system of solitary species occupying large home ranges (Erofeeva and Nadenko 2011) where females are dispersed widely, and males actively seek them. Feral cats fit this category. Depending on author, the mating system of free-ranging cats has usually been classified as polygynous or promiscuous (Leyhausen 1973, Liberg 1983, Natoli and De Vito 1999, Say *et al.* 1999), but any such distinction seems influenced more by circumstance than biological directive. Cats generally fail the polygyny test, which requires one or more of the following: (1) intense male–male competition, (2) pronounced sexual dichotomy as an evolutionary consequence of male–male competition, (3) higher susceptibility of males to starvation, (4) comparatively higher and earlier male mortality, (5) pre-copulatory mate guarding, (6) aggregation of females in stable groups, and (7) territoriality (Clutton-Brock 2009, Clutton-Brock and Lukas 2012).

A free-ranging male cat might be called polygynous if his home range overlaps those of two or more females and he mates with them exclusively, a situation less likely in cities and suburbs than rural areas where fewer cats reside. However, defining a mating system based solely on circumstances seems an aleatory decision; after all, the pattern of dispersion in which a cat finds itself (Chapter 2) is inevitably circumstantial and therefore not the basis of a species-defining mating system. Where females are dispersed in clumped patterns (e.g. urban strays) and males extend their normal home range boundaries to find mating partners, the system just appears more promiscuous. In either case – rural or urban – an estrous female attracts adult males from all around, which might always be the same individuals depending on the circumstances of their distribution.

I therefore contend that population density can affect mating *patterns* of free-ranging cats without altering the mating *system*. The first depends purely on dispersion and population density; the second is a fact of biology. During a French survey, nine microsatellite loci were typed to compare paternity in cats from a rural and an urban location (Say *et al.* 1999). The sample size was large, 312 offspring of 76 mothers and 65 putative fathers. The rate of multiple paternity was 70–83% in the urban population, but only 0–22% in cats occupying the rural area, indicating that a rural male could father a female's entire litter. These results are not necessarily what the authors claimed them to be; that is, evidence of different "tactics" used by males to enhance reproductive success. The more parsimonious explanation: fewer cats and wider dispersion result in less variable paternity. A behavioral factor must also be considered. Estrous females are perhaps less selective when courted by fewer males (Ishida *et al.* 2001). With a reduced number of mating opportunities, a male's so-called "tactics" and female choice (Section 4.7) are both restricted. A higher incidence of multiple paternity is expected where food is abundant and clumped and cats aggregate to feed. Permanent residency is not a requirement: paternity at a waste-disposal site on the central highlands of New South Wales, Australia, was shared by resident and nonresident males alike (Denny *et al.* 2002).

A pervasive notion in the literature has free-ranging male cats seeking to "monopolize" matings (e.g. Kerby and Macdonald 1988), the consequence of an evolutionary imperative for an individual to spread its genes through a population at the expense of rivals. He accomplishes this by "defending" the estrous female against other males and fighting them for exclusive access to her. This gives a misleading impression of what generally seems like a peaceful clustering of males each waiting with mild disinterest to take his turn (but see Liberg 1983). Estrous females are perceived as passive "resources" when their home ranges overlap that of a male (Macdonald 1983), at which point they become available or not to the male depending on its "dominance" status (Natoli and De Vito 1991). Such a picture distorts the domestic cat's mating system.

When food is clumped, so is the occupancy pattern of cats exploiting it. The home ranges of female strays are often contained partly or wholly within that of a male. Not surprisingly the number of females living inside a male's home range often correlates positively with its size (Say and Pontier 2004). Nonetheless, 30 kittens (28%) born to a group of strays in Lyon between 1996 and 1998 were fathered by males having home ranges that *did not* overlap with the mothers (Say et al. 1999). Clearly, the mating pattern of males is *not* to monopolize individual females but rather mate with them and move on, impregnating as many as possible even if only siring some kittens of every litter instead of all of them (Sandell 1989).

A subtle teleology threads through some of these reports, as if proof of multiple parentage is too weak to support without induction. For example, Say and Pointier (2004: 178) wrote, "Female cats behave in ways that promote conflict between males." No supporting data were offered. Instead, the authors cited references mostly describing the mating behavior of elephants, ground squirrels, and wallabies, leaving an impression that females in general help to undermine what might otherwise be more stable mating systems. Even in the unlikely case that this is true of cats, why the implication that a male cat's evolutionary "preference" is to defend a single estrous female against other males, mate with her monogamously, and sire all the kittens? Consequently, the male is left little choice except to mate promiscuously: "When the number of male candidates increases, the cost of defending the female becomes too high. ... " This makes no sense considering the many successful promiscuous and polygynous species and the incompatible observation – occasionally by the same authors – that male cats in estrous groups rarely contest for matings.

Clutton-Brock (1989) implied that bonds form between polygynous males and the females with which he mates, although the nature of these attachments was not explained. Some degree of familiarity might be expected, but whether male cats meet Clutton-Brock's additional criterion of mate defense, either directly by pre-copulatory guarding of the female or indirectly by guarding the space she occupies, is doubtful. Male cats, as emphasized, make no overt attempt to "monopolize" matings. As to guarding space, they also fail to meet the requirements of territoriality as traditionally defined (Chapter 2). Neither are free-ranging female cats polyandrous and restricted to mating with the same group of males during successive reproductive cycles, unless influenced by circumstance (i.e. the number of adult males is limited and they happen to live in the vicinity), in which case the *appearance* of polyandry, like that of polygyny, is an artifact of dispersion and population numbers. Everything considered, the domestic cat's mating system strongly favors promiscuity over polygyny or polyandry. The last two categories, if referred to at all, should be qualified as facultative.

The factors discussed so far, combined with a general lack of aggression among male cats in estrous groups (Section 4.6), reflect the mating behavior of the thirteen-lined ground squirrel (*Spermophilus tridecemlineatus*), another asocial, promiscuous species (Davies 1991). Males do not guard estrous females and abandon them after mating to seek other females. The females ordinarily are scattered instead of clumped, a pattern similar to that seen in feral cats, and males are forced to search widely. As in cats, putative signs of male dominance is a poor predictor of mating success, and aggression among males is rare, something else the species have in common (Section 4.6).

The first squirrel to mate with a female sires ~75% of her offspring, making it advantageous to mate with her just once then seek another partner. Male cats often behave this way. In thirteen-lined ground squirrels the frequency of multiple paternity within litters varies between reproductive cycles, from 0 to 50% (Schwagmeyer and Brown 1983, Schwagmeyer and Parker 1987). Schwagmeyer and Woontner (1986) classified this squirrel's mating system as "scramble competition polygyny," but calling it promiscuous seems more accurate. Female promiscuity in cats might be explained as an evolutionary mechanism for enhancing fertility; that is, a hedge against teratospermia and pseudopregnancy. Mating with several males could increase the chances of receiving adequate spermatozoa of high quality, but this is simply conjecture.

4.5 Female mating behavior

Mating behavior of domestic cats has been well documented, and my description is a composite from the literature (e.g. Beaver 1977; Stabenfeldt and Shille 1977; Whalen 1963a, 1963b; Young 1941). The female emits a characteristic vocalization when in estrus, variously termed a "mating," "heat," and "rutting" call. This has been described as a "curious, low vocal sound" (Stabenfeldt and Shille 1977: 520); "low growling cries" (Green *et al.* 1957: 508); and like "the crying of a human baby" (Shimizu 2001: 88). The male responds with a similar vocalization, and the two call back and forth. According to Shimizu's (2001: 88) description, "Both males and females produced it as a series of several calls when walking or pausing." Dards (1983: 137) described the male's call as having a "prau" sound, which is "a short cry, relatively high-pitched with a rapid rise in frequency, and is emitted repeatedly." Neither mating vocalizations nor those serving other putative functions are universal. At some locations the females are silent (Dards 1983), and Konecny (1987b) recorded feral cats vocalizing just once during 502 field days in the Galápagos Islands.

Certain proestrous behaviors (Section 4.2) continue to be displayed: rubbing the head, neck, and chin on objects or the ground, and rolling repeatedly on the ground (Green *et al.* 1957, Rosenblatt and Schneirla 1962) either in the presence or absence of a male (Green *et al.* 1957). She also exhibits lordosis, sometimes to the extent of moving with her forelegs collapsed, chest and abdomen pressed to the ground, and hindquarters elevated. In the male's presence she crouches down and makes stepping movements by alternately lifting her hind legs. Her vocalizations become low-pitched and more frequent. If the male tries to mount prematurely she might turn on him, claws bared, causing him to step back. A female's disposition while in heat varies by individual, ranging from docile to vicious (Green *et al.* 1957). Coitus ends when the female emits a yowl (the mating cry) and attempts to disengage from the male (Johnson and Gay 1981b, Wildt *et al.* 1980).

The male's behavior influences the female's receptivity. Whereas an aggressive male might attack and subdue a resisting female, a shy male can be discouraged if the female rebuffs him too abruptly (Green *et al.* 1957). A female ready to mate lowers her head, crouches, and lifts her pelvic area when touched on her back, shifting her tail to one side. The male grips the skin on the back of her neck in his teeth and mounts her (Fig. 4.9). As the male withdraws his penis she pulls forward, yowls a "mating cry," and turns aggressively toward the male, forcing him to back away (Fig. 4.10). The

Fig. 4.9 Cats mating. The female assumes the position of lordosis; the male grasps her in a neck grip and mounts. Source: © Thomas Graversen | Dreamstime.com.

Fig. 4.10 After copulation the male dismounts quickly because the female often attacks him aggressively. Source: © Thomas Graversen | Dreamstime.com.

cause of this might be pain from the anteriorly pointed cornified papillae of the male's penis (Section 4.3) against her vaginal surfaces during withdrawal. Having freed herself from the male she rolls on the ground, rubbing her head and neck as during proestrus (Section 4.2) and then licks herself – especially her genitals – at times almost frantically. Afterward she might lie on her side in seeming contentment with eyes partly closed, opening and closing her forepaws and purring. Time until the next bout of receptivity and mating varies among individuals, but it can be frequent, several times in an hour (Section 4.2).

4.6 Male mating behavior

Males in the laboratory display sexual behavior starting ~11 months and within the range of 8–15 months (Rosenblatt and Schneirla 1962). Hormones are important factors in the sexual behavior of male cats, but so is mating itself. Previous sexual experience increases the chances that a castrated male will mate, although sexual behavior even in these cats declines over subsequent months (Rosenblatt and Schneirla 1962). Experienced cats also displayed more sexual behavior after injection with testosterone proprionate.

At least in captive conditions, it is sometimes necessary that the male is in a familiar location for him to mate (Green *et al.* 1957, Michael 1958, Scott 1955, Scott and Lloyd-Jacob 1955, Stabenfeldt and Shille 1977), although not necessarily inside his home space (Beaver 1977). In the laboratory this means acclimating the male alone in the space for several days. Males placed in new controlled environments sometimes ignore estrous females or act aggressively toward them (Michael 1961), according to some investigators because they have not yet established territories (Goodrowe *et al.* 1989). A more likely reason is unfamiliarity with novel spaces. Breeding is often more successful when a female is moved to a male's established location (Goodrowe *et al.* 1989).

The male investigates the area in the vicinity of the female, spraying nearby objects, and rubbing against places not yet sprayed (Fox 1975), behavior thought to increase familiarity of space (Beaver 1977). On encountering an estrous female he vocalizes with her using his own "rutting" call (Yamane 1998) and sniffs her perianal area. When the female is receptive she crouches, especially on feeling his forelegs on her back. He bites down on the back of her neck and mounts by straddling her, first with his forelegs and then his hind legs, sometimes emitting a "chirping sound" (Rosenblatt and Schneirla 1962, Stabenfeldt and Shille 1977: 520). He makes stepping movements with the hind limbs until in the copulatory position, at which point his hindquarters shift forward and his back arches (Fig. 4.11). He gives several strong pelvic thrusts to

Fig. 4.11 At intromission the male hunches forward and arches his back. Source: © Saherli | Dreamstime.com.

achieve intromission and stops thrusting with onset of ejaculation. The sequence takes only a few seconds (Section 4.2), and then the penis is withdrawn at once (Beaver 1977). The male quickly dismounts, avoiding the female's claws as she hisses and throws him off her back (Fig. 4.9). He sits down, licks his penis and forepaws, and either leaves or waits for the female to become receptive again.

During courtship, several males might surround a single estrous female (i.e. form an estrous group). Larger males have been seen to get nearer the female and copulate more often (Liberg 1983, Yamane et al. 1996), but these observations have not been documented elsewhere with any consistency (Ishida et al. 2001). Liberg (1983), Yamane (1998), and Yamane et al. (1996) wrote that male body mass correlates positively with mating success. Larger males occupied positions closer to the female and copulated with her more often. Liberg (1983) claimed, in addition, that "dominant" males monopolized copulations, but this was not observed by Natoli and De Vito (1991). According to Ishida et al. (2001), mating success of males on Ainoshima Island did not correlate positively with body mass, age, or persistence, and these cats were from the same aggregation of strays that Yamane and colleagues had studied earlier. That females do not mate consistently with males appearing to be the "winners" in male–male competition could also be seen as indirect evidence of female choice (Ishida et al. 2001).

Natoli and De Vito (1991) described a linear hierarchy among 14 male strays during the breeding season, but their data and results showed questionable confirmation. A male labeled GI was the "victor" in 36 of 38 agonistic encounters with other males, yet interacted with only 8 of the others. No male filled all cells of the frequency table (Natoli and De Vito 1991: 236, Table II) by interacting with each of his 13 counterparts, making a pattern impossible to discern. "Victories" correlated positively with threats given and received and cheek-rubbing against objects (Chapter 1). The first assessment seems inconsistent (i.e. that a putatively dominant male would need to consistently display threat behavior that in turn requires a reciprocal response). What, if anything, cheek-rubbing signifies is unknown. "Dominance" (as measured by "victories" against a paired antagonist) did not correlate with distance from the estrous female, frequency of urine spraying on objects in the habitat, interfering with another male attempting to copulate with the female, number of "vocal duels" (undefined), number of mounts (whether successful or not), or with crouching or fleeing (both considered submissive responses). In fact, GI's behavior seems mostly incompatible with the notion of dominance (Chapter 1), considering he allowed "subordinate" males to copulate with the female being courted and that his own copulations were interrupted by "lower ranked males." Natoli and De Vito (1991: 233) wrote: "Sometimes these interferences were successful, the copulation between GI and the female was interrupted, and GI did not react aggressively."

Although agonism undoubtedly occurs between males during the breeding season, its importance is often misrepresented. Ishida et al. (2001: 248), for instance, gave no supporting evidence for the statement, "Aggressive encounters of male cats are frequently observed during the breeding season." Ordinarily, male strays in estrous groups seldom interact aggressively, even taking turns copulating with the female (Natoli and De Vito 1991; Say et al. 1999, 2001; Yamane et al. 1996). The consequence of promiscuous mating is a high incidence of multiple paternity; in one case 80% of litters were fathered by at least two males (Say et al. 1999), and males of putatively high dominance rank did not sire more offspring than those in the next group below them (Say et al. 2001), making it difficult to understand how males that copulate more often are proven recipients of greater "mating success" as claimed by Yamane *et al.* (1996, 1997).

As already emphasized, much of the literature would have us believe that attempting to "monopolize" estrous females and subsequently "ensure paternity" is a male cat's evolutionary mandate (e.g. Kerby and Macdonald 1988, Natoli and De Vito 1991, Say and Pontier 2004, Say *et al.* 1999, Yamane 1998). Male "dominance" is treated as an exclusionary factor in which some individuals deny their supposed rivals (lesser males) access to "resources," specifically estrous females (Sandell 1989). Dominance then serves as implicit agent and *cause* of the inequitable distribution of resources. Even if this were true, confirmation would offer limited explanatory power. Any measure of dominance necessarily includes the scores of all agonistic encounters, most of which are about resources of one kind or another.

Descriptions of how males really interact in an estrous group paints a contradictory picture, one of general peace and mutual tolerance, not of fierce competition for mating privileges. Fights among males in these circumstances are rare (Dards 1983, Green *et al.* 1957). Any directed aggression is likely to be by a female toward an approaching male while she is not yet receptive (Dards 1983). Males neither "guard" nor "defend" estrous females, but apparently try to mate with as many females as possible (Say and Pontier 2004, Say *et al.* 2001). Social rank among adult males – if it even exists – is therefore a dubious factor in male reproductive success (Say *et al.* 2001).

For example, Liberg and Sandell (1988: 88) stated, "According to our basic assumption, males compete for access to females." Why assume this? Competition is the condition of competing. It connotes rivalry, elements of contention, even combat. Before accepting the premise of male–male competition for females, we should examine the evidence. As discussed previously (Chapter 1), dominance can be defined in either a structural or a functional context (Bernstein 1981). Structural definitions are descriptive, reporting and quantifying interactions observed by the investigator but opaque to explanation. They are data-driven, the objective being to seek patterns in a collection of observations that could indicate dominance (Hinde and Datta 1981). Functional definitions, in contrast, are theory-driven. They too incorporate descriptions of dominance behavior but allow interpretations to be derived from its apparent function. Structural theories crowd the free-ranging cat literature; far fewer are theory-driven, offering insight into the evolutionary role of dominance.

I found no convincing structural descriptions, much less functional ones, of male cats guarding an estrous female or attempting to monopolize copulations with her, only weak and inconsistent observations that certain males are more likely than others to mate. Sometimes – but not always – they might be bigger and more aggressive. However, if an individual was truly dominant and intent on keeping the female for himself, it seems in his best interest to stay around until her estrus ended, fending off competitors and mating at every opportunity. Doing so would make him the "fittest" by eliminating any risk of sperm competition from rival males and the possibility of multiple paternity. But this view of fitness is incompatible with both behavioral observations and the reality of multiple paternity. Male–male and male–female interactions precede – and ultimately dictate – the subsequent meeting of ova and spermatozoa.

According to traditional theory, a male can enhance his fitness by monopolizing matings with as many females as possible while simultaneously preventing his rivals from mating. A female should mate promiscuously to assure fertilization. A cat that proves itself dominant in a majority of dyadic interactions with other males is not always the winner in terms of number of matings, perhaps because "winner" in this context is applied incorrectly. Dominant males are identified through consistent asymmetry of dyadic contests, which supposedly reveals that "winners" gain mating opportunities while denying or limiting access to females by the "losers." However, the outcome of round-robin dyadic contests offer limited predictive value in this regard. The pattern among male cats is to *not* monopolize matings. Males thought to be dominant seldom attempt to interfere when subordinate males mate (Leyhausen 1973). The Roman stray designated GI was the winner of 36 dyadic encounters and a loser only twice, although other males mated with a female while he courted her and other "lower ranked" males interfered while he copulated (Natoli and De Vito 1991). And still he did not react aggressively toward these presumed rivals.

Males are rarely aggressive toward estrous females, and most interaction is amiable, especially if individuals are familiar (Dards 1983). A male encountering a receptive female has two choices. He can stay and copulate with her repeatedly or wander off seeking other prospects. Even a male with a home range overlapping an estrous female's is not guaranteed monogamy through her estrus, but this is an unlikely problem. Other roaming males are apt to find and mate with her too. According to Sandell (1989: 175), "If the most rewarding tactic for the dominant male is to roam, it is impossible for other males to have exclusive ranges." This presumes dominance to be mitigating, which it doubtfully is. By occupying home ranges instead of defending territories (Chapter 2), free-ranging male cats are free to seek mates wherever they might be without encountering the aggression of other males, and one male's range during the breeding season is likely to overlap that of another also seeking females (Turner and Mertens 1986). This is hardly a suitable "strategy" for monopolization, nor do promiscuous mating systems conform with traditional notions of fitness seen in terms of minimizing sperm competition among putative rivals.

4.7 Female choice

Animal mating systems, driven by the urges of individuals, are not the products of species evolution (Clutton-Brock 1989). Variation in a female cat's receptivity can be caused by several factors, including degree of sexual experience (Dards 1983) and behavior of the male (Rosenblatt and Schneirla 1962). Females, although commonly mating with multiple males, appear to show some degree of choice, occasionally preferring certain individuals over others (Leyhausen 1973), notably previous partners (Rosenblatt and Schneirla 1962). A male's aggressiveness and how he approaches the female can also affect her receptiveness (Section 4.5).

Clutton-Brock and McAuliffe (2009: 5) defined *female choice* as instances in which "females show an active preference for mating with particular categories of males, whether or not matings lead to conception." This requires that females recognize certain attributes, or traits, in males that make one individual more appealing than another. Among mammals generally, these includes such factors as maturity and dominance or social status, in addition to qualities indicating degree of fertility, symmetry, coloration, vocal or olfactory display, degree of relatedness or familiarity, or previous mating success. Most mammals, as mentioned, are polygynous. If female choice occurs in the domestic cat, its nuances are subtle.

A salient feature of polygyny is that males are usually larger than females, and they use their greater size and strength to overcome female mate choice (Clutton-Brock and Parker 1995). Male cats are ~30% larger than females at adulthood (Chapters 3 and 5), not a big disparity compared with other polygynous species. Nonetheless, sexually experienced, aggressive males frequently overpower estrous females and mate with them (Section 4.5), effectively constraining any opportunity they have to choose a mate. Here the male dominates the female during an act-specific situation. However, because cats are asocial, adults of the opposite sex interact mostly during reproduction, and these acts of dominance are therefore not incidences of *social* dominance. In social species, females often select dominant males in preference to subordinates, and female choice is then based on a male's status, not simply the extent of his aggressiveness.

Female choice in social mammals is also constrained when opportunities to choose a mate are restricted by males competing to monopolize access to them (Clutton-Brock and McAuliffe 2009), an unlikely situation in estrous groups of domestic cats. Not all attempts at mounting are accepted, and female choice is perhaps expressed by rejection. For example, 8 females monitored on Ainoshima Island were courted by 9–19 males but mated with only 3–9 (Ishida *et al.* 2001). Females having shorter durations of estrus and smaller body mass were least selective, as were those courted by fewer males. Females reject unwanted suitors either before or after mounting but before copulation by pawing and hissing at them, running away, or refusing to assume the mating pose (Ishida *et al.* 2001). According to Beaver (1977), an estrous female does not necessarily mate with the male appearing to be the most dominant of her suitors. Female choice might explain the lack of aggression among males of an estrous group: why risk aggressive encounters when mate selection is not entirely their choice (see Smuts 1981)?

Other differences are apparent in female–male interactions. Some laboratory males achieve successful intromission more often with certain females than with others (Whalen 1963a), and male–female interaction is what ultimately determines the frequency and success of intromission. In addition, males that have become sexually exhausted from mating with the same female could experience renewed arousal when presented with a new partner. The reverse was not true, and a sexually exhausted female remained so when offered a new male. This suggests that, in males, performance must be based partly on the rewarding properties of mating and its reinforcement by repetition (Whalen 1963a).

Although mating behavior is innate and hormonally driven, its efficiency is learned. Naïve females in the laboratory have been seen resisting males until after the first intromission (Whalen 1963b), indicating an incomplete mating pattern. Receptivity then increases progressively. After ~20 intromissions the females accepted strange males readily. Whalen (1963b: 463) wrote: "Hormones induce the appropriate postural responses, yet mating experience seems necessary to condition behavioural receptivity."

5 Development

5.1 Introduction

Generalities regarding postnatal development in cats are confounded by differing rates of ontogenesis between individuals and even from one litter to another (Martin and Bateson 1985a). Males do not provision the female during gestation or while rearing her young, nor does he assist with rearing in any way. His absence, combined with the severe physiological drain caused by lactation, places the female in a vulnerable position, especially if food is hard to procure.

5.2 Intrauterine development

After mating, progesterone remains at baseline levels (<1 ng/mL) until the LH surge, when it rises with onset of ovulation at 40–72 h after LH spikes (Schmidt *et al.* 1983, Wildt *et al.* 1981). Goodrowe *et al.* (1989) and Wildt *et al.* (1981) showed that progesterone peaks at ~40–90 ng/mL 9–23 d following ovulation where it remains until day 41 or so, at which time the concentration decreases to baseline (Fig. 4.3c). Transport of zygotes through the oviduct and into the uterus takes 4–5 d after coitus and is accompanied by estrogen sensitization and stimulation by progesterone (Herron and Sis 1974). Implantation occurs at days 12–15 (Stabenfeldt and Shille 1977). Radiographs first show uterine enlargement at day 21, and by day 35 the swelling is visible externally (Stabenfeldt and Shille 1977). Confirmation of pregnancy can sometimes be obtained radiographically as early as day 17 post-coitus and with more certainty on days 21–25, as seen by embryonic swellings in the uterus and its visible enlargement and displacement (Boyd 1971, Tiedemann and Henschel 1973). Partial ossification is often apparent at day 40, and ultrasound diagnosis can be made at 30–35 d (Stabenfeldt and Shille 1977). The mammary glands feel turgid by day 45 (Stabenfeldt and Shille 1977); by day 50 the fetuses are about 100 mm long, and pregnancy can be diagnosed by palpating them 26–30 d following coitus (Stabenfeldt and Shille 1977). Palpation must be gentle so as not to induce abortion (Christiansen 1984: 277).

Fetal growth in length is linear (Fig. 5.1), and gestation through diestrus lasts ~63 d with a range of 56–71 d (Concannon 1991, Dawson 1946, 1952; Foster and Hisaw 1935; Nelson *et al.* 1969; Schmidt *et al.* 1983; Scott 1955; Scott and Lloyd-Jacob 1955; Sojka *et al.* 1970; Stabenfeldt and Shille 1977; Verhage *et al.* 1976), similar to the domestic dog. During gestation, prolactin rises from 4–12 to 20–45 ng/mL from days 40 to 60, then to 35–55 ng/mL 2–3 d before parturition (Banks *et al.* 1983). It stays high for the first 4 weeks of lactation, diminishes during weeks 5 and 6, and falls to baseline 1–2 weeks after weaning. Some pregnant cats display signs of estrous behavior during gestation (Beaver 1977), but this is atypical. When it occurs, the timing is usually between days 21 and 24 of pregnancy (references in Christiansen 1984: 275).

Free-ranging Cats: Behavior, Ecology, Management, First Edition. Stephen Spotte.
© 2014 John Wiley & Sons, Ltd. Published 2014 by John Wiley & Sons, Ltd.
Companion Website: www.wiley.com/go/spotte/cats

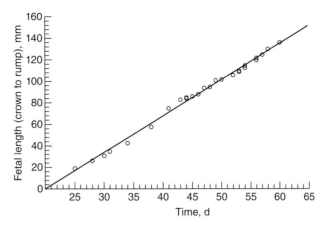

Fig. 5.1 Linear regression plot of the crown–rump lengths of fetal cats measured *ex utero* from day 25 after coitus to parturition ($n=33$). Parturition occurred at 145 mm (cumulative days after coitus not stated). Growth and time demonstrate tight correlation ($R=0.997$, $F_{1,30}=5431.1$, $\beta=-68.53$, $p=0.00$, $n=32$). Source: Data from Boyd 1971, table 1, p. 502.

Neither LH nor progesterone fluctuates much after parturition, but estradiol-17β surges episodically at intervals of 10–18 d concomitant with follicular activity (Goodrowe et al. 1989). Relaxin can be detected around day 25 of gestation (Stewart and Stabenfeldt 1985). It spikes at 4–11 ng/mL from days 40 to 50, diminishing by half from then until parturition when it drops abruptly, becoming undetectable a day after the kittens are born.

The domestic cat is *polytocous*, producing several offspring at a time, each group called a *litter*. Adult females in a group of laboratory cats kept under constant long-day conditions (14L : 10D) gave birth to an average of 9.1 kittens in 2.1 litters annually over 10 y (Robinson and Cox 1970), and some highly domesticated breeds (e.g. the Angora) might go into estrus three times annually (Liche 1939). Forbush (1916: 19) stated (without citing evidence), "Cats are known to have from two to four broods yearly. ... " Free-ranging cats usually have 2 litters per year, but some can produce 3 (Turner and Mertens (1986). Feral cats in southeastern Australia averaged about 2 litters annually (Jones and Coman 1982a).

When reproduction is bimodal the first litter is typically born in spring, the second in summer or early autumn (e.g. Jones and Coman 1982a, Nutter et al. 2004). Not uncommonly, the second peak is lower (Fig. 5.2). Females at Macquarie Island in the Southern Hemisphere sometimes had 2 litters per year with peaks in November–December and February–March (Brothers et al. 1985). Pregnant females were found only between 6 October and 26 April, lactating females between 9 December and 26 May. In similar fashion, females at the Crozet Islands (46°06′S, 50°14′E) reproduced from October–May with bimodal peaks in November and April, austral spring and autumn (Derenne and Mougin 1976). Females in the Kerguelen Islands (49°25′S, 69°58′E) also showed a bimodal pattern, giving birth in September–October and January–February (Pascal 1980). Derenne (1976) reported Marion Island (46°52′S, 37°51′E) cats reproducing from September–March, but did not state whether the pattern was unimodal or bimodal. Urban strays in Jerusalem evidently displayed unimodal reproduction, the females coming into estrus in January and February (Mirmovitch 1995). Feral cats in an agricultural area of northern Italy gave birth in May (Genovesi et al. 1995); those in Scotland bred throughout the year (Corbett 1978). According to Apps (1983), on Dassen Island, South Africa (33°25′S 18°5′E), births of feral cats occurred 8 months of the year with a major hiatus in winter (May–July) and a shorter one in summer (December).

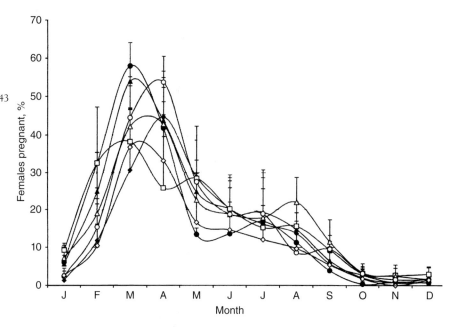

Fig. 5.2 Percentage by month ($\bar{x} \pm$ SD) of female stray cats in the United States that were found to be pregnant (years 1993–2004). Total n (males + females) = 103 643 of which 53.4% ($\pm$2.2 SD, range = 50.9–56.5%) were intact females. Of these, 15.9% (~55 345 cats) were pregnant. Note that most pregnancies occurred in spring regardless of location (Hawai'i to North Carolina). Source: Wallace and Levy 2006. Reproduced with permission of Sage Publications Ltd.

5.3 Dens

Free-ranging cats do not excavate dens but select ready-made sites offering shelter and minimal disturbance (Deag et al. 1988, Izawa and Ono 1986). Feral cats give birth in holes and depressions at the bases of trees and stumps, in hollow trees and logs, in crevices among fallen logs, and in dense thickets (Fitzgerald and Karl 1986, Forbush 1916: 27, Harper 2004, Jones 1989). Pregnant females at Macquarie Island used rabbit burrows as dens (Brothers et al. 1985), as do feral cats throughout Australia (Jones 1989). The sizes of rabbit burrows are sometimes borderline: Norbury et al. (1998b) reported that two radio-collared feral cats at New Zealand's South Island became stuck in the entrances and died.

Tree hollows offer not just protection from rain, snow, sun, and wind, but a stable temperature. Stains (1961) monitored temperatures inside a hollow tree used by raccoons (*Procyon lotor*) in Illinois. The den's entrance was 14 cm high, 11.5 cm wide, and faced north. The entrance opened into a cavity of 74-cm diameter enclosed in a wall of living wood 7.6 cm thick. Mean temperature inside the hollow changed 0.32°C per 1.75°C of mean change outside.

Urban and suburban strays and feral cats establish birth sites in locations that can be hidden from dogs and other predators and are protected from the weather, such as inside boats and vacant buildings, underneath bushes, and on cultivated land (Izawa and Ono 1986). The birth site is not necessarily permanent. Lactating females often move their kittens among shelters (Baerends-van Roon and Baerends 1979: 12; Fitzgerald and Karl 1986), especially after a disturbance (Macdonald et al. 1987: 25; Smucker et al. 2000).

5.4 Parturition

Parturition consists of four phases divided arbitrarily: contractions, emergence of the fetus, delivery of the fetus, and passing of the placenta (Baerends-van Roon and Baerends 1979: 11–12, Fox 1975, Rosenblatt and Schneirla 1962). A mother cat seeks a secluded location (Cooper 1944). Parturition is preceded by restlessness, inappetence, and decline in rectal temperature that becomes lowest 24–48 h prior to delivery (Stabenfeldt and Shille 1977). Other behavioral changes can include heightened "nesting" behavior (e.g. turning around in the birth area, clawing at the bedding, "covering up" the bedding as if after urination or defecation), and aggression (Cooper 1944, Liberg 1980).

With onset of parturition the vulva slackens and expels clear mucus. The mother squats or lies on one side, contractions begin, and a little more of the fetus protrudes with each. Secretions become heavier. Delivery of a kitten varies from a few minutes to an hour or so (Concannon 1991, Cooper 1944), although intervals of 11–24 h between births are not uncommon (Concannon 1991). Fetal membranes are ingested postpartum (Fox 1975, Stabenfeldt and Shille 1977). The mother cat licks herself, the kittens as delivered, and the surrounding area, ridding it of fluids (Rosenblatt 1971).

As defined by Millward (1995: 94), *growth* is "irreversible structural change. ... " Its principal feature involves interactions among linear extension of bone, protein deposition in skeletal muscle, and dietary intake of protein (Millward 1995). Kittens at birth are ~90–125 g (Hall and Pierce 1934, Nelson *et al.* 1969) and grow at a mean rate of ~13 g/d over the following 56–70 d (Martin 1986; Wichert *et al.* 2009, 2011), although rates >30 g/d have been reported (Morris 1999). Kittens attain 700 g at weaning, 1 kg at 10–12 weeks, and 2.0–2.5 kg by 20 weeks (Martin 1986, Scott and Scott 1957). The comparative growth of males versus females varies. Neonate males and females are similar in size with females tending to be slightly heavier, but males catch up and eventually pass them (Latimer and Ibsen 1932, Rosenstein and Berman 1973), as depicted (Fig. 5.3). Another study (Deag *et al.* 1987) found the opposite: males were significantly heavier than females at first, but during week 1 the females gained weight faster, catching up with the males until in weeks 2–3 the difference disappeared. Martin (1986) reported that males outgrew female littermates from birth onward and were significantly heavier by 4 weeks despite not nursing more often. Males grew at 13.8 g/d, or significantly faster than females (11.7 g/d).

Despite differences in early growth pattern, males at maturity are consistently ~1 kg heavier than females (Yamane *et al.* 1994), but a true size disparity usually does not appear until week 8 or so (Deag *et al.* 1987). Feral and stray males, like house and laboratory cats, are usually heavier than females. Read and Bowen (2001) collected a series of free-ranging cats at Roxby Downs in the arid northern part of South Australia with these results: males ($\bar{x}$ = 3.9 kg ± 0.10 SEM, n = 263); nonpregnant females ($\bar{x}$ = 2.8 kg ± 0.07 SEM, n = 217); pregnant females ($\bar{x}$ = 3.8 kg ± 0.28 SEM, n = 15). One male was 7.3 kg.

Nutrition naturally affects growth rate. Cat milk, like that of dogs and some primates, contains components that facilitate development. One is bile-salt-activated lipase, which aids digestion of lipids and stimulates growth (Adkins *et al.* 1997; Wang *et al.* 1989). Kittens fed a milk-replacement formula containing purified human milk bile-salt-activated lipase grew twice as fast as kittens fed the formula without it (Wang *et al.* 1989). Milk composition varies naturally over the course of lactation (Adkins *et al.* 1997; Keen *et al.* 1982). Protein decreases from days 1–3 before rising steadily until lactation ends; alternatively, it rises steadily and significantly from the start (Table 5.1) from <4 to ~7.5% at day 43+ of lactation. Differences at the start can probably be attributed to noise in the data. The ratio of nitrogen in casein compared with whey rises slowly from 40/60 in the colostrum to 56/44 by the end of lactation. Fat also rises significantly throughout, from 1.4% at the start to ~5% late in lactation.

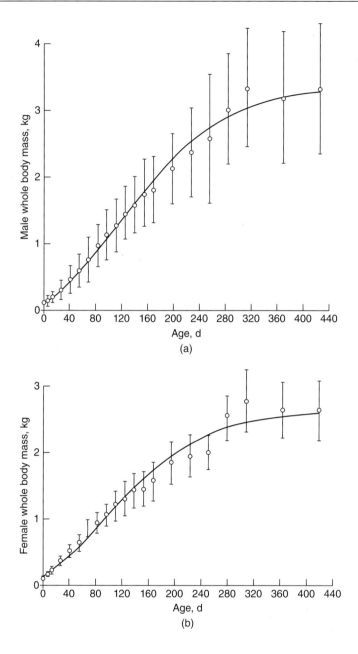

Fig. 5.3 Changes in whole body mass with age ($\bar{x} \pm 1$ SD) of cats reared in a captive group. Sample size unknown, but >500. Curves are least-squares fits to the data using a form of the Gompertz growth equation: (a) males, (b) females. Source: Rosenstein and Berman 1973, figures 1 and 2, p. 576. Reproduced with permission of the American Veterinary Medical Association.

Amino acids also show changes over time (Table 5.2). Total lipid concentration in colostrum was indistinguishable from that of milk, but energy content of the milk dropped from ~5442.8 to 3558.78 kJ/L on day 3 and rose thereafter through lactation. Carbohydrate changed little, holding at ~4%. Lactose increased from a concentration of 29.9 g/L in colostrum to 39–42 g/L in milk.

Litter size varies from 1 to 9 (Forbush 1916: 19). The relationship between litter size and birth size varies. Body mass at birth is apparently affected little by the number of kittens born, although the heaviest birth masses cluster at litter sizes of 3–6 in litters

Table 5.1 Changes in the proximate composition of cat milk during lactation ($\bar{x} \pm$ SEM), n in parentheses.

Days of lactation	0-2	3-7	8-14	15-21
Protein	3.97 ± 0.64 (5)	4.36 ± 0.31 (17)	4.89 ± 0.25 (15)	5.53 ± 0.39 (14)
Fat	3.39 ± 0.74 (5)	3.49 ± 0.38 (15)	3.68 ± 0.42 (15)	4.80 ± 0.32 (14)
Carbohydrate	3.57 ± 0.15 (6)	3.69 ± 0.18 (16)	3.62 ± 0.19 (15)	3.85 ± 0.18 (14)
Days of lactation	22-28	29-35	36-42	43+
Protein	6.49 ± 0.56 (11)	6.16 ± 0.34 (11)	6.55 ± 0.45 (10)	7.46 ± 0.41 (6)
Fat	5.08 ± 0.28 (11)	5.86 ± 0.47 (11)	5.47 ± 0.43 (10)	5.31 ± 0.9 (6)
Carbohydrate	3.44 ± 0.17 (11)	3.91 ± 0.15 (11)	4.10 ± 0.22 (9)	4.29 ± 0.24 (6)

Source: Adapted from Keen *et al.* 1982.

Table 5.2 Mean amino acid composition of protein in domestic cat milk during lactation (μmol/g protein; ± SEM omitted).

Amino acid	Day 1	Day 3	Day 7	Day 14	Day 20	Day 42
Alanine	539.3	570.9	529.9	508.1	504.1	473.6
Arginine	352.3	400.3	400.1	402.6	395.2	394.2
Asparagine + aspartic acid	691.0	799.0	768.6	771.1	802.6	770.3
Cysteine	130.4	149.4	122.0	116.9	111.1	93.4
Glutamine + glutamic acid	1474	1516	1581	1583	1598	1622
Glycine	240.8	222.3	198.9	193.8	202.2	186.0
Histidine	193.2	198.4	205.7	214.2	226.7	240.9
Isoleucine	290.9	277.0	291.4	308.2	307.2	318.4
Leucine	987.8	972.0	996.6	981.0	950.4	952.8
Lysine	490.8	525.6	516.4	539.8	547.7	548.1
Methionine	213.3	225.4	206.2	219.1	2050	195.8
Phenylalanine	251.3	233.7	223.8	219.2	222.4	217.0
Phosphoserine	22.8	30.6	31.6	29.7	33.4	31.4
Proline	1073	909.7	929.3	883.7	829.8	847.0
Serine	546.2	482.2	467.4	458.6	480.2	481.3
Taurine	52.5	90.0	128.3	147.5	75.8	75.0
Threonine	517.4	479.3	479.8	471.5	476.9	478.5
Tryptophan	94.1	95.2	94.1	100.0	102.2	107.0
Tyrosine	278.2	274.1	275.3	285.6	300.0	307.9
Valine	414.8	364.9	375.5	373.4	348.2	353.4

Source: Adkins *et al.* 1997. Reproduced with permission of the American Veterinary Medical Association.

ranging in number from 1 to 8 (Hall and Pierce 1934). Hall and Pierce (1934: 113, Table 1) listed mean birth mass in grams for kittens from various litter sizes. Using their data I performed a simple linear regression and found that number of kittens in a litter did not affect birth mass significantly: $R = 0.49, F_{1,5} = 0.12, p = 0.92$. In assessing 169 litters ($n = 662$ kittens), Nelson et al. (1969: 188, Table 5) found minimal differences in body masses of kittens born into litters of 1–4; in larger litters (5–7 kittens) birth mass correlated inversely with litter size (a decrease of ~5%). Factors controlling litter size are unknown (Deag et al. 1987, 1988), but primiparous mothers tend to have smaller litters (Robinson and Cox 1970), and the effect, at least in one study, was independent of age (Connelly and Todd 1972). According to Dawson (1952) and Hall and Pierce (1934), small females often have larger kittens (Fig. 5.4).

Wallace and Levy (2006) reported the average prenatal litter size of stray cats in the United States to be 4.1. The average litter in US laboratory cats is ~4 (Hall and Pierce 1934, Nelson et al. 1969, Robinson and Cox 1970, Stabenfeldt and Shille 1977),

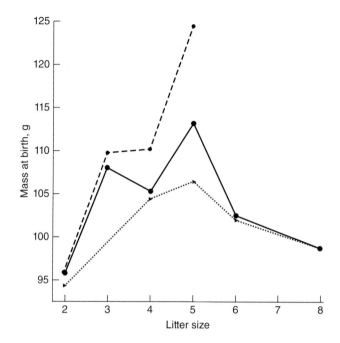

Fig. 5.4 Relation of mean mass at birth to litter size based on 33 litters. Dashed line ($n = 39$ kittens, mothers 2.38–3.45 kg); dotted line ($n = 48$ kittens, mothers 3.54–6.00 kg); solid line represents all kittens combined ($n = 128$). The regression equation is $y = 3.4004x - 68.5322$: $y =$ crown-to-rump length (mm), $x =$ time (d). Source: Data from Hall and Pierce 1934, figure 1, p. 116.

tending to be smaller in very young and very old females (Dawson 1952, Sojka et al. 1970). Generally, more males are born than females, but the difference is seldom significant. In a study conducted over several years (Nelson et al. 1969: 184, Table 1), 286 litters produced 1058 kittens (576 males, 482 females). Seventy-one litters of laboratory cats in Scotland averaged 4.4 kittens (Deag et al. 1987). Free-ranging house and farm cats in central Illinois, produced 1.6 litters/year, a litter averaging 4.4 kittens (Warner 1985). Each reproducing female averaged 7.1 kittens/year, with births of litters distributed as follows: 8% (December–February), 47% (March–May), 34% (June–August), and 11% (September–November). Prenatal litter size of free-ranging cats trapped or shot in southeastern Australia was also 4.4 (Jones and Coman 1982a) and ~4 in the southern region of Australia's Northern Territory (Strong and Low 1983). Litter size of Jerusalem's urban strays was 2.1 for 18 litters, but most kittens were not seen until ~4 weeks old and some mortality was probably missed (Mirmovitch 1995). Data on litters of strays at Ainoshima Island ranged from 2 to 5 ($\bar{x} = 3.9$) but also did not account for early postnatal mortality (Izawa and Ono 1986). Litters in the Kerguelen Islands averaged ~4 (Pascal 1980), those at Marion Island ~3 (Derenne 1976). The number of embryos in cats collected at Macquarie Island by Brothers et al. (1985) ranged from 1 to 9 ($\bar{x} = 4.7$). In a large study conducted by Nutter et al. (2004) using North Carolina strays, 334 pregnant strays averaged 1.4 litters/y. Prenatal litter size ranged from 1 to 6 ($\bar{x} = 4.2$) and was significantly larger than postnatal litter size (Nutter et al. 2004) in a similar group of cats ($n = 50$ litters). A few cats had 3 litters/y, but ordinarily when all kittens in one of the earlier two litters had died before weaning.

5.5 Early maturation

Kittens are altricial at birth (Fig. 5.5): blind, deaf, nearly helpless, and relatively immobile for the first 21 d or so (Kolb and Nonneman 1975), but capable of making forward groping movements by the end of week 3 (Leyhausen 1979: 81). The eyes usually open

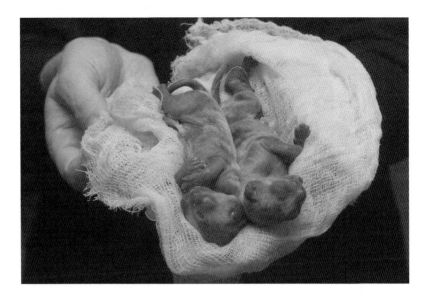

Fig. 5.5 Domestic cats are born altricial, unable to see, hear, and barely able to move. Source: © Igor Dmitriev | Dreamstime.com.

~7–13 d postpartum (Dawson 1952, Ewer 1959, Kolb and Nonneman 1975, Rogers 1932, Rosenblatt 1971, Villablanca and Olmstead 1979) but might not open fully until day 21 (Dawson 1952). Vision continues to mature at a slow pace and is not completely developed until ~ day 60 (Villablanca and Olmstead 1979). The cat's early maturation has been reported several times. Here I mainly follow interpretations of Villablanca and Olmstead (1979), but also see Baerends-van Roon and Baerends (1979: 14–34, 67–68). Kittens stand and walk imperfectly by 31.4 d (± 4.4 SD, $n = 27$), and walk and run normally by 44.4 d (± 5.4 SD, $n = 28$) with a remaining slight ataxia of the forelimbs. The external auditory canal matures in linear fashion from days 1 to 15, opening at 12 d (± 3 SD, $n = 30$), at which point its depth is ~15 mm. The adult pinna is formed by 31 d (± 4 SD, $n = 27$), although movements of the pinnae are evident from birth. Kittens orient to auditory stimuli by 5 d (± 1.5 SD, $n = 37$), but differential responsiveness to classes of stimuli are not apparent until 20 d (± 2, $n = 26$) and not clearly so until 24.6 d (± 4.5 SD, $n = 26$). Rudimentary olfactory responses are in place at birth. Neonates respond to unpleasant odors by initial sniffing followed by "escape" movements of the head, wrinkling the muzzle, vocalization, and tachypnea (rapid breathing). When the kittens are a little older these responses can include *grimace*.

Young kittens are thigmotactic, huddling together and often clustering in a heap (Fig. 5.6) during times when the mother leaves briefly to eat or drink (Rosenblatt and Schneirla 1962). Neither the invoking stimuli nor the function of this behavior has been determined. Hypotheses include olfactory, haptic, and thermal (body heat conservation) factors acting singly or together (Rosenblatt and Schneirla 1962). The last is perhaps most likely. Body temperature is regulated by the anterior hypothalamus (Adams 1963). Stability of core temperature is acquired gradually, rising from a mean rectal value of 37°C at 5 d to adult levels (~38°C) at ~28 d (Olmstead *et al.* 1979). Thermoregulation develops over the first 45 d, measuring ~37.0°C at 5 d and increasing to adult values (38.2°C) at ~7 weeks (Olmstead *et al.* 1979). When kittens were removed from the home area at ages up to ~14 d and placed at ambient temperature (23–25°C), body temperature declined at ~0.02°C/ min. From day 1 kittens showed the capacity

Fig. 5.6 Kittens often sleep in close contact, perhaps to conserve body heat or to satisfy a need to be thigmotactic. The behavior becomes less common with age. Source: © Dreamframer | Dreamstime.com.

to move toward warmth along a thermal gradient. After 45 d they responded to cold by shivering and piloerection, and panted when overheated. Survival of kittens living outdoors during cold or hot periods can be largely a matter of age. At ~15°C, body temperature drops at 0.2°C/min at age 10 d, declining to 0.1°C/min at age 40 d. Body temperature of heat-challenged kittens rises at 0.05°C/min at 10 d, declining to 0.02°C/min at 45 d.

For the first 4 weeks kittens limit movements to the small area around the birth site. They can walk at 3 weeks but not well until ~4 weeks (Rosenblatt and Schneirla 1962), and adult-like walking and running is apparent at 6–7 weeks (Villablanca and Olmstead 1979). Spatial orientation develops during the first 21 d (Rosenblatt 1971, Rosenblatt and Schneirla 1962).

The mother's nutritional status has a profound effect on the development of her kittens. During pregnancy, her gain in weight is attributable to accumulation of body reserves in addition to the fetuses, their placentas, and associated tissues (Wichert et al. 2009). Loveridge and Rivers (1989) monitored changes in body mass of 75 laboratory cats during gestation and lactation. Cats in the study gave birth to litters of 1–5 kittens and reared them successfully. Gain during pregnancy was nearly linear, the loss after parturition less so (Fig. 5.7). Mean gain between mating and parturition was 1.2 kg (±0.31 SD), a 39% increase. Litter size and pre-mating body mass were unrelated. However, mean gain rose with number of kittens in a litter, although not proportionately because even cats producing only 1 kitten gained 32%. The regression equation is:

$$y = 106x + 888.9 \qquad (5.1)$$

where y = gain in body mass (g), and x = number of kittens in a litter. Notably, the slope of the equation (106.5 g) equaled the mean individual birth mass of the kittens (106.2 g). The cat is unusual in that most weight is gained linearly, not accelerating steadily until just before parturition. Instead, it peaks by the end of the second

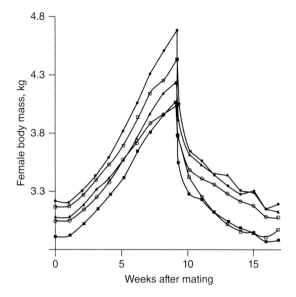

Fig. 5.7 Body masses of females during gestation and lactation producing litters of different sizes: 1 kitten (■), 2 kittens (○), 3 kittens (Δ), 4 kittens (□), 5 kittens (●). Source: Loveridge and Rivers 1989. Reproduced with permission of Cambridge University Press.

trimester, serving as an energy store to be tapped in late pregnancy for fetal growth, and afterward declines (Loveridge and Rivers 1989).

Free-ranging feral and stray cats often live where sources of food are unreliable or deficient in critical nutrients. Laboratory experiments show that kittens born to malnourished mothers suffer behavioral and developmental deficiencies seemingly serious enough to reduce longevity or otherwise impair fitness. Gallo *et al.* (1980) fed pregnant test cats a diet containing increasingly less protein beginning after week 6 of gestation: 30% protein starting day 43, 24% starting day 50, and 18% starting day 57. Controls were fed a diet containing 36% protein. Test cats gave birth to more underweight kittens, compared with controls. Those pushed away from the nursery area by larger littermates were not retrieved by the mother and died within 4 d postpartum. Underweight survivors vocalized more than controls, were often ignored by their mothers, and their postnatal gain in body mass lagged behind the controls. In a similar experiment (Gallo *et al.* 1984), test kittens showed defective locomotor development and increased loss of balance. Overall, results indicate that kittens born to undernourished mothers are at a disadvantage, suffering emotional distress, disrupted maternal attachment, and aberrant or retarded growth. This is a potential problem for stray cats scavenging on garbage deficient in protein, but unlikely in feral cats except during times of low prey abundance (Chapters 7 and 9).

Pregnant laboratory cats were purposely malnourished from the midpoint of gestation through weaning, and the kittens then fed commercial cat food *ad libitum* (i.e. were "rehabilitated") from weeks 6 to 16 (Smith and Jansen 1977a). The objective was to monitor changes in growth and certain brain components, and to identify effects of having been undernourished early in life. Weight gain of food-restricted mothers (50% less food than controls fed *ad libitum*) was 25% that of the controls from mid-gestation to parturition. Starved mothers tended to be poor caregivers. They proved less tolerant of their kittens, oriented less toward them, did little to facilitate nursing, often rebuffed or attacked kittens, and even allowed them to die of hypothermia (Smith and Jansen 1977a, 1977b).

Kittens of food-restricted mothers were slightly smaller at birth (~97 g vs. ~102 g) and remained so until ~4 weeks after start of the *ad libitum* diet. A discontinuity was then apparent in the controls, which failed to gain weight during the first week after weaning. This brief interruption in growth at weaning is typical of kittens raised under normal conditions (Deag *et al.* 1987, Smith and Jansen 1977b). Mean body mass of kittens of underfed mothers was ~55% of the controls at 6 weeks when food restriction stopped. Weight gain afterward was similar in both groups. When the experiment ended at 16 weeks the mean body masses ($\pm$ SEM) were: control males (1414 g $\pm$ 90), rehabilitated males (1371 g $\pm$ 119), control females (1257 g $\pm$ 61), rehabilitated females (1159 g $\pm$ 83).

About 73% of adult brain cells form during the first 6 weeks of life (Smith and Jansen 1977b), and kittens of malnourished mothers showed substantial changes in brain composition in addition to delayed growth, impaired learning, and reduced motor skills (Simonson *et al.* 1972; Smith and Jansen 1977a, 1977b). At 6 weeks their cerebrum, cerebellum, and brain stems were significantly smaller, and certain measures of brain chemistry (e.g. cholesterol, protein/DNA ratio, galactolipid) registered in abnormal ranges (Smith and Jansen 1977b). At 16 weeks, kittens fed normal rations post-weaning (i.e. "rehabilitated") nonetheless appeared clumsy, their coordination compromised. The test females ran about aimlessly and climbed less than the female controls reared normally; males jumped more than the male controls (Smith and Jansen 1977b). Also at 16 weeks, while males reared normally spend more time at object play (Chapter 6) and less time interacting with other cats, the males of malnourished mothers preferred to engage in aggressive social play (Smith and Jansen 1977b). Compared with controls, "rehabilitated" kittens of both sexes explored less and spent more time stalking others.

Of the long-chain polyunsaturated fatty acids (PUFAs), arachidonic acid (20:4n6) and docosahexaenoic acid (22:6n3) are crucial lipid components of the brain and retina (Green and Yavin 1993), making up 20–25% of the brain's dry mass (Bourre *et al.* 1991). These are essential fatty acids (EFAs) obtained only from food. General signs of EFA deficiencies in cats include fatty liver degeneration, excessive fat deposition in the kidneys, dystrophic mineralization of the adrenal glands, testis degeneration, and hyperkeratosis of the skin (MacDonald *et al.* 1984b). Dietary deficiencies of linolenic acid, eicosapentaenoic acid, 20:5n3, and perhaps other PUFAs in pregnant and lactating rodents and ungulates can result in offspring with compromised vision or brain function, manifested as disabilities in learning and emotional activity (Duvaux-Ponter *et al.* 2008). Whether this could occur in kittens during development is unknown. Docosahexaenoic and eicosapentaenoic acids are derivatives of α-linolenate (18:3n3) and discussed in Chapter 7. The cat's capacity to synthesize fatty acids from glucose or its precursors and thus "spare" amino acid oxidation is beneficial when high-protein foods are available only sporadically (Silva and Mercer 1985), which is possible for feral and stray cats.

5.6 Nursing

Kittens begin nursing within 30 min or so of birth, or even before younger littermates have been born (Cooper 1944), and almost always within 1–2 h (Ewer 1961). The cat has 4 pairs of mammary glands, 2 pairs on the thorax, the remaining pair on the abdomen. They rarely lie exactly opposite (Schummer *et al.* 1981: 493). Kittens quickly develop specific teat, or nipple, preferences (Deag *et al.* 1988; Ewer 1959, 1961; Rosenblatt 1971; Rosenblatt and Schneirla 1962). The behavior is learned, starts within a few hours of birth, is established by the end of day 3 (Ewer 1959, Rosenblatt and Schneirla 1962, Rosenblatt et al. 1961), and eventually attenuates, partly because as kittens age they nurse less often as a group (Ewer 1959). During the first 2 weeks the female spends ~35% of her time nursing (Rosenblatt and Schneirla 1962). She approaches the kittens and arouses them, lies on her side, and presents her mammary glands. Neonates nurse almost continuously during the first 24 h (Dawson 1952). Kittens recognize their general location as early as day 4, and home orientation begins ~day 5, as shown when they try to crawl back to the natal area after having been moved 45 cm away (Rosenblatt 1971).

Nursing behavior from birth to weaning is well described (e.g. Fox 1975, Rosenblatt 1971, Rosenblatt and Schneirla 1962, Rosenblatt et al. 1961). For the first 2–3 weeks the mother initiates nursing. She visits the kittens often, lying down with them in the rearing area and allowing them to attach to her nipples. She grooms them by licking their fur and also licks the anogenital regions of each kitten, causing it to urinate and defecate reflexively, and then ingests the waste materials, which keeps the rearing area clean. Without this stimulation they would hold back excretory products and soon die (Rosenblatt Schneirla 1962). About halfway through week 3 the kittens start to move away from the rearing area intermittently, and the mother retrieves them. Meetings to nurse gradually involve mutual approaches by kittens and mother, and nursing can extend from the rearing area to spaces nearby (Ewer 1961). On becoming ambulatory the kittens often initiate nursing (Rosenblatt et al. 1961, Rosenblatt 1971). In the last stage of nursing they approach the female increasingly more often, following her if she moves away. The mother then begins leaving them for longer periods. Starting at week 5 or so the kittens can consume foods other than milk, and maternal care diminishes further (Rosenblatt and Schneirla 1962). By the end of week 5 a mother cat is nursing her kittens <20% of her time (Schneirla et al. 1963) and delivering dead prey to them instead.

Behavioral estrus in the mother is ordinarily not evident during lactation (Foster and Hisaw 1935, Schmidt et al. 1983), although some cats have been reported to enter estrus while still nursing, in which case polyestrus commences 4–6 weeks into lactation (Scott 1955, Scott and Lloyd-Jacob 1955). Thus the current litter is usually – but not always – weaned before onset of the next estrus (Scott 1955).

Folliculogenesis ordinarily begins after weaning, causing estradiol and LH to rise above baseline and triggering the onset of estrus within 17 d (Schmidt et al. 1983). If ovulation occurs the rise in progesterone is delayed, and spikes during the first week after mating are half as large as those of cats that have not just reared a litter (Schmidt et al. 1983).

Kittens in small litters have a nutritional advantage of larger proportions of milk available, hastening their growth potential (Fig. 5.8). In a study by Deag et al. (1987), kittens >8 weeks grew at 7.3 g/d in litters of 7 and 8 compared with 13.7 g/d for kittens in litters of 2. Mother cats increase milk production with larger litters, although not in proportion to the number of kittens.

5.7 Weaning

Weaning produces substantial changes in a kitten's nutritional requirements and behavior and culminates in complete independence (Martin 1984a, 1986). Lactation extends 50–60 d, or roughly 6 weeks (Concannon 1991), although at 4–6 weeks a kitten can take solid food (Dawson 1952). Consumption of solid food usually commences at 4–5 weeks when kittens begin to wean themselves (Deag et al. 1987,

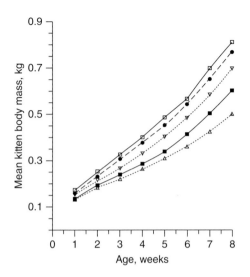

Fig. 5.8 Relationship between mean kitten body mass and age for 71 litters ($\bar{x}=4.4$ kittens/litter) of different sizes ($n = 312$ total kittens). Litter sizes (number of kittens): □ = 2, ● = 3, ▽ = 4, ■ = 5 and 6, △ = 7 and 8. Note that growth discontinuities in week 5 are not apparent, concealed by individual differences. In this study they become visible only when individual litters were examined. The mean age at discontinuity of growth was day 31.6, similar to the age of 29.7 d reported by Bateson and Young (1981). Source: Deag et al. 1987, figure 3, p. 161. Reproduced with permission of John Wiley & Sons.

Martin 1986, Rosenblatt and Schneirla 1962, Schneirla et al. 1963), complete independence taking place ~4 weeks later (Jones 1989, Rosenblatt et al. 1961). Most kittens start eating solid food at 27–30 d (Hall and Pierce 1934, Martin 1986), and weaning is complete by day 48 or so (Martin 1986). Martin (1984a) considered weaning to be a process, or phase, not an event, during which parental investment drops sharply. Its progression can be charted in the declining volume of milk transferred from mother to young. Until weaning commences a kitten is completely dependent on its mother to supply all nutrition needed for physiological functions (e.g. basal metabolism, thermoregulation, growth, locomotor activity).

Weaning tends to begin earlier in smaller kittens, presumably because the mother's supply of milk has become inadequate, and heavier mothers are able to delay the process (Deag et al. 1987). Some think that mother cats can adjust weaning to environmental conditions (Bateson 1981). Survival of feral and stray kittens likely demonstrates a positive correlation with body mass at weaning. Deag et al. (1987) speculated that infanticide of her own offspring by a free-ranging cat might be a consequence of large litters delivered during times of food deprivation, its occurrence associated with the mother's body mass at parturition and not just her overall physical condition. Nutter et al. (2004) also mentioned a possible incidence. When infanticide does occur the victims are probably not yet weaned, and documenting their disappearance could often be difficult.

Weaning does not necessarily coincide with the appearance of adult dentition. Milk teeth are apparent between days 11 and 35, and permanent teeth replace them at 4–7 months (Dawson 1952). Of these, canines and incisors break through first, the second upper premolars last (Dawson 1952). Weaning requires correct timing of changes in the mother's behavior and physiology with the developmental stage of her kittens (Koepke and Pribram 1971). Through the nursing period the mother makes herself increasingly less available to be suckled. Her behavior during this time ranges from actively encouraging her kittens to nurse immediately postpartum to gradual disinterest over the next few weeks and ultimately hostility and avoidance (Koepke and Pribram 1971, Martin 1986, Panaman 1981, Rosenblatt and Schneirla 1962). In

Fig. 5.9 A stray cat nurses her kittens, which are nearly weaning age. Source: © LohKH | Dreamstime.com.

between, mother and offspring begin a process of approaching each other less often (Martin 1986). Weaning inhibits lactation, and the mammary glands run dry by 5 d post-weaning (Dawson 1952). Kittens near weaning are nearly as big as their mother (Fig. 5.9).

5.8 Survival

Even in laboratory-reared cats the survival of every kitten is not assured. Some are stillborn; others die between birth and weaning (Robinson and Cox 1970). Development is an unpredictable sequence from pre-conception to weaning of a live kitten. Not all follicles mature into corpora lutea and ovulate (Chapter 4). For example, from an average of 4.9 CL detected per female, 67.4% ovulated and produced a kitten (Schmidt et al. 1983). Casual observation often misses early mortality because kittens might not emerge from the den until full weaning at ~8 weeks (Denny et al. 2002, Jones 1989, Jones and Coman 1982b). Aborted fetuses and stillbirths are easily overlooked if the female eats them (Scott 1955), and intrauterine mortality and resorption of embryos is impossible to monitor. Females occasionally consume their own newborns accidentally with the fetal membranes (Cooper 1944). Finally, all individuals of a large litter (>8) seldom survive to weaning (Dawson 1952).

Survival of stray cats can be brief even if ample food is available (Page et al. 1992). Survival of kittens to 10 months at Ainoshima Island was 9.5% (Izawa and Ono 1986) and 16% among urban strays in Jerusalem aged <6 months (Mirmovitch 1995). Mortality at this location was probably higher because most kittens were not seen until at least 4 weeks old. Turnover in feral cats (emigration and mortality) living completely outdoors at a location in northern Italy was 92% over 2 y (Genovesi et al. 1995), an estimate based on individuals seen one year but not the next. Nutter et al. (2004) monitored 169 stray kittens. Their mothers had been given supplementary food, but 127 (75%) nonetheless died ($n = 87$) or disappeared ($n = 40$) prior to age 6 months. The mother's experience (i.e. maternal parity) was not a significant factor. Most deaths were trauma-induced (e.g. dogs, encounters with motor vehicles).

The survival of young appears to be higher in low-density, expanding populations, but this is not necessarily a trend. At the time of van Aarde's (1978) publication, the cat population at Marion Island had been expanding at 23.3% annually for 26 y. Nonetheless, the early mortality of kittens was 42%. Dispersal from the natal areas and survival of juveniles were higher in the Kerguelen Islands when the adult population was small (Devillard et al. 2011). Climate is a factor. The 5-month survival (January–June 1980) of kittens born into the favorable Mediterranean climate of Dassen Island, South Africa, was 85.7% (Apps 1983). The sample size was small (~3 litters), but in June 1980, 51–56% of all cats at this location were <1 y old. The feral cat population in semi-arid Hattah-Kulkyne National Park, Victoria, southeastern Australia, was maximum in summer when the age profile of the population was 57% adults, 11% adolescents, and 33% juveniles (Jones and Coman 1982b). Few adolescents and juveniles survived into winter. Compare this with the situation on Marion Island where the climate is far harsher. During the 1970s the combined neonatal and preweaning mortality was an estimated 42% and 37.9% for cats aged 1–12 months (van Aarde 1984). The overall annual mortality of 79.9% in the first year of life left ~20% of kittens born surviving to potential reproductive age. The survey revealed 54% of the population to be <2 y and 71% of cats were <3 y. However, cats that lived 3–4 y had a survival rate of 93.7%, and some of these attained maximum longevity of 9 y (females) and 8 y (males).

Availability of food is a crucial factor affecting survival of young cats. Juvenile mortality restrains population growth during times when prey is limiting (Jones and Coman 1982b). At Stewart Island/Rakiura, New Zealand, this was clear from growth of the cat population following rat plagues (Harper 2004). At Macquarie Island the estimated survival to 6 months was <43% (Brothers et al. 1985). Most mortality occurred in winter when prey (rabbits and burrowing petrels) were scarce. Feral cats at Grand Terre, Kerguelen Islands, experienced highest mortality during the first 3 months of austral winter, or May–July (Devillard et al. 2011).

House cats survived longest among free-ranging cats monitored for a year in Caldwell, Texas, followed by strays given supplementary food. Feral cats, which by definition are not fed, had the poorest survival (Schmidt et al. 2007). Although sample sizes were small, feral cats also had fewer litters per year ($\bar{x} = 1.0$) than strays ($\bar{x} = 1.6$). Average litter size of feral females was 3.5 of which reproductive success was 1.75. Strays had the same average litter size (3.6) but higher reproductive success (2.75), a difference of 36%. Reproductive success was defined as number of kittens surviving ≥12 weeks. Horn et al. (2011) projected that 50% of feral cats in an eastern Illinois farming region would die within 392 d, but that 92% of house cats would still be alive after 596 d. Despite better care, free-ranging house and farm cats in central Illinois, although fed daily, rarely survived past 5 y, and <1% lived to be 7 y or older (Warner 1985).

Still other factors affect survival of free-ranging cats. Sterilization, for example, can increase longevity (Levy and Crawford 2004), and West (1979) made the anecdotal comment that adult male cats sometimes kill young free-ranging kittens. This has been documented (Macdonald et al. 1987: 24–25), but the behavior is rare. West did not speculate about whether the killings were acts of predation. The kittens recorded by Macdonald and colleagues were evidently not eaten. Adult male cats usually ignore kittens where ample food is available (e.g. Izawa and Ono 1986). How these incidences affect populations is unknown.

Individual cats vary in their motivation and skill to hunt. Being an indifferent or poor hunter poses no liability to a pampered house cat. A feral cat unable to kill faces starvation. Cats also vary in the learning time needed to hunt competently, and in the path taken to achieve hunting prowess. Influential factors include early experience with prey, time spent in the company of littermates, and behavior of the mother (Adamec et al. 1980a; Baerends-van Roon and Baerends 1979: 44; Caro 1980a, 1980b, 1980c; Kuo 1930). Laboratory kittens exposed to prey when the mother was present demonstrated increased killing rates, and this and other predatory behaviors persisted into adulthood (Caro 1980c, Kuo 1930).

An upbringing of impoverished experience is not always detrimental as an adult (Caro 1980a, 1980b, 1980c, Leyhausen 1979: 90). Play experience (Chapter 6), for example, seems unnecessary for development of the basic predatory movements used by domestic cats (Martin and Caro 1985), although object play between 4 and 12 weeks is apparently required, or at least helpful, in developing normal predatory behavior (Caro 1980c). Ontogenetic buffering evens out and lessens variation in experience during development, as embodied in *equifinality,* a concept positing that maturation is flexible and organisms can arrive at the same final status in different ways despite what we perceive as handicaps or disadvantages. Thus a kitten can still achieve adult competency despite its mother's aberrant or indifferent behavior, competitive littermates, or restricted play opportunities. These and other factors sometimes influence adult competency, but a determinant pathway to maturation is not always essential. Poorly performing kittens often recover to become effective adults (Baerends-van Roon and Baerends 1979: 58, Caro 1979).

Transformation from dependent kitten to skilled predator is delayed until sensory and mental maturity permits escape from ontogeny's pull. One necessary development is *object permanence* (Gruber et al. 1971: 9), the knowledge that an object exists in continuous time and space, "that, despite its disappearance from view in a variety of circumstances, it can easily be recovered; and that, when recovered, it will be the same object in all its physical attributes as when it disappeared." As a predator a cat spends part of its day or night stalking and pursuing prey, and on occasion the animal it stalks or chases disappears. The hunting cat thus needs a way to predict where its prey will reappear based on the location where it vanished. In laboratory tests, kittens at 16 weeks can visually track the trajectory of a disappearing object, if distractedly. Object permanence is not completely developed until ~24 weeks (Gruber et al. 1971), or about the time of full independence from the mother.

5.9 Effect of early weaning and separation

Cats weaned prematurely sometimes demonstrate abnormal behavior, including insomnia, excessive displays of affective behavior, impaired capacity to learn, increased likelihood of gastric ulcers, and other disorders (Martin 1986, Seitz 1959). Weaning even slightly earlier than normal can also have pronounced effects on the play behavior of kittens (Martin 1984b, Martin and Bateson 1985b, Wyrwicka 1978).

Some free-ranging cats no doubt become malnourished during lactation. Laboratory tests show that depriving the mother of food results in rapid loss of body mass and changes in her behavior (Martin 1986). She becomes intolerant of her kittens, often avoiding them and rebuffing their attempts to nurse. This can escalate to open aggression when they try to approach. The kittens become distressed, stop playing, and lose weight. If early weaning then takes place it happens abruptly and agonistically. These responses are reversible if the mother is again provided adequate food.

Female cats reared by humans and having no contact with other cats displayed strong aggression as adults toward their human caretakers and also males with which they were later paired (Mellen 1992). Seitz (1959) reported marked differences starting at age 9 months in kittens separated from their mothers at 2 weeks (Group 1) compared with others separated at ~6 weeks when weaning occurred (Group 2) or left with their mothers until 12 weeks old (Group 3). As adults, Group 1 cats were the most randomly active; the others displayed more goal-directed movements, such as jumping onto a ledge. Overall, they showed more restlessness and anxiety, greater fear and hesitation in novel situations, and were slowest to recover after stressful events. Their fear was persistent, they vocalized the most, were least inclined to explore, and had difficulty solving problems. In feeding competition tests, Group 2 and Group 3 cats consistently reached the food first, and those in Group 3 spent the most time eating (interpreted as being dominant). Paradoxically, Group 1 cats were the most aggressive in these contests, pushing, pawing, biting, growling, and hissing at the others, which nonetheless kept them from the food. Cats from groups 2 and 3 were willing to share food rather than behave aggressively, but not those of Group 1. When faced with a frustrating or novel situation, Group 1 cats were slowest to learn. Throughout their lives Group 3 cats were the most docile, calm, and friendly. Seitz attributed these results to the Group 1 cats having been traumatized by early weaning, an experience from which they never recovered. He posited that learning and experience through social interaction with their mothers and littermates had given Group 2 and Group 3 cats greater tolerance and confidence that made them calmer and better adjusted as adults.

5.10 Early predatory behavior

Cats are adapted to capture living prey. Early experience helps (e.g. Berry 1908, Caro 1980c), but is not required (Baerends-van Roon and Baerends 1979: 35–66, Yerkes and Bloomfield 1910), and experience with one species of prey does not necessarily improve later adult predatory skills with unfamiliar species (Caro 1980c). In fact, early experience has little general effect on later predatory behavior, especially on the ability to kill rats, although it does make the killing of recognized small prey (e.g. mice, birds) more efficient (Caro 1980c).

The building blocks of predation are evident at a young age and indistinguishable from certain elements of play (Chapter 6). Rudiments of approach to prey (rats and mice) begin ~day 22 and are developed by day 52 (Adamec et al. 1980b). Withdrawal from prey first appears ~day 27 and is developed by ~day 37. Withdrawal is seen mainly from rats, not mice, which usually do not intimidate kittens of any age (Adamec et al. 1980b). Both approach and withdrawal appear prior to predatory attack behavior. Motor patterns involved in predatory attack have been identified. Instead of being fixed they are modulated by such factors as motivation, early experience, peer competition and facilitation, and size and response of the prey (references in Adamec et al. 1980a). An adult cat's decision to attack or defend – especially when confronting a large rat or other potentially dangerous antagonist – is influenced strongly by early rearing conditions. These can include the mother's participation in learning by delivering live but injured prey at a critical juncture in development, a kitten's state of hunger when presented with live or dead prey, and the extent to which littermates compete for prey brought by the mother. Results vary. Kittens without any previous exposure to prey can start killing mice as soon as 4 weeks (Yerkes and Bloomfield 1910). Aggressive competition suppresses the development of predatory attack and consummatory behavior in some kittens and facilitates these responses in others (Adamec et al. 1980a).

6 Emulative learning and play

6.1 Introduction

As Seligman (1970: 407) pointed out, any organism is designed for a task it performs naturally. In experimental situations, "It brings specialized sensory and receptor apparatus with a long evolutionary history which has modified it into its present appropriateness or inappropriateness for the experiment." Restated in a slightly different context, behavior that is species-specific arises from endogenous components of ontogeny; individual behavior depends on experience of the individual (see Ewer 1961). Simply put, an animal that evolved without fingers is unlikely to ever play the violin.

6.2 Emulative learning

Cats, both as kittens and adults, can learn to solve problems or perform a task by watching another cat and then emulating its behavior (Adler 1955, Berry 1908, Chesler 1969, Herbert and Harsh 1944, John et al. 1968, Rosenblatt and Schneirla 1962), but success is more assured when the expected response already comprises part of the natural repertoire of cats (Adler 1955, Seligman 1970). *Any such task or problem, having been demonstrably accomplished or solved by one animal, is potentially within the realm of success of any naïve conspecific working on its own.*

Learning of this kind has traditionally been called *observational learning,* but closer inspection of its working components inspired a change in terminology. The process still involves one animal watching another, but investigators studying social learning in primates (e.g. tool use) noticed that achievement can proceed along more than one pathway. Humans are apparently unique in how we learn and transmit knowledge. Young children placed in experimental situations focus not just on the reward given after successful completion of a task but also the process by which the reward can be obtained, a form of *imitative learning* that *copies the process* and is largely beyond the capabilities of other animals (Tennie et al. 2009). While observing, even chimpanzees use *emulative learning* during which the naïve animal watches the accomplished one and *copies the product* (i.e. the effect, or outcome) while ignoring the action, or process (Call et al. 2005, Tennie et al. 2009, Tomasello 1996). The chasm separating imitation and emulation is enormous. Call et al. (2005) found that human children focus mainly on the demonstrator's actions (imitation, or process copying), but chimpanzees focus on the outcome (emulation, or product copying).

Cats are the same as other animals in relying on emulation instead of imitation. Picture two cats in separate but identical cages, each occupant in clear view of the other. Cat *A* has been trained to pull a ribbon, an act that releases a small food pellet. Cat *B* has never done this. It watches as Cat *A* engages the ribbon with the claws of

---

Free-ranging Cats: Behavior, Ecology, Management, First Edition. Stephen Spotte.
© 2014 John Wiley & Sons, Ltd. Published 2014 by John Wiley & Sons, Ltd.
Companion Website: www.wiley.com/go/spotte/cats

a forepaw and gives a tug. It eventually solves the puzzle, but perhaps by using its teeth to pull the ribbon instead of its claws. Cat A has emulated B by focusing on the product. The point is, emulation requires reinvention; imitation does not. The capacity to imitate allows humans to design and build computers and rockets and microscopes because information gained at each step supports the next advance. For other animals, every bit of knowledge transmitted is duplicated only after trial and error.

One means of validating whether emulative learning has occurred is by testing if animals learn a task more quickly or efficiently from watching another perform the same task or solve a problem successfully compared with deciphering it on their own (Herbert and Harsh 1944). We know that after a suitable number of trials laboratory cats can emulate other cats to obtain food by pulling a ribbon, pushing a lever, or rotating a turntable. Emulative learning necessitates the untrained cat watching a trained individual perform the task, then acquiring appropriate skills needed to manipulate objects, modifying its own behavior to duplicate the trained cat's results. Cats differ considerably in temperament and are easily distracted. In general, they learn poorly by observation under experimental conditions, and trial and error produces more reliable results, meaning that control cats working alone often learn the task at a steadier rate than cats that have watched others perform it (Adler 1955), evidence they learn by emulation, not imitation.

Emulative learning nonetheless makes free-ranging kittens more effective hunters as adults, perhaps because even laboratory cats seem keenly interested in watching a fellow test subject eat (Winslow 1944a). This convinced Adler (1955: 174) that "cats tend to pay attention to each other's behavior, especially if food is involved." The amount of time kittens demonstrate awareness of prey delivered to them by the mother cat, spend time watching it, and interacting with it increases significantly with age (Caro 1980a, Caro 1981b). In weeks 4–5 they pay little attention, each member of the litter watching a live prey animal independently of the others (Caro 1980a). In weeks 6–8 they interact with the prey simultaneously. At >8 weeks they watch prey alone more than expected by chance.

A mother's participation tracks the ontogenetic development of her kittens, presumably sustaining their progress by providing opportunities for them to emulate her behavior and subsequently reinforce their own successes. But learning by observation does not explain how a mother cat exploits the nascent predatory traits of her kittens (Caro 1980a). This requires, at minimum, that she temporarily alter her own behavior to accommodate her offspring. Perhaps, as Caro (1980a) suggested, descriptions of mothers returning with live prey, presenting it to their kittens, and recapturing it when it tries to escape focuses their attention; that is, her behavior *facilitates* their learning. Successful emulative learning by kittens was once thought to depend on their capacity to focus on the adult (Chesler 1969), which increases with age, rising from 26% of the test time at age 22 d to 65% at day 119 (Adamec *et al.* 1980b). However, increased attention paid to the adult is doubtful if the main learning pathway is emulation rather than imitation, in which case the focus is actually on the reward instead of the process required to obtain it.

Caro's experiments revealed that mothers interact with live prey more often and longer when their kittens do not, and kitten–prey interaction occurs more often following the mother's contact with live prey than if contact had been made by a littermate. Kittens were also more inclined to intervene in the mother's interaction with prey than to interfere with littermates. Moreover, the interest and persistence of interaction

Fig. 6.1 Influence of mother cats on the attention of their kittens toward prey. Ordinate shows ratio of the percentage time any kitten in a litter watched or interacted with live prey simultaneous with the mother watching or interacting with it. The plot shows the median ratio of 7 litters against increasing kitten age and demonstrates a negative correlation ($R = -0.57$, $p < 0.02$, $n = 17$). Vertical bars indicate interquartile ranges. Source: Caro 1980a. Reproduced with permission of Brill Academic Publishers.

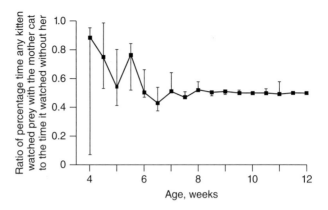

by kittens with prey was more intense in the mother's presence than in the presence of just siblings. If her offspring failed to kill prey delivered to them their mother might perform this task, but leave it for the kittens instead of eating it herself even if they had lost interest. These observations suggest a superior role for the mother in the predatory development of her young. However, if their learning is indeed emulative then paying more attention to the mother than to littermates might simply be less competitive, offering greater latitude to interact with the prey uninterrupted.

A mother cat's window of influence over the developing predatory behavior of her offspring is narrow, starting when the kittens are ~4 weeks old and tapering off rapidly ~3.5 weeks later (Fig. 6.1). The maximum latency when mothers did not kill prey in the presence of kittens peaked during week 6. A mother's interaction with live prey included approaching, sniffing, pawing, biting, and carrying it, behaviors that happened infrequently after the kittens reached 7 weeks. A mother cat emits "chirp" vocalizations when approaching her young (Martin 1986: 244). Vocalization while approaching and carrying prey – a signal thought to alert and condition the young to impending food – rose abruptly when the kittens were ~5.5 weeks and fell just as quickly by 6 weeks (Caro 1980a). Overall, a mother cat demonstrates an inverse association of her own predatory behavior with that of the kittens, but her involvement in their development as predators declines when they reach 4–6 weeks.

The sequence involved in exposing kittens to prey follows this general pattern (Baerends-van Roon and Baerends 1979: 36–37, 68–81; Caro 1980a). A mother cat takes dead prey to the den and eats it in the presence of her kittens starting when they are ~4 weeks old. The mother's behavior in the presence of prey objects presumably excites and conditions them to focus (Rosenblatt and Schneirla 1962). Later she returns with dead prey but does not eat it. Next she brings live prey to the kittens for them to manipulate in the manner of object play (see later). If the prey tries to escape she retrieves it. Eventually the kittens capture and eat the prey without her assistance. After week 5 the mother spends little time with the prey and seldom eats it. The mother's impending latency to kill prey increases, and her behaviors attendant to predation (e.g. sniffing, pawing, biting) decline concomitantly. Up to 8 weeks the mother interacts with prey mainly if her kittens become disinterested. Her intervention, along with vocalizing, are believed to revive flagging interest.

Ontogenetic progress by the kittens, in combination with the mother's participation, is important: kittens exposed to mice in the mother's presence were better mousers when tested as adults than kittens presented with mice only in the mother's absence (Caro 1980c). Furthermore, it is doubtful whether mother cats can respond to any differing developmental skills of their individual offspring. All kittens of a litter are subjected to the same duration of time in which to become effective predators. The process just described involves a progressive diminution in the mother's predatory behavior (Ewer 1969). She first must delay eating the prey until it has been taken to her offspring. Prey animals delivered later are then killed, but not eaten, and finally she brings back animals alive, retrieving them if her young fail to kill them. The pattern – a progressive sequence – is instinctive, and learning results when it coincides with ontogenetic development of the kittens.

That a cat or any other species of animal actively teaches its young to hunt is unlikely (Ewer 1969). In my opinion (Spotte 2012: 232), no animal can teach another if the term is used in a conventional sense of one individual instructing for the receiver's benefit (but see Hoppitt et al. 2008). Caro and Hauser (1992: 151) argued that teaching need not involve "active participation of instructors. ... " They speculated that teaching in the animal world does not require sensitivity by the instructor "to the pupil's changing skills or knowledge, and the instructor's ability to attribute mental states to others ... are not necessary conditions of teaching ... because guided instruction without these prerequisites could still be favored by natural selection." Maybe so, but it seems like wishful thinking for any observation when nothing more specific than natural selection is offered as a causal hypothesis (Gould and Lewontin 1979). Evidence points to teaching – in particular, verbal instruction – as the pivotal force behind cumulative culture, which then excludes all species except humans (Tennie et al. 2009). A more parsimonious perspective on Caro and Hauser's theme has animals incapable of teaching conspecifics while retaining some capacity to learn from watching another's behavior. Intent never enters the learning sequence. As Ewer (1969: 698) put it, a mother cat "does not necessarily bring back live prey *for* [italics added] her young to kill: she brings it back and they kill it."

Emulative learning has two pragmatic components for investigators: (1) to demonstrate its occurrence in the test species; (2) if confirmed, establish its importance in the animal's real world. The capacity to learn by watching conspecifics might be best suited to species needing the skill to survive. To capture prey a feral cat must become an effective hunter, but chimpanzees lacking the skill to "dip" for ants by poking sticks into anthills are unlikely to starve considering that ants are only incidental dietary items. Tests of learning by observation in primates have yielded mixed results, except in the case of humans. There its effectiveness is undeniable. Dean et al. (2012: 1115), compared knowledge transmission in human children, chimpanzees, and capuchin monkeys. The children were 3–4 y, the apes and monkeys mixed juveniles and adults. The results were starkly different, demonstrating that "teaching, communication, observational learning, and prosociality all played important roles in human cultural learning but were absent (or played an impoverished role) in the learning of chimpanzees and capuchins."

Although Caro and Hauser's (1992: 153) definition of teaching fails by excluding more parsimonious explanations, it still has merit: "An individual actor A can be said to teach if it modifies its behavior only in the presence of a naive observer, B, at some cost or at least without obtaining an immediate benefit for itself." This implicitly accounts for the mother's hormonal status working in parallel with the ontogenetic development of her kittens and rightly excludes a self-motivated laboratory cat tugging a string to receive food while remaining oblivious to nearby conspecifics watching its behavior. The wording carefully – and importantly – avoids the pitfalls of intentionality and attribution of mental state, two human characteristics as yet undemonstrated conclusively in any other animal.

The act of a mother cat bringing a mouse to her kittens and leaving it for them instead of eating it herself consistently meets the required modification of her own behavior in the presence of naïve observers, and at personal cost; that is, she denies herself food by not eating the prey, thus losing the nutrition it provides. Laboratory chimpanzees, in contrast, often steal food from their offspring (Dean et al. 2012).

Caro and Hauser (1992) gave the following example of *not* teaching. During weaning a mother changes her behavior when she starts rejecting her infant's attempt to nurse. Consequently, the infant learns the appropriate times and circumstances when nursing is permitted. The mother remains "sensitive" to its infant's needs while simultaneously gaining an immediate energetic benefit by reducing milk production. Such situations fail to consider modifications in the mother's behavior and physiology postpartum and prior to weaning, which conform with Caro and Hauser's proposed definition of teaching but make little sense in that context. The interval between birth and weaning involves extensive behavioral changes by a mother cat while she nurses a litter of "naïve" kittens, and it includes those physiological consequences of energy loss through lactation she routinely tolerates until stopping milk production. These are substantial (Chapter 8). For mammals in general, energy represented by the total biomass of offspring (e.g. a litter) equals energy transferred during only 1–2 d of lactation (Martin 1984a), which can be especially detrimental to malnourished mothers (Bateson et al. 1990). Such factors surely fall outside the purview of teaching.

"Cultural transmission" occurs in many kinds of animals (e.g. van de Waal et al. 2013 and references), particularly between mother and young. There can be little doubt that a mother cat's presence and behavior influences the foraging behavior of her kittens. After adult females were trained to eat bananas or mashed potatoes in the presence of meat pellets using these unnatural foods as rewards following hypothalamic stimulation, 18 of 22 kittens in the process of weaning emulated her behavior (Wyrwicka 1978). Emulation began at ≥ 35 d, and in the majority (10 of 18) at 49–56 d. Habituation extended over post-weaning at 9–27 weeks. Therefore, emulation of the mother affected later food preferences in the mother's absence and not just at the critical weaning period. In other experiments 12 of 16 kittens of 6–8 weeks consumed jellied agar, which is odorless and tasteless, when their mother had been conditioned to eat it (Wyrwicka 1979). When two plates of agar were presented they first ate from the mother's, then from both plates. In another experiment, one group of kittens (aged 30–39 d) was trained to eat canned tuna, the other cereal (cooked cream of wheat with addition of 3% vegetable oil and broth flavor) (Wyrwicka and Long 1980). Each group was subsequently divided into two subgroups, one offered the new foods in the mothers' presence (previously conditioned to eat them), the other in the absence of their mothers. Kittens with mothers nearby started to eat on days 1 or 2 (mean delay of 0.2~d). In the other groups the delay averaged 4.8~d whether the food was tuna or cereal, indicating the importance of the mothers' presence. As before, effects of early experience persisted. Kittens conditioned to eat cereal later preferred it to meat pellets; those conditioned on tuna later refused cereal even when hungry.

That food preferences develop from experience would hardly surprise cat owners. Laboratory cats of 4–5 months kept together since weaning showed food preferences by litter, but no such pattern was evident in tests of littermates living in different households (Bradshaw 2006). The conclusion (Bradshaw 2006: 1930S) was that "litter-specific (presumably maternally influenced) food preferences may persist into adulthood only when the feeding experiences of the littermates remain virtually identical." Common preferences were retained in littermates at 7 weeks, before kittens were distributed to different homes. In still another test a female that had eaten fish almost exclusively refused to eat the test diet, and so did 3 of her 4 kittens. The overall conclusion was that "preferences of pet cats are initially influenced by the feeding habits of the mother, but are subsequently modified by the range of feeding experiences made available by the owner." A house cat and the human with which it cohabits eventually complete a mutual training program, after which the cat's feeding preferences vary less "as the purchasing habits of the owner come into dynamic equilibrium with what cats will and will not eat."

6.3 Play

What should we make of animal play, and how does it factor into the lives of free-ranging cats? Is play an appropriate descriptor of specific behavior or simply a flaccid construct tethered within the constraints of human bias? Loizos (1966) considered play a human-centered concept appropriate for describing activity that appears to be the opposite of work. Logically, if animals do not work, neither can they play. No wonder many investigators think of animal play as useless or frivolous, having no functional endpoint or obvious survival value yet bearing an occasional resemblance to utilitarian behaviors displayed in adulthood. Were play serious – that is, had a real function – then it would not be play. But suppose play really is serious. To Ghiselin (1974: 261), "animals at play behave very much as if they were at work." This is the theme I intend to develop.

A structural–functional distinction was raised in how dominance is defined (Chapter 1), and the same concept is useful when considering play behavior (Fagen 1974, Martin 1984b). In this case a structural approach concentrates on the form and temporal patterns of play and the motor acts involved. Function emphasizes play's biological consequences. Both approaches have shortcomings. Definitions conceived with structure in mind are inconsistent if observers describe the same motor patterns differently. Functional definitions lack a factual foundation because the function of animal play – if any – is unknown (Martin and Caro 1985, West 1979).

Martin and Caro (1985: 60) defined *function* as "the particular consequences of a behavior pattern which *currently* increase the individual's chances of survival and reproduction in the natural environment and upon which natural selection acts to maintain that behavior pattern." For clarification, they noted that during incubation the shell of a bird's egg warms and expands, which is not a "function" of incubation, merely a result of it. Note the emphasis they placed on "currently," which makes the concept more testable by eliminating all aspects of a function's historical origin.

A benefit, real or presumed, is not necessarily synonymous with a function, but because tests of benefits are easier to falsify they might be more reliable variables than functions *sensu stricto*. Then again, ignoring the possible benefit of a behavior can appear to simplify things, at least superficially. A functional distinction is often presumed by claiming an animal at play derives no immediate benefit from its activity (Bekoff and Byers 1981, Martin 1984b, Martin and Caro 1985, Mendoza and Ramirez

1987), or at least none obvious. Because this is still difficult to test (i.e. benefits must be imagined and then tested), many observers have settled for structural descriptions interpreted functionally; that is, analyses of motor patterns corresponding with the behavior itself accompanied by explanations of how they potentially serve function. The approach involves listing and then describing motor patterns (e.g. arch-back, piloerection, chase, paw) and devising functional definitions that seem appropriate. Any remaining behaviors are implicitly relegated to "not-play," the alternative category. A definition of play by Bekoff and Byers (1981: 300) is often cited despite several unmentioned shortcomings, including difficulty of falsification: "*Play* is all motor activity performed postnatally *that appears to be purposeless* [italics added], in which motor patterns from other contexts may often be used in modified forms and altered temporal sequencing."

At least three problems are apparent. First, this definition is purely structural with a vague function ("purpose") that seems impossible to falsify. Even were "purposeless" properly limited the remainder is subjective: what *appears to be* purposeless to me might not be to you. Second is "purposeless" itself, which is teleological, never mind its presence or absence. Third is the restriction that behavior not demonstrably "purposeful" constitutes play, which then encompasses all motor activity not linked with a purpose, an animal's every movement being either play or not-play.

Martin and Caro (1985: 65) proposed what seems meant as a function-based definition: play is "all locomotor activity performed postnatally that appears to an observer to have no obvious immediate benefits for the player, in which motor patterns resembling those used in serious functional contexts may be used in modified forms." Here several similar but slightly different problems arise. The first repeats the failure of Bekoff and Byers to account for observer bias by allowing personal impressions (what constitutes a "benefit" in this case) to stand as part of the definition. Behavior appearing as play to an observer, much less its possible benefit, is ultimately irrelevant anyway. A properly designed experiment incorporates measures of error and bias to strengthen and refine the falsifiable statement of hypothesis and accompanying definition, provided the definition is clear and free of contingencies. This one is not. Applied in concert these factors allay confusion by excising all potential functions except the one to be tested. Requiring an observer to then distinguish play from "serious" behavior seems a dubious undertaking, keeping in mind our present inability to explain why an adult cat sometimes kills a mouse and then "plays" with it. The predatory act was presumably serious to both parties, especially the mouse. Second, although "serious" tacitly means the opposite of play it contains no specific contextual meaning. Third, if "serious" alludes to "function" then the process becomes circular by introduction of a term ("serious") that itself is part of the definition, and by assuming prior understanding of it. Fourth, obvious circularity haunts any assumption that readers will fully understand the meaning of "functional contexts." This is impossible, considering the actual function of animal play has never been ascertained.

Conceptual problems with function are difficult to surmount. As Martin (1984b: 74) pointed out, "In its weakest sense, the function of a behaviour pattern is simply a consequence which appears beneficial – of which there may be many." Conversely, function in a strong sense suggests advantageous consequences affecting survival and reproduction dependent on sustained selection for that specific pattern. A behavior, however, could have multiple functions, and today's fitness consequences might not accurately reflect earlier occurrences in an organism's genetic history. Thus any beneficial consequences of play might be out of alignment with current function, which in turn could differ from that for which a playful behavior was selected originally (Martin 1984b). This makes Martin and Caro's restriction of function to just those behavioral patterns that *currently* increase an individual's chances of survival and reproduction difficult to accept and, in the end, equally difficult to falsify.

Structurally, play can be placed in one of three categories (Spotte 2012: 209). *Locomotor play* (also called *self-play*) is usually performed by young animals when alone and involves activities like tail-chasing, jumping, and so forth. Early play consists mostly of solitary locomotor patterns (locomotor play) and evolves into social and object play. It is neither specifically social nor directed at objects but simply involves movement (Martin and Bateson 1985a). *Social play* incorporates direct interaction between or among conspecifics or other animals, including us. West (1974) identified eight structural patterns of social play in kittens (Table 6.1). During *object play* the animal paws, bites, chews, chases, bats, paws, or pulls an object such as a stick or ball (Fig. 6.2).

From a structural perspective it seems appropriate to first determine how play might be distinguished from other motor activities, not a simple task in the absence of a

Table 6.1 Structural categories (motor patterns) of object and social play in kittens. Range in age (d) when first observed in brackets. Some elements of object play (West 1979) are without age ranges. Sources: Data from West 1974 and West 1979. Reproduced with permission of John Wiley & Sons.

Structural category	Description
Belly-up	Lies on back with paws and belly up; makes treading movements with hind legs, pawing movements with front legs. Mouth is open, teeth exposed. Usually another kitten straddles the prone one, and all paws of prone individual contact underside of standing kitten. Sometimes an object is used instead of another kitten. [21–23]
Stand-up	Stands near another kitten or straddles it, head oriented toward the other's. Mouth open and bites prone kitten. Sometimes raises a paw at prone kitten. Sometimes an object is used instead of another kitten. [23–26]
Wrestle[a]	Combination of belly-up and stand-up except contact can be even closer, the kittens clasp, sometimes biting each other and tumbling around.
Face-off	Sits near another kitten, body hunched forward, tail moving back and forth, paws at head of other kitten, which might respond in kind. [42–48]
Vertical-stance	Sits and rocks back on hindquarters, lifts front paws off ground and stretches them out perpendicular to body; might partially stand in bipedal position. Directed at an object or other kitten. [35–38]
Horizontal-leap	Stands laterally to another kitten with back arched (arch-back), tail curved upward and leaps off ground. [41–46]
Side-step	Similar in posture to horizontal-leap, but kitten walks sideways and stiff-legged, often with exaggerated movements, around another kitten or object. [32–34]
Pounce	Crouches with head-low, back legs tucked in, tail straight back and sometimes moving laterally back and forth; moves hindquarters back and forth and creeps forward, thrusting suddenly with hind legs and propelling itself toward another kitten or object. [33–35]
Chase	One kitten chases another, which runs away, or chases an object. [38–41]
Scoop	Picks up an object with front paw by curving paw under object and grasping it with claws.
Toss	Releases object from mouth or paw with sideways shake.
Grasp	Holds object between front paws or in mouth.
Poke/bat	Makes contact with an object using either front paw directed from the vertical (poke) or horizontal (bat).
Bite/mouth	Places object in mouth and closes and opens its mouth around object.

[a] My addition.

Fig. 6.2 Object play, a kitten with a ball. Source: © Andrey Medvedev | Dreamstime.com.

clear and falsifiable definition of the term itself and our general inability to distinguish "playful" from "serious" activities, assuming they differ (see later). The dilemma has led to some curiously disjunctive statements. West (1979: 179), for example, introduced her article on play in domestic kittens by stating that although play "is apparently easy to recognize," theories that might explain it remain untested "partly because of the difficulties of studying a behavior we cannot define. ... " The opaqueness of this and similar juxtapositions has led some to conclude that defining difficult concepts is not worthwhile, the best approach being to avoid definitions altogether (Mendoza and Ramirez 1987). Such beliefs subsume science in metaphysics. Cloaking problems in the guise of "semantic quibbling" (e.g. references in Martin and Caro 1985: 61) fails to recognize that clear-eyed empiricism is unable to proceed until important variables have been defined in falsifiable formats and their meanings clearly delineated. Some investigators (e.g. Mendoza and Ramirez 1987) appear to have avoided the problem, although Martin and Caro (1985: 61) acknowledged that "an inability to define play must cast considerable doubt on the degree to which it is understood." Without knowing play's function, a functional definition is impossible. I see only one choice at present: tests of hypotheses must be based exclusively on structural definitions, which has been the standard applied so far. This *might* provide limited insight into function (Bekoff and Byers 1981), assuming play has a function. Structurally, no single character (i.e. behavioral phenotype) defines the class of behaviors we call "play," thus rendering it a polythetic concept.

Bekoff (1976) and Loizos (1966) offered several patterns they believed distinguish play from the "serious" structural sequences of similar behaviors. These can include reordering a sequence of motor patterns, interrupting such a sequence by irrelevant activities with later resumption, and exaggerating or repeating movements within a sequence or failing to complete them. None has been tested rigorously using domestic cats as subjects. Tests of potential function are rare for any species, and of those suggested none has strong empirical support. For reasons outlined later, I dispute the claim of Bekoff and Byers (1981: 298), even at a structural level, that "play is readily characterized and is different from other behaviors that it resembles." Also questionable is their comment on social play of young animals being clearly distinguishable from adult behavior. Such thinking discounts the influence of ontogenetic progression and the possibility that juvenile play and adult activities are actually the same, not merely similar.

Some play is clearly of doubtful importance to mammals. Martin and Caro (1985: 84) considered it "unlikely to be a crucial ingredient of normal behavioral development." In their opinion, "A more plausible view is that play is a facilitatory developmental determinant whose benefits outweigh its costs in some favorable environments but not in others." I interpret this to mean there are no major benefits of play, just structural descriptions. These reveal how extrinsic factors affect play, the ontogenetic progression of play behavior, and why kittens raised under optimal conditions are more playful than others encountering difficult and dangerous conditions, although the difference in terms of normal development is probably irrelevant. Once again, *why* animals play is unknown (Loizos 1966, Martin and Caro 1985, Mendoza and Ramirez 1987, West 1974), but Section 6.4 outlines some popular functional hypotheses, none yet validated (Bekoff and Byers 1981). An even shakier hypothesis purporting to show that play is costly states that it subtracts time a young animal could spend doing something more constructive (Martin 1982), although what this might be and whether the time would be so used are unknown.

6.4 Ontogenesis of play

Play is typically characterized by youth, the outward expressions of animals still under parental care, and its origins are in movement. Loizos (1966) listed six arbitrary ways through which patterns of movement change during the transition to play: (1) the sequence might be *reordered*, (2) movements comprising the sequence can be *exaggerated*, (3) certain movements within the sequence can be *repeated*, (4) the sequence is sometimes *fragmented* (i.e. interrupted by other activities) and resumed later, (5) movements can be both *exaggerated and repeated*, (6) movements can be *incomplete*. The activities we think of as play are restricted mostly to young animals that are not hungry, thirsty, tired, sick, too cold or too hot, or otherwise unduly uncomfortable or suffering (Adamec et al. 1980c). Play also appears to be a voluntary activity: an animal can be invited to play, but not coerced.

In kittens, uncoordinated play is sometimes evident as early as day 12 (Adamec et al. 1980b). Villablanca and Olmstead (1979) briefly outlined the ontogenetic progression of play, also reporting a first glimmer of play behavior as soon as the second week when chance interactions (e.g. incidental body contact) result in orientation toward the source and ineffectual attempts at biting. Incidences like these increase from days 10 to 30, pawing becoming especially prominent after day 25. Starting in week 5, with encounters still undirected, kittens act as if "chasing butterflies." In Villablanca and Olmstead's (1979: 116) words, "Although occasional leaps were directed at specific identifiable objects such as the mother's tail, the edge of the cardboard lining of the whelping cage, etc., the target was usually not readily identifiable." Play behavior becomes more focused in weeks 6–7, is directed at individual littermates, and in large litters preferences for play partners form. Play with littermates, however rudimentary, therefore occurs at 10–15 d, and ludic behavior is well developed by 35–40 d.

Locomotor play in cats begins ~3 weeks of age and social play ~2 weeks later (West 1974), both developing in step with maturing sensory and locomotor capacities (Baerends-van Roon and Baerends 1979: 93–98, Kolb and Nonneman 1975, Villablanca and Olmstead 1979, West 1974). All three structural categories are expressed by ~6 weeks (West 1974).

Object play follows the appearance of social play, becoming common at 6–8 weeks (Barrett and Bateson 1978, Moelk 1979). From 4–12 weeks some aspects of social play (e.g. *stalk, wrestle*) increase in frequency (Fig. 6.3); others, such as direct cat-to-cat contact and arch-back, diminish (Barrett and Bateson 1978). At 7–8 weeks the frequency of object play (e.g. pawing and biting objects) rises noticeably (Barrett and Bateson 1978). Social play peaks ~12 weeks (Caro 1981a, Guyot *et al.* 1980, Mendoza and Ramirez 1987) and then diminishes (Caro 1981a, Moelk 1979, West 1974), perhaps in response to the break from maternal care, dispersal, or simply the disinclination for social contact characteristic of adult cats (Mendoza and Ramirez 1987; West 1974, 1979).

These changes in play coincide with separation from the mother, as shown by experiments designed to test effects of premature weaning. For unknown reasons, early weaning increases the frequency of both object and social play (Bateson and Young 1981), is often manifested as hyperactivity (Wyrwicka 1978), and the time spent in play peaks at an earlier age (Koepke and Pribram 1971). Separating kittens from their mother ~day 35 slightly hastens the play sequence transition (Bateson and Young 1981), as does rearing kittens with nonlactating females and feeding them by stomach tube (Koepke and Pribram 1971). Early weaning also can weaken emulative learning (Section 6.2) because kittens pay less attention to their mothers (Wyrwicka 1978). Injecting females with bromocriptine to briefly interrupt lactation when their kittens are age 35 d has the same effect, sharply increasing the frequency of object play 2–3 weeks later (Bateson *et al.* 1981). These different studies point to ontogeny as an

Fig. 6.3 Wrestling is one form of social play in kittens. Source: © Tony Campbell | Dreamstime.com.

important primer of play behavior and not just the mother's increasing absence with impending weaning. Stated differently, whatever play's function, its components proceed in a sequential pattern that seems deterministic and necessary to finish in the proper order. A kitten weaned early is forced to take care of itself at a younger age. To quote Bateson *et al.* (1981: 844): "Therefore, it could be important for the early-weaned kitten to accomplish as much play as possible while it still had the chance." Interaction with littermates in general becomes less frequent from ~16 weeks, when feral and stray mothers typically abandon their litters (West 1974). Object play might continue beyond 12 weeks (West 1974) and extend into adulthood (Hall and Bradshaw 1998, Hall *et al.* 2002).

Play behavior changes substantially ~4 weeks following parturition. The incidence of object play rises sharply in parallel with increased motor activity (Barrett and Bateson 1978); simultaneously, elements of social play that had peaked earlier diminish in frequency (Barrett and Bateson 1978, Bateson and Young 1979). These events coincide with weaning (Hall and Pierce 1934, Rosenblatt and Schneirla 1962) and mark the important switch from complete dependency on maternal care to increasingly independent activity and predation. If, as some believe, play's function is to acquire survival skills (e.g. Bekoff 1978) then it should progress in a predictable way.

Social play in domestic cats starts and ends over a few weeks early in life. Most social activity coincides with decline of maternal care (Rosenblatt *et al.* 1961) when the mother becomes less interested in playing with her kittens (Rosenblatt *et al.* 1961). Additional factors are increasing interest in sexual behavior, becoming independent from the mother, and dispersal of littermates (West 1979). The presence of other kittens often stimulates object play, and mother and littermates have an additive effect on development of predation (Egan 1976). Whereas the mother displays appropriate skills and becomes a model for emulative learning, littermates, by competing, stimulate rapid prey capture. Thus mother and kittens act in concert to establish what Eagan (1976: 165) called "an important learning unit, where knowledge of local prey and expertise in dealing with it are transmitted from one generation to the next. ... " Only the first part of Eagan's statement is correct: evidence of cumulative culture has not been found for any animal except humankind, not even chimpanzees (Dean *et al.* 2012).

Social play ought to occur prior to weaning in those species producing more than one offspring, so that playmates of the same age and stage of development are available (Mendoza and Ramirez 1987). Single kittens attempt more interactions with the mother than kittens in litters of 2, but they spend more time alone and play less than kittens with a littermate (Mendl 1988). Object play, which does not require playmates, should peak post-weaning when the mother brings home prey, dead and alive, which are objects too. These situations apply favorably to the predictable and rapidly maturing behavior of kittens. As Bateson and Young (1981: 179–180) argued, "The skills involved in dealing with other cats are acquired ... at a time when siblings are readily available; those involved in dealing with the inanimate environment and prey are acquired after weaning when the kittens have moved out of the den."

However, as the authors pointed out, these events need not be causal, merely ontogenetic. As a test they simulated early weaning by gradually removing from their mothers 2 kittens from 7 litters of 4 until the separation process was complete at 35–39 d. The separated siblings were kept together; the 2 control kittens from each litter remained with their mothers. All kittens were then observed for frequency of social play and object play. From days 38–49 the separated kittens performed more social play than the controls and more object play during days 50–61. Over the whole period they also vocalized more. The kittens weaned early displayed greater activity, perhaps because, unlike the controls, they spent no time nursing. Their growth lagged behind the controls, not catching up until 33 d following separation. The overall result for the separated kittens was not the predicted lowering of age at which social play diminished and object play increased, but a rise in the frequency of both categories of play at developmental stages when they normally reach their highest levels anyway. Thus the time of weaning directly affects play. Whether this has a survival function is unknown, but kittens weaned early must naturally start foraging sooner, and some studies (e.g. Tan and Counsilman 1985) have shown that kittens weaned early become effective predators sooner and at a younger age.

Guyot *et al.* (1980) assessed social and object play of kittens reared in pairs or with a littermate to determine the effect of interactions. Both groups were kept with their mothers from ages 4 to 12 weeks and tested individually, either alone or in the company of a strange "test" kitten of the same age. When assessed in isolation, kittens reared alone except for their mothers demonstrated more object play than kittens reared with a littermate. They bit objects, displayed piloerection more often, and vocalized less. When placed individually with a "test" kitten, those reared with a littermate showed a greater frequency of object play. In the social situation (i.e. in the presence of a "test" kitten) the kittens reared with littermates displayed fewer incidences of piloerection, and they bit objects less often than those reared as single kittens. Kittens raised singly were more aggressive, demonstrating significantly more slapping, chasing, wrestling, ears-flat, stand-sideways, and approach. Their claws seemed extended more often than kittens reared with a littermate. Arch-back was not mentioned and evidently not observed. The overall interpretation (Guyot *et al.* 1980: 325) was that kittens deprived of a littermate "did not appear to know how to play, or that they were not aware of play versus aggressive signals."

That sick or malnourished animals are less inclined to play (Martin 1984c) could be interpreted from an adaptive standpoint as play's marginal benefits being less than its costs (Martin and Caro 1985), but a mother's health and nutritional status affect the play of her unweaned kittens indirectly by modulating her nursing behavior. Kittens reared by mothers on rationed food (80% of the energy ingested by the control group mothers fed *ad libitum*) showed greater frequency of object play at age 21–84 d than control kittens, but there was no difference between the groups in incidence of social play (Bateson *et al.* 1990). Both feeding regimens had been implemented 1 d following birth of the litters. The kittens of rationed mothers were not more active; instead, their behavior seemed associated with the greater time their mothers spent away from them during the first 18 d postpartum. Although these kittens were "buffered" nutritionally (i.e. insulated against abnormal growth and development), for unknown reasons they spent more time from days 3 to 18 nuzzling their mothers' nipples. Play usually diminishes as an animal becomes malnourished, but kittens of the rationed mothers had unlimited access to food after weaning, and the subsequent increase in object play was not a result of malnutrition postweaning. These findings probably mirror true-to-life situations in stray and feral cats if the mother's access to food is restricted, the food source undependable, or her litter too large. In such situations, free-ranging mothers are thought to wean their kittens early (Deag *et al.* 1987, 1988; Martin 1986).

Barrett and Bateson (1978) monitored kittens at play. Frequency of object play, wrestle, and stalk increased significantly from weeks 4–7 to weeks 8–12. Contact with other cats (social play) and arch-back decreased significantly over the same period, as did neck-flex (associated with arch-back) and *rear* (kitten rears on its hind legs). Sexes did not differ in incidences of social play, neck-flex, stalk, and wrestle, but males displayed arch-back more often than females at age 4–7 weeks, and females displayed rear more than males at 8–12 weeks. Differences were slight in both cases. Mendoza and Ramirez (1987), in contrast, reported arch-back failing to decline in frequency past weeks 12–16 and then increasing, not peaking until weeks 20–24.

Locomotor and social play progress at the same rate in male and female kittens. The frequency of object play by males increases sharply at 7–8 weeks. Similar behavior is not seen in all-female litters (Bateson and Young 1979), a persistent trend at least through 12 weeks (Barrett and Bateson 1978). However, no sex differences have been seen in any aspect of later predatory behavior (Caro 1980c, 1981b). The probability of an all-female litter is <4%, indicating that even if important the absence of the extra experience of play with males affects only ~4 of every 100 females (Deag *et al.* 1988). Laboratory experiments showed that females having at least one brother played significantly more with objects (pawing, batting, or biting a toy) than females without a brother, and frequency of object play by those with brothers was indistinguishable from that of males the same age (Bateson and Young 1979). The short-term presence of a brother is evidently not the stimulating factor because this trend persisted even when males were absent (Bateson and Young 1979). The presence of a sister had no effect on either males or females. Caro (1981a) showed how the number of brothers has a graded effect on male-type behavior of a female kitten's play: females that play with boys evidently behave like tomboys.

When data for object play were subtracted from this mix, none of the other correlations among play behaviors proved significant. Barrett and Bateson (1978: 118) wrote, "Moreover, the positive associations between the measures of play in the 4–7 week period fell away significantly in the 8–12 week period … as though some kind of behavioural differentiation had taken place and, as this happened, the various measures of play came increasingly under separate types of control." They proposed as an untested explanation the transition to adult behaviors, retention of arch-back perhaps becoming associated with agonism, that of object play with exploration and capturing prey.

Play sequences shorten as kittens mature, typically ending in growling, hissing, and similar agonistic behavior, play itself initiated by a pounce and ending with a chase (West 1979). Structural elements of play (Table 6.1) are sometimes repeated instead of happening sequentially, and some might occur more than others during a play bout (West 1974, 1979). *Pounce,* for example, is an opening pattern of most play bouts and is used in play solicitation (i.e. to decrease physical distance and initiate contact). Chases and horizontal leaps often terminate play. Other patterns (e.g. belly-up, stand-up can occur anywhere in a play bout, singly or in sequence).

Changes in daily activity become noticeable starting at ~16 weeks (West 1974). The amount of time spent sleeping increases, while that of locomotor and social play declines. Also increasing is the amount of time spent quiet and alert but not moving. Allogrooming among kittens declines from age 5 to 16 weeks. Kittens are often away from littermates and their activities less synchronized. If given the opportunity house cats at this age are outdoors more often, and play activities are displaced by exploration and hunting. Also about this time feral mothers abandon their kittens and move to another part of the home range. By age 16 weeks a feral kitten has left the den, separated from its littermates, and is living on its own. At ~18 weeks male sexual behavior becomes evident. Males might try to mount females, grasp the backs of their necks, or sniff their perianal areas (Chapter 4), but the females ignore or avoid them until becoming sexually mature at ~22 weeks.

6.5 What is play?

So what is play? Does it even exist? As Bateson (1981) pointed out, thinking that play must be "playful" and not provoke annoyance or aggression from the playmate or cause pain to either party places limitations on how we interpret it. Barrett and Bateson (1978) identified several "play" behaviors of kittens (e.g. arch-back with neck-flex, rear, stalk, wrestle). The arch-back and neck-flex combination is usually associated with agonism, but the other behaviors seem indistinguishable from those expressed during hunting and prey-killing. When wrestling, for example, the cat on the bottom rakes with its hind feet. During prey-capture this injures the prey if the claws are extended (Chapter 9). Meanwhile, the opponent's upper body is clasped between the forelegs while neck bites are delivered (Fig. 6.4). And few would dispute that stalking and hunting are closely associated. The possible misinterpretation could be accepting that such activities are "practice" for adulthood or somehow superfluous when they might be simply as they appear. Hall *et al.* (2002) used object play and predatory play as synonyms. Structurally, object play by adult cats resembles predation (Egan 1976),

Fig. 6.4 Two cats at "play," except that biting down on an opponent's neck, wrapping the forelegs around its upper body, and raking its abdomen with the hind paws are identical to behaviors a cat would use when attempting to subdue a prey animal its own size. Source: © Lucaturati | Dreamstime.com.

Fig. 6.5 A hunting cat pounces on a small animal. Source: © Wenbin Yu | Dreamstime.com.

and Caro (1979) saw a similarity between predatory behavior and certain aspects of social play. The stimulus properties of the objects eliciting play are often similar in size and texture to rodents and other typical prey animals, leading to the hypothesis that play and predation have a common motivational basis (Hall and Bradshaw 1998) and not simply that predation is motivated behavior whereas play is not. I find this lack of distinction important. Pounce has an obvious function in adult predatory behavior (Fig. 6.5).

Rearing with the forelegs extended simultaneously releases the claws of the forepaws from their sheaths (Chapter 9), a prelude to pouncing on a mouse. From the play-as-reality viewpoint, a kitten stalking another is actually hunting it; a cat batting its prey, whether alive or dead, is at the culmination of a hunting sequence (Fig. 6.6). A wrestling match is not merely a simulation of techniques and patterns used to subdue prey, but a very real attempt by each combatant to subdue another animal it fully intends to eat. If valid, this hypothesis could explain why the killing bite develops later than these other behaviors (see later). After a play bout among kittens, behavior shifts back to standard species-recognition signals; that is, predation is set aside and predictable interactive responses are elicited by the pertinent stimuli (e.g. a slap elicits a hiss).

My take on the subject is straightforward. It seems to me there is no such entity as play. It does not exist, and purported evidence of what we call play representing anything besides predation, copulation, or some other essential function is invention. Things are exactly as they are inside this artificial playpen we construct for animals, not as they seem, our odd failing being an inability to understand that what we observe and interpret is not always sufficient to make it real.

Admittedly, this is a radical view, but doubters have lots of other choices. Alternative suggestions of play's function are legion (Martin 1984b), starting with believing that play exists but ascribing to it no function. Bekoff and Byers (1981) condensed play's possible functions into three groups: motor training, cognitive training, and socialization. All presume play to somehow prepare young animals for adulthood. However, as

Fig. 6.6 Adolescent cat bats a rodent it has killed or is about to kill. The function of such "play" behavior is undetermined. Source: © Ivanov Arkady | Dreamstime.com.

Martin (1984b: 81) pointed out, "The developing animal is not merely a half-built version of the adult. ... " Spotte (2012: 212) suggested that "we could delude ourselves into thinking of play as practice for the adult activities it resembles, a teleological presumption that ontogenetic stages 'exist for the sake of adult life.' " (The internal quotation is from Ghiselin 1974: 259.) To Martin and Caro (1985: 88) it is "highly misleading to regard [youthful] behavior as no more than an incompetent version of adult behavior." Adaptation, they emphasized, is necessary at every ontogenetic stage. Loizos (1966) pointed out that play is not necessary to practice. As to play providing information about the environment, so does every other activity, exploration in particular. She wrote (Loizos 1966: 5), "Quite simply, it is not necessary to play in order to practise: there is no reason why the animal should not just practise."

Everything described above seems logical had it been said while expelling the concept of play itself from the discussion and seeing structure and function in a different light (see later). Locomotor, social, and object play are broadly inclusive of *what* animals do when we think of them playing, but not of *why* they do it. Several theories have attempted to remedy this deficiency, although none has so far survived a rigorous empirical challenge. I shall consider three of the most popular four. Play in young animals is (1) practice for later social interaction and predation (Baerends-van Roon and Baerends 1979: 37, 79; Caro 1979; Mendoza and Ramirez 1987; West 1974, 1979), (2) training of muscles and motor functions for later adult activities like hunting and agonistic encounters (Fagen 1974, 1976; Martin and Bateson 1985a; West 1974), (3) training through exploration and familiarization with novel objects and environments (Martin 1984b, West 1974), or (4) because it appears to the observer to be "fun" or pleasurable (Bekoff and Byers 1981; Martin and Bateson 1985a; West 1974,

1977) and therefore self-rewarding (Loizos 1966). To my knowledge, this last idea has never been tested, and I have nothing further to say about it except this. The stipulation that play should include only behaviors devoid of immediate benefit (Bekoff and Byers 1981) or obvious "endpoint" similar to exploration (Hall *et al.* 2002: 264) seems unnecessarily stringent, if not peculiar, under the circumstances of function-seeking, especially because some "behaviors" given space in ethograms describe animals in static positions not doing much of anything except breathing. This raises questions of what those immediate benefits might be other than sustaining autonomic functions. In any case, exploration and manipulation seem rewarding, if not pleasurable, in themselves and fully active without derived incentives (Miles 1958).

Movement is a natural component of ontogenetic development, leading some to think of play as essential for the maturation and functioning of somatic tissues and coordinated locomotion (Fagen 1976, West 1979). Movement requires energy, the presumed link being that physical fitness enhances evolutionary fitness, and a behavior persisting through natural selection retains benefits outweighing the energetic cost of sustaining it (Martin and Caro 1985). In theory, a play bout should last long enough to work muscles to near fatigue and then stop, and data for the length of time this takes should display a nonrandom distribution. An experiment using California ground squirrels (*Spermophilus beecheyi*) as test animals produced randomly scattered data, forcing acceptance of the null hypothesis of no effect (McDonald 1977). In addition, play is affected profoundly by food (Bateson *et al.* 1990, Egan 1976) – its availability, distribution, and composition. Similarly, animals tend to play less if hungry or malnourished (Martin and Caro 1985).

Altricial species do not play early in life. According to Fagen (1976: 201), this is because the cost of play, defined as "the decrease in fitness resulting from diversion of one unit of time or energy from feeding and temperature maintenance is greater than the benefit of play, defined as the fitness increment resulting from the allocation of one unit of this resource to play." Cost of play then diminishes more rapidly over time while its advantages rise concomitantly, lowering the cost/benefit ratio as ontogenesis proceeds. Reinforcing this argument is the observation that young animals tend to play more than adults (Loizos 1966; West 1974, 1977, 1979), which still fails to explain why adults sometimes play too and what its functional advantage could be (Loizos 1966). In any case the effect is probably minor despite statements to the contrary (e.g. Bekoff and Byers 1981): costs and subsequent benefits in most mammals (including kittens) are probably <10% of the daily energy expenditure (Martin 1982, 1984c; Martin and Caro 1985).

The "play as exercise" hypothesis usually includes discussions of immediate costs versus later benefits and the trade-off between risk of injury (Bekoff and Byers 1981) and advantages of enhanced adult conditioning. The point is to explain why adults play comparatively little (rigorous adult activities supposedly preclude any need), but the general reasoning is circular: compared with juveniles, adult animals play less by being adults.

Some think playing is practice for later adult life – mating, agonistic encounters, predation, evasion of predators – because play and some of these adult activities are similar structurally; that is, they look the same (e.g. Forbush 1916: 15). Supporting evidence is weak (West 1979) and the logic dubious: behaviors with similar structure can have different functions (Martin and Caro 1985). Equating object play with hunting, for example, seems misplaced. As West (1979: 185) wrote, "Play with objects resembles activities with captured not uncaptured prey." Stated differently, "hunting" suggests not-capture. Stealth used by adult cats while hunting, and which involves stalking and pouncing, is seldom seen in object play, and if, as West (1979) and others have suggested, hunting might be the most effective practice for hunting, I see little to distinguish one from the other. Unlike the "serious" patterns it resembles, play seems to occur largely in the absence of context, which I find hardly surprising. Falsifiable experiments comparing youthful play against adult performance require an empirical method of effectively isolating the test animal from itself, an impossible feat but theoretically necessary if experience is to be dissected away from effects imposed by ontogenesis. West (1979: 191) suggested that when seeking function in play we might emphasize "requirements of the young rather than … the accomplishments of the adult." This, of course, presumes that play as a self-serving entity actually has a function.

To be viable the "play as practice" hypothesis would need to overcome still another hurdle by explaining why experienced adult cats indulge in object play using prey animals either alive or dead. No evidence shows that adults deprived of play as kittens are necessarily inferior hunters, although it can be demonstrated that hunting experience at 4–12 weeks improves predation efficiency on the same species of prey at age 6 months (Caro 1980c). In fact, structural activities manifested in the play of kittens have little apparent impact on the adult cat's efficacy as a predator (Caro 1979, 1980c). Experience also affects preferences, and early predation on mice, for example, can lead to preferential killing of mice in later life.

What I shall call the "play now, pay later" hypothesis posits that play has immediate functions and benefits and delayed consequences (Bekoff and Byers 1981). The notion that routine play is dangerously risky or energy-intensive (Bekoff 1976, Bekoff and Byers 1981, Loizos 1966) has little empirical support (Martin 1982). Such thinking also disregards potential short-term advantages of play by discounting any immediate benefits it could have (Martin 1984b). The "pay to play" hypothesis posits the opposite. It claims that play grants long-term benefits at the same potential costs (e.g. loss of energy, risk of injury or predation) paid by the young with advantages accruing into adulthood (Martin 1984b, Martin and Caro 1985), the implication being that a playful adolescence, even if expensive, confers later fitness benefits. This could be true only if later benefits exceed future costs, which is always doubtful, and whether the observer is correct that all benefits are future ones after play has been largely extinguished by maturation (Martin and Caro 1985). Neither hypothesis has much to offer. Both claim play to be costly and therefore must obviously bestow benefits of some kind, a questionable presumption.

Suggestions of social play increasing longevity or otherwise affecting survival have not been tested in free-ranging cats. What Martin (1984b: 78) called "a family that plays together stays together" offers the advantage of keeping a litter of kittens at the den site instead of wandering off to starve or be killed (West 1974, 1979), which would certainly be beneficial. These and other ideas still lack unequivocal evidence of play's positive value, including its importance during normal development. Whether early play directly helps any species of adult animal seems doubtful (Bekoff and Byers 1981, Martin 1984b). Some evidence shows that kittens reared by their mothers but without littermates were deficient in certain forms of interaction (Guyot et al. 1980). Any adverse effect on fitness is unknown.

Sometimes a cat "plays" with prey instead of killing and eating it (Fig. 6.5). This happens if the species is unfamiliar (Caro 1980c), but under other conditions too, such as play with an inedible object. As Hall and Bradshaw (1998: 149) reported, "On the assumption that object play in non-hunting cats is actually predation which has not been refined by experience to enable the cats to respond to only live prey with specific characteristics [e.g. small size], it is not surprising that the expression of object play is influenced by hunger in the same way as predation in hunting cats." Their conclusion that object play is not practice for predation by inexperienced, non-hunting cats but actual predatory behavior directed at toys is eminently sensible. So is the link between object play and hunger (Bateson *et al.* 1990, Hall and Bradshaw 1998, Martin and Bateson 1985b). Biben (1979) reported cats least likely to play with their prey when hungry and the prey is small, and most likely when satiated and the prey is small or the prey is large and difficult to kill and the cat is hungry. Kuo's (1930) experiments demonstrated that kittens are reluctant to attack large rats until having attained sufficient age and size to do so successfully. As mentioned, starving animals are seldom playful (Martin 1984b), although even hungry cats sometimes play with a mouse or rat before killing it (Biben 1979). Overall, prey size and state of hunger exert a strong combined effect on the predatory response (Biben 1979, Kuo 1930).

I reject all four previously offered hypotheses for why animals play. I can think of no "reason," ontogenetic or otherwise, for play to exist and prefer banishing the concept altogether as confusing and irrelevant. I believe the behaviors we call play (or playful) are deadly serious activities at every stage of development. Can we claim with certainty that a kitten chasing a ball of inedible paper is not hunting or that two jousting young chimpanzees are not fighting? In both examples, how do we know that function (or call it "motivation") is not being fully applied as we watch, the only unapparent elements being experience, skill, and context? Perhaps trying to distinguish "play-hunting" from "hunting" or "play-fighting" from "fighting" reduces these concepts and others like them to artifacts seemingly separable as hypotheses to us, but experienced as a continuum to the subjects themselves.

In cats, all structural behaviors required in adult hunting appear during week 5. Pertinent motor skills are crude and clumsy at first, but by week 8, according to Rosenblatt and Schneirla (1962: 454), kittens "can be seen stalking, chasing, climbing, following, leaping at one another, and performing various manipulative activities involving their forepaws." Everything is there except the killing bite, which according to Leyhausen (1979: 81) "seems to need a special form of elicitation." This could be stimulation by movements of the prey, excitement prompted by the mother's behavior, or competition from siblings (Leyhausen 1979: 90). Then again, Leyhausen never tested any of his suggestions, so the causal factor could be something else entirely. Motivation to capture prey varies by individual, and some house cats reared normally never develop into predators (Leyhausen 1979: 91). Still others, having no exposure to live prey until 9 weeks, kill without the need of experience or any recognizable stimulus (Leyhausen 1979: 92). Kittens reared in homes or laboratories and not presented with prey by their mothers between weeks 6 and 20 might kill later, although often in a slow, uncertain way (Leyhausen 1979: 92).

The same argument can be restated in slightly more formal language. Kittens, like other young mammals, confront the environment by *learning*; that is, by processing specific extrinsic stimuli and adjusting their behavior in response. This is accomplished partly through conditioning during times when the same or closely similar stimuli are experienced repeatedly. Conditioned reactions can also be "unlearned" if the prolonged absence of a stimulus causes extinction of the response. Responses sometimes thought of as new are often modifications of existing ones. Separating them is a stringent empirical task. Take, for example, those sequences of play resembling adult hunting behavior. To be falsifiable, any presumed dichotomy branching from the common phenotypic stem must first invoke and then demonstrate dissimilar motivations while simultaneously controlling for changing ontogenetic responses.

If play is indistinguishable from work – manifested as hunting in a free-ranging cat – its function has a more parsimonious explanation than proposed above: *every structural aspect of play we can identify is simultaneously functional*. In basic form this is not a new idea (see Egan 1976), nor is it conceptually tautologous if the behavioral repertoire of a species is actually restricted to functional activities, their structural aspects simply a matter of context regulated by factors like stage of development and hormonal status. Then a kitten at play "hunts" just as an adult hunts, its behavior real, not "practice" for the killing moment or for honing its motor skills to become a coordinated predator in the future. *It is already a hunter,* the presence or absence of prey having no consequential bearing on ontological status, which remains unchanging through all the dynamics of ontogeny. If true, there can be no difference between "play" and the "serious" adult behaviors that play resembles. The killing of something, often cited as how predation and play are distinguished (Martin and Caro 1985, but see Biben 1979), then becomes incidental and superfluous. Following this reasoning, what we call "play" is motivated behavior performed efficiently at a moment in development (Loizos 1966), with this important distinction: *it is not play*.

7 Nutrition

7.1 Introduction

Here I describe some basic aspects of cat nutrition. I also discuss why certain dietary needs of cats make them different from other animals and how commercial diets fed to house and laboratory cats vary in composition from the items captured and scavenged by feral and stray cats. Domestic cats have the same requirements whether free-ranging or not and suffer the same consequences when failing to meet them. The extent to which a free-ranging cat is able to maintain body mass depends on its food supply and hunting skills. The body mass of one Macquarie Island male live-trapped 12 times fluctuated between 2736 and 3450 g (Brothers *et al.* 1985). Another male went from 3100 to 4250 g in <1 month. Changes in another five cats (males and females combined) was <600 g.

Feral cats feed exclusively on animal tissues, and Felidae comprises the only family of carnivorans in which all members are truly carnivores; that is, have nutrient requirements that can be satisfied completely by eating animal flesh (Bradshaw *et al.* 1996, MacDonald *et al.* 1984a). Put simply, every substance cats are unable to synthesize is available to them in a carnivorous diet, and a free-ranging cat that feeds exclusively on other animals should rarely be malnourished if adequate numbers of prey are available. Not discussed are the needs and functions of different inorganic elements (e.g. calcium, iron) because I could not find examples of deficiencies in free-ranging cats.

In a meta-analysis of 27 reports used to derive 30 dietary profiles and based on 6666 stomach, gut, and scat samples, Plantinga *et al.* (2011) showed that feral cats obtain 52% of their daily energy intake from crude protein, 46% from crude fat, and only 2% from carbohydrate (Fig. 7.1), ratios offering ample proof of carnivory (Morris 2002). That animal tissues contain little carbohydrate and high concentrations of protein relative to energy is a situation to which cats have adapted during their evolution. Because the animals likely to be eaten have a similar composition the only difficulty confronting a feral cat would be a shortage of prey or suitable animal matter to scavenge. As Hirsch *et al.* (1978: 287) succinctly stated, "The carnivore's nutritional problems are essentially solved once food is captured." In this regard the following example should quell any doubt about the cat as opportunist: feral cats on Guadalupe Island, México, were seen attempting to steal milk from lactating northern elephant seals (*Mirounga angustirostris*) resting on the beaches. In the words of Bradford *et al.* (2006: 3), "Curiosamente, también se han observado gatos robando leche a los elefantes marinos lactantes. ... ").

Nearly everything we know about cat nutrition comes from the laboratory, and while such information is important to scientists, cat owners, and manufacturers of

Free-ranging Cats: Behavior, Ecology, Management, First Edition. Stephen Spotte.
© 2014 John Wiley & Sons, Ltd. Published 2014 by John Wiley & Sons, Ltd.
Companion Website: www.wiley.com/go/spotte/cats

138 Chapter 7

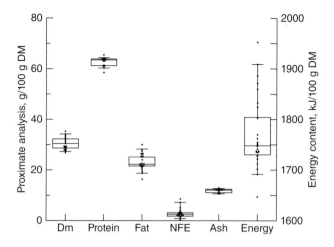

Fig. 7.1 Box and whisker plots showing proximate composition of the natural diets of feral cats. Upper and lower limits of boxes represent 75th and 25th percentiles. Bands inside boxes are medians. Whiskers bracket 5% and 95% confidence intervals. Means (g/100 g ± SEM): DM (30.5 ± 0.4), protein (62.7 ± 0.3), fat (22.8 ± 0.5), carbohydrate (2.8 ± 0.3), ash (11.8 ± 0.1); E (1770 kJ/g DM ± 13). "As fed" values are recalculated in the text. Source: Plantinga *et al.* 2011. Reproduced with permission of Cambridge University Press.

pet foods, it often has little direct application when assessing the diets of free-ranging cats. However, because laboratory findings make up our largest reservoir of knowledge, I shall dip into them often when attempting to infer how cats living on their own manage to fulfill nutritional needs that seem ordinary in some respects and peculiar in others (MacDonald *et al.* 1984a, Morris 2002, Zaghini and Biagi 2005), and how commercial cat foods can fail to meet the dietary requirements of confined cats or even fed strays if their diets are served exclusively by these sources. As I use the term, "nutrient" is restricted to substances contributing energy rather than *contributing to* energetic processes, thereby excluding individual elements (e.g. calcium, iron) and vitamins.

7.2 Proximate composition

The rough composition of a sample of food assessed in the laboratory is commonly expressed as its proximate analysis, or proximate composition. It comprises dry matter (DM) after removal of the water and includes everything the DM contains: protein, fat, carbohydrate, fiber, and ash (the largely inorganic residue remaining after a sample has been combusted and the organic content burned off). Fat content typically correlates inversely with water and ash and directly with the amount of protein. Energy content is sometimes given too, although energy is not part of a proximate analysis. Energy was previously expressed in kilocalories (kcal) but more recently as kilojoules (kJ). To convert, 1 kcal = 4.184 kJ. Results of the other parameters are usually expressed as percentage DM on a mass basis (equivalent to g/100 g DM) or as g/kg DM. Crude protein (or simply protein) is a measure based on ash-free lean dry mass and determined using an analytical method that measures its nitrogen (N) content. The result is then converted to crude protein by N × 6.25.

If the proximate analysis of a natural food is listed "as is" or "as fed" it sometimes means the water has been included in the results. The sample, in other words, was dehydrated as usual for laboratory analysis of DM, then "reconstituted" by replacing the water arithmetically for tabular purposes. Doing this is useful because the composition of natural foods (e.g. prey animals) can then be compared directly, something difficult to visualize when proximate analyses are expressed as DM components.

Table 7.1 Mean proximate composition of laboratory mice (*Mus musculus*) and rats (*Rattus norvegicus*) at different ages and sizes, called "developmental stages" here ($\bar{x}\% \pm$ SD, sample sizes in parentheses). Means of components have been recalculated as percentage wet mass. Values are approximate because summed percentages of protein, fat, and ash from the original source do not always equal total percentage dry mass in the original data.

Developmental stage	Body mass, g	Length, cm	Water	Protein	Fat	Ash
Mice						
Pinkie (8)	1.7 ± 0.5	3.0 ± 0.4	82.0 (7)	11.56 (7)	2.74 (7)	1.75 (7)
Fuzzy (7)	3.8 ± 1.1	3.7 ± 0.4	70.8 (6)	12.20 (6)	13.64 (6)	2.45 (6)
Crawler (12)	6.5 ± 1.3	4.7 ± 0.4	69.0 (6)	14.45 (6)	14.45 (6)	2.63 (6)
Small (12)	7.4 ± 1.0	5.2 ± 0.4	71.3 (5)	16.01 (5)	6.72 (5)	3.84 (5)
Medium (12)	14.8 ± 3.0	7.0 ± 0.7	69.7 (6)	17.75 (6)	6.03 (6)	5.15 (6)
Large (6)	36.2 ± 8.8	8.2 ± 0.9	62.9 (6)	16.69 (6)	11.91 (6)	5.04 (6)
Rats						
Pinkie (10)	10.0 ± 3.6	5.2 ± 0.6	79.2 ± 19 (5)	12.04 (5)	4.93 (5)	2.54 (5)
Small (5)	54.1 ± 9.9	12.6 ± 1.3	70.0 ± 35 (5)	11.22 (5)	5.5 (5)	2.96 (5)
Medium (5)	117.2 ± 4.5	14.3 ± 1.3	69.8 ± 11 (7)	18.15 (7)	7.25 (7)	4.80 (7)
Large (5)	256.9 ± 39.0	19.9 ± 1.7	65.2 ± 38 (5)	17.85 (5)	12.21 (5)	3.06 (5)

Source: Douglas *et al*. 1994. Reproduced with permission of Elsevier.

A freshly killed mouse eaten by a cat "as is" includes tissue water, which might be 70% of its total mass. A proximate analysis would typically list the mouse's composition in terms of the different fractions of its DM, which makes up the remaining 30%. A cat eating it ingests the identical amount of protein in either case. The same situation applies when comparing "as is" canned ("wet") commercial cat foods and "dry" foods, or *kibbles*. Both are consumed "as is," but the canned food might be ~70% water, the kibbles 4–12%.

Several factors affect proximate composition, one being age of the animal from which the tissues being tested were obtained. From dromedary camels (*Camelus dromedaries*) to mice and even some insects, the tissue water as percentage of live body mass varies inversely with age while protein, fat, and ash increase (Ademolu *et al*. 2010, Dawood and Alkanhal 1995, Fedyk 1974, Kaufman and Kaufman 1977, Myrcha and Walkowa 1968, Sawicka-Kapusta 1970, Xiccato *et al*. 1999). Old-field mice (*Peromyscus polionotus*), for example, are 71.5% water at age 7 d and 64.0% at age 42 d, and within this interval protein rises from 14.3% of fresh mass to 16.95%, fat from 4.0 to 11.9%, and ash from 2.4 to 3.7% (Kaufman and Kaufman 1977). On a mass basis, the nutritional value of a mouse is thus determined partly by its age (Table 7.1).

7.3 Proteins

As mentioned briefly in Chapter 6, protein is the main dietary component affecting growth (Millward 1995). Kittens require a minimum of 19–21% dietary protein by DM (Rogers and Morris 1982), or ~1.5 times more than growing pigs (*Sus scrofa*) despite pigs depositing higher proportions of the nitrogen from food in their own bodies, and adult cats need 2–3 times the protein of adult omnivores and herbivores for tissue maintenance (Morris 2001, 2002). Adolescent cats (0.5–1.0 kg) gained body mass at the rate of 11 g/d on a diet of 23% protein, and doubling the protein content to 46% effectively doubled the growth rate to 21 g/d (Mercer and Silva 1989). Rogers and Morris (1982) reported that kittens grew at 20–30 g/d on 22% dietary protein (Fig. 7.2).

Fig. 7.2 Effect of feeding increasing amounts of dietary nitrogen on weight gain of kittens fed purified diets containing free amino acids or casein. Mean of two experiments, one using amino acids, the other casein as major nitrogen sources. Each experiment was a balanced Latin square 6 × 6 for both males and females ($n = 24$). Source: Rogers and Morris 1982, figure 4, p. 526. Reproduced with permission of John Wiley & Sons.

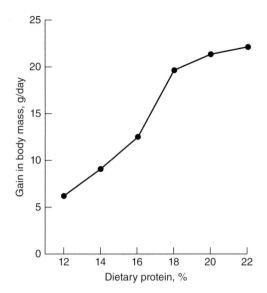

Adult animals need protein to replace amounts lost in hair, skin, digestive enzymes, and from the breaking down, or catabolizing, of amino acids (Rogers and Morris 1982). Young animals require protein for both tissue maintenance (nitrogen loss and replacement) and tissue growth (nitrogen retention). Amino acids, the structural constituents of proteins, participate in several functions including glucose synthesis, or *gluconeogenesis* (Millward 1995, Moundras *et al.* 1993), which supplies energy (Chapter 8). In cats, amino acids serving this purpose and to synthesize proteins are not oxidized completely (Beliveau and Freedland 1982). Some proteinaceous breakdown products go to form creatine and other necessary compounds; others are oxidized to ketone bodies and carbon dioxide. Nitrogen freed by catabolism is assimilated into urea and excreted or else incorporated into ammonia and uric acid.

The *essential amino acids* (*EAAs*) are the ones an animal is unable to synthesize (Table 7.2). *Dispensable amino acids* are those that when omitted from the diet do not cause appetite suppression and weight loss, nor does their absence prevent weight gain. Protein requirements of domestic cats published by the NRC (2006) are based on food

Table 7.2 Amino acid requirements of the cat.

Essential amino acids	Insufficient synthesis	Adequate synthesis
Arginine	Asparagine	Alanine
Histidine	Taurine	Aspartic acid
Isoleucine		Cystine
Leucine		Glutamic acid
Lysine		Glutamine
Methionine		Glycine
Phenylalanine		Proline
Threonine		Serine
Tryptophan		Tyrosine
Valine		

Source: Rogers and Morris 1982, table 4, p. 525. Reproduced with permission of John Wiley & Sons.

containing a maintenance energy (*E*) component of 16.75 kJ/g (4 kcal/g). From this criterion the minimum crude protein allowance for kittens (NRC 2006: 364, Table 15–9) was established as 180 g/kg of food (18.0 g/100 g DM, or %DM), broadly consistent with an earlier estimate of ≤300 g/kg DM (Hammer *et al.* 1996). The minimum allowance for adults (NRC 2006: 366, Table 15–11) was 160 g/kg DM; for gestating and lactating cats, 170 and 240 g/kg DM, respectively and summarized here in Table 7.3. These standards were subject to amino acid profiles meeting the minimum requirements for adult cats (Table 7.4).

Aside from sugars (see later), cats can taste what they eat, responding positively to some tastes and negatively or not at all to others. Cats have adapted to a diet high in animal flesh, leaving them limited metabolic latitude for subsisting on foods of variable protein concentration. Evidence of their extreme carnivory starts with taste buds in the facial nerve that are keenly sensitive to amino acids. They react positively to the amino acids proline, cysteine, ornithine, lysine, histidine, and alanine, are inhibited by tryptophan, isoleucine, arginine, and phenylalanine, and generally unresponsive to the carbohydrates mono- and disaccharides (Boudreau 1974, Bradshaw *et al.* 1996, White and Boudreau 1975). Monophosphate nucleotides accumulate in decaying meat, and they too inhibit the amino acid receptors of cats, perhaps allowing them to distinguish food quality (Bradshaw *et al.* 1996).

In addition to these adaptations, most of a cat's high-protein need is predicated on tissue maintenance, not growth (Rogers and Morris 1982), and culminates from the limited capacity to regulate catabolic enzymes of amino acid metabolism (MacDonald *et al.* 1984a). These are the aminotransferases used in metabolizing dispensable nitrogen, and the urea cycle enzymes (Morris 2001). A high-protein diet keeps nitrogen catabolic enzymes in the liver functioning efficiently (Rogers *et al.* 1977), which alone permits little latitude in how far dietary protein can decline. As MacDonald *et al.* (1984a: 534) pointed out, "These enzymes are nonadaptive, so the obligatory nitrogen loss is high even when cats are fed low-protein diets. … " Thus cats fail to conserve amino acids whether feeding on diets of either high or low protein (Hammer *et al.* 1996, Rogers *et al.* 1977), a condition traceable to rigid resistance when confronted by changes in aminotransferases. Consistent with this pattern, their nitrogen losses during periods of starvation exceed those of omnivores.

The amount of urea generated from protein catabolism increases with dietary protein. Most is excreted, resulting in a small fraction recycled to the gut for anabolic functions (Russell *et al.* 2000). The pattern varied little when laboratory cats were fed a diet of either moderate amounts of protein (MP, 20%) or one containing high protein (HP, 70%), both by DM. On the MP diet, 15% of urea was recycled compared with 12% on the HP diet (Russell *et al.* 2000). The respective amounts diverted into the ornithine cycle were 57% and 59%, leaving 40% from both regimens to be excreted as fecal nitrogen. The amounts remaining for anabolism were low (6.41% for the MP diet and 4.79% for the HP diet). Russell *et al.* (2000: 597) concluded: "These results show that cats operate urea turnover, but at a lower rate, and with less nutritional sensitivity than has been reported for other species."

The general lack of enzymatic adaptation has been thought to place cats in the catabolic fast lane with hepatic enzymes tuned permanently to metabolizing high concentrations of protein, meaning that acceleration is essentially invariable, not adjusting smoothly to lower levels of protein (Rogers and Morris 1982). As a profligate user of

Table 7.3 Minimum requirements or adequate intake of adult cats for protein and amino acids by dry matter (kg DM), energy (kJ DM), and metabolic body mass ($M^{0.67}$).

	Gestating cats			Lactating cats		
	Amount/kg DM	Amount/4186.8 kJ DM	Amount/kg $M^{0.67}$	Amount/kg DM	Amount/4186.8 kJ DM	Amount/kg $M^{0.67}$
Protein, g	170	43	5.90	240	60	12.9
Amino acids, g						
Arginine	15.0	3.75	0.52	15.0	3.8	0.81
Histidine	4.3	1.08	0.15	7.1	1.8	0.39
Isoleucine	7.7	1.93	0.27	12.0	3.0	0.64
Leucine	18.0	4.50	0.63	20.0	5.0	1.07
Lysine	11.0	2.75	0.38	14.0	3.5	0.75
Methionine	5.0	1.25	0.17	6.0	1.5	0.32
Methionine + cystine	9.0	2.25	0.31	10.0	2.6	0.56
Phenylalanine	—	—	—	—	—	—
Phenylalanine + tyrosine	15.3	3.83	0.53	15.3	3.83	0.52
Taurine	0.42	0.105	0.015	0.42	0.105	0.015
Threonine	8.9	2.23	0.31	10.8	2.7	0.58
Tryptophan	1.9	0.48	0.07	1.9	0.48	0.10
Valine	10.0	2.50	0.35	12.0	3.0	0.64

Source: National Research Council (U.S.). Ad Hoc Committee on Dog and Cat Nutrition 2006. Reproduced with permission of the National Academies Press, Copyright 2006, National Academy of Sciences.

Table 7.4 Minimum requirements or adequate intake of adult cats neither gestating nor lactating for protein and amino acids by dry matter (kg DM), energy (kJ DM), and metabolic body mass ($M^{0.67}$).

	Amount/kg DM	Amount/4186.8 kJ DM	Amount/kg $M^{0.67}$
Protein, g	160	40	3.97
Amino acids, g			
Arginine	7.7	1.93	0.19
Histidine	2.6	0.65	0.06
Isoleucine	4.3	1.08	0.11
Leucine	10.2	2.55	0.25
Lysine	2.7	0.68	0.07
Methionine	1.35	0.34	0.03
Methionine + cystine	2.7	0.68	0.07
Phenylalanine	4.0	1.00	0.10
Phenylalanine + tyrosine	15.3	3.83	0.38
Taurine	0.32	0.08	0.008
Threonine	5.2	1.30	0.13
Tryptophan	1.3	0.33	0.03
Valine	5.1	1.28	0.13

Source: National Research Council (U.S.). Ad Hoc Committee on Dog and Cat Nutrition 2006. Reproduced with permission of the National Academies Press, Copyright 2006, National Academy of Sciences.

nitrogen the cat's urea cycle proceeds unabated, perhaps a necessary situation to preclude ammonia toxicity after ingestion of a high-protein meal (Green et al. 2008), and to provide the carbon substrates for driving gluconeogenesis (Chapter 8). Instead of being recycled into dispensable amino acids, as might be the case with other animals, ammonia generated through deamination is shunted to the urea-forming pathway to be excreted in urine, accounting for up to 86% of urinary nitrogen (Thrall and Miller 1976).

Feed a cat a low-protein diet and it still loses nitrogen rapidly (Rogers et al. 1977), explaining in part why obese cats shed weight far more quickly on diets low in protein than obese dogs, rats, pigs, and humans, all of which are omnivores. In effect, cats need large amounts of protein to continue excreting high concentrations of nitrogen independently of their protein status, and their high protein need is actually a requirement for dispensable nitrogen (Morris 2001). Some salutary features of this situation are nonetheless retained. As Morris (2002: 155) wrote, were the enzymes catalyzing the *initial* degradation of EAAs not controlled, "cats would have a high requirement for essential amino acids as well as for N. ... " The additional benefit cats derive from slack regulation of overall nitrogen metabolism by the aminotransferases (Morris 2001, Rogers and Morris 1980) allows them to maintain blood glucose stability during starvation, which food-deprived omnivores do less well.

This model is in conflict to some extent with other studies showing that cats have flexibility in how they modulate protein metabolism (e.g. Kettlehut et al. 1980; Russell et al. 2000, 2002, 2003; Silva and Mercer 1985). Suppression of lysosomal degradation of proteins could be an important means of nitrogen conservation in cats forced to consume low-protein diets (Silva and Mercer 1991), which could readily happen to scavenging strays. Using 5 laboratory cats, Russell et al. (2003) demonstrated that diets in which most energy came from both high (70%) and moderately high (20%) protein concentrations could balance protein breakdown and synthesis, thus maintaining nitrogen balance as measured by comparing rates of ureagenesis and urea excretion. Both proteolysis and urea excretion ramp up with protein intake, tripling between 20% and 70% (by DM) protein consumption. However, these measures are not completely effective, leaving intact a basic view of the cat as a user of outsized quantities of protein necessary to offset a high obligatory nitrogen loss. The cat's partial latitude to control these processes might simply be narrower than that of other species (Hendriks et al. 1997). Furthermore, no work has yet accounted for the strange absence of rate-limiting amino acid control over body protein catabolism based on momentary metabolic needs.

Without a tight rein on its liver enzymes, how does a cat modulate the effects of a high-protein intake? More specifically, what mechanisms maintain plasma amino acid and nitrogen concentrations at the micromolar and millimolar level, respectively, even during times of rising protein consumption? One possibility is by enlargement of the liver, which also speeds up amino acid catabolism. Another is by increasing the rate at which amino acids are degraded as the level of protein increases, which naturally affects whole-body nitrogen turnover.

As to the second, aside from arginine the cat's EAA requirements for growth are not inordinately great (Rogers and Morris 1979), but the amount needed for tissue maintenance is something else, as assessed in adult cats. Newer findings have shown that the cat's enzymatic system is more flexible than previously thought. Revisiting the experiments of Russell et al. (2003) is useful here. They tested 5 adults at two protein concentrations, medium (MP = 20% DM) and high (HP = 70% DM) using a cross-over design and following full-body nitrogen turnover for 48 h in urine and feces after administering a single intravenous dose of [^{15}N]glycine. Nitrogen flux increased with protein concentration (56 mmol N/kg M/d at MP vs. 146 mmol N/kg M/d at HP). Protein synthesis also showed a direct association (6.6 mmol N/kg M/d at MP vs. 75 mmol N/kg M/d at HP). Protein breakdown demonstrated the same trend, being higher at HP (72 mmol N/kg M/d) than at MP (44 mmol N/kg M/d), the equivalent of 6.3 vs. 3.9 g protein/kg M/d. Finally, on a comparative basis using Kleiber's scaling exponent for body mass (g/kg $M^{0.75}$/d), cats turned over whole-body protein more slowly than herbivorous and omnivorous species (e.g. rat, rabbit, sheep, human, cow), for which average rates are $\geq$12.6 mmol N/kg M/d versus those of the cat at 4.9 mmol N/kg M/d (MP) and 9.7 mmol N/kg M/d (HP).

Of the EAAs (Table 7.2), asparagine is necessary for maximum growth in newly weaned kittens, but only for a few weeks (MacDonald et al. 1984a). Cats are unable to synthesize arginine (Morris 2002). Kittens in particular might not obtain enough dietary arginine and lose weight in addition to developing cataracts (Adkins et al. 1997). Arginine deficiencies would be exceedingly rare in free-ranging cats (Morris 2002), but quick death is the result of no arginine at all for cats of any age. Arginine is the most immediately essential of the EAAs: just one arginine-free feeding and a cat is doomed.

Cystine "spares" ~35% of the methionine requirement in kittens (Rogers and Morris 1982, Schaeffer et al. 1982). Dietary leucine in excess is growth-suppressing, a condition reversible by adding isoleucine and valine (Hargrove et al. 1988). However, blending excessive amounts of leucine into experimental diets (100 g/kg of food) does not hold down appetite (Hargrove et al.1994), and cats are not particularly sensitive to large quantities of dietary branched-chain amino acids (Hargrove et al. 1988). For the cat, we must also include taurine among the EAAs, which prevents central retinal degeneration and myocardial failure (Hayes and Carey 1975, Pion et al. 1987) despite some capacity for its conservation and therefore resistance to depletion.

Among mammals, taurine is an EAA only for cats and perhaps human infants. It functions in cats as a conjugate for bile acids (Morris 2002, Morris *et al.* 1994) , but its presence in animal tissues means that shortages are never a problem provided the diet is carnivorous and the protein source not exclusively casein. More than half an adult cat's taurine requirement never enters the metabolic system but is directed instead toward replacing that degraded by intestinal microorganisms (Morris *et al.* 1994).

The dog conserves taurine by employing glycine conjugation when dietary taurine is in short supply. Cats must use the alternative pathway. Taurine is a metabolite of cysteine. Synthesis from cysteine is inefficient, however, making exogenous sources necessary (Glass *et al.* 1992). Under natural conditions, cats capitalize on cysteine catabolism to yield pyruvate, which can be oxidized and used as energy. When dietary taurine is ingested directly the advantages gained through these pathways disappears: any amount generated would be excreted with the urine, its potential energy value lost (Morris 2001, 2002). As predators of small animals, cats typically consume their prey whole (Chapter 9), ingesting a range of tissues varying in composition and ordinarily satisfying taurine requirements. Doing so obviates any need for synthesis.

The situation is different for confined house and laboratory cats, and even strays if they too rely exclusively on human-produced foods. Here *digestible protein* (i.e. the fraction actually used and not excreted) and the manner in which it has been processed become important. Taurine present in animal tissues is an adequate dietary source, but the situation changes if vegetable matter has been included, which happens when a cat eats either unsorted and unprocessed human garbage or commercial cat foods (Section 7.5). If the manufacturing or cooking process has yielded Maillard products, microbial degradation of taurine, once ingested, accelerates (Morris 2002). Obviously, both canned cat foods and kibbles must have enough dietary taurine for confined cats unable to hunt or scavenge, but canned products need to contain double the minimum amount in "dry" foods for maintenance of normal plasma concentrations (Morris 2002). Glass *et al.* (1992) considered a normal dietary concentration to be 1 g/kg. Excess is excreted harmlessly with the urine (tauriuria).

As mentioned again later, cats offered food *ad libitum* choose to eat several small meals over the diel cycle, and at least part of this behavior is based on functional needs tied to nitrogen assimilation and excretion. A confined cat of normal weight fed once or twice daily is usually unable to ingest enough bulk without returning to eat again. If the food is picked up when the cat first walks away, its intake might become limiting over time. Bradshaw *et al.* (1996: 206) called this requirement of multiple meals "nibble eating," noting that food available at all times reduces the postprandial alkaline tide. They speculated, in addition, that cats lack an "effective mechanism to avoid struvite [magnesium ammonium phosphate] crystal formation because the problem would be unlikely to develop under natural feeding patterns." Because urolithiasis, urethral obstructions, and related maladies are common in confined house cats (e.g. Barker and Povey 1973, Jones *et al.* 1997, Lekcharoensuk *et al.* 2001, Segev *et al.* 2011), an etiological study assessing whether feral cats are less afflicted might be interesting.

The amino acids ingested by a cat that has been fasted overnight are quickly deaminated to provide energy (Morris 2002). Ammonia, an end product of deamination, is diverted continuously into the urea cycle instead of relinquishing nitrogen to synthesize dispensable amino acids. The hours without food – even if just overnight – will have depleted plasma arginine along with other intermediate compounds vital to anaplerosis and closure of the urea cycle (Morris 2001). Were the cat not a cat but some other mammal the ammonia would be converted to urea as usual, but a night without food induces a drop in blood amino acid levels, especially arginine. As Morris (2001: 189) wrote: "When cats ingest a meal containing all essential amino acids, arginine has an anaplerotic effect on the urea cycle, and allows normal disposal of the ammonia from deaminated amino acids." But fasting for several hours has depleted the remaining arginine stores in the liver. Under extreme conditions in which no arginine is taken in for several hours the result is diminished ammonia clearance forcing its retention. In such situations cats lose body mass rapidly (Morris *et al.* 1979). Deamination does not cause a rise in ammonia in other mammals (including humans) because they have the capacity to synthesize intermediate compounds for injection into the urea cycle (Carey *et al.* 1987).

Hyperammonemia is therefore at the zenith of extreme arginine deprivation in cats. Some of its clinical signs are excessive salivation (frothing), neurological dysfunction, emesis, moaning, extended claws and limbs, hyperesthesia, ataxia, hyperglycemia, coma, and death (Morris and Rogers 1978a, 1978b). As Morris (2002: 158) wrote, "Cats are exquisitely sensitive to arginine deficiency, for there is no other example in a mammalian species where consumption of a single meal lacking an essential nutrient can lead to death."

The idea that cats somehow sense a nutritional deficiency and pick out foods to correct it (Bradshaw and Thorne 1992, Church *et al.* 1996) has mixed empirical support. Cats do not select for sweetness (Section 7.5), nor for sodium even if sodium-deficient (Yu *et al.* 1997). Rogers *et al.* (2004) assessed choices made by laboratory kittens offered diets deficient in threonine or methionine compared with others containing a balanced array of amino acids. Kittens require threonine at 6 g/kg of food (Hammer *et al.* 1996), but were allowed to pick from diets that contained 325 g/kg amino acids including 0, 4, or 6 g/kg threonine or a protein-free diet. The diet with amino acids contained all the EAAs except threonine. The tendency was to eat more of the diet containing amino acids as opposed to the one devoid of protein, but the threonine concentration seemed irrelevant. The kittens lost weight. They apparently selected for taste (texture was controlled) and evidently were not under physiological pressure to detect and correct this specific deficiency. In contrast, when the experiment was repeated in a slightly different configuration using adult cats with methionine (0, 2, or 4 g/kg of food) substituted for threonine, the diet containing 2 g/kg (the apparent minimum requirement) was selected most often. Selection for methionine was confirmed when other experiments showed palatability not to be a factor.

Palatability might take precedence over physiological need, similar to the effect of junk foods on humans. Some fats improve taste (Kanarek 1975). However, fat content alone (i.e. independent of flavor or texture) has little effect on dietary preferences (Kane *et al.* 1987). Cats prefer protein hydrolyzates and meat extracts (Boudreau 1974, White and Boudreau 1975).

Cats are notable for being "finicky" and rejecting foods that are too powdery or have unusual textures, and not controlling for texture can be a confounding variable in experiments devised to evaluate food selection (Cook *et al.* 1985, Hirsch *et al.* 1978, Kane 1989, Rogers *et al.* 2004). Cats can also become habituated to foods, even unusual ones, if exposed to them early in life (Chapter 6). Palatability involves aroma, taste, texture, and consistency (Bradshaw *et al.* 1996, Bradshaw and Thorne 1992, Kane 1989).

Bradshaw *et al.* (1996: 206) proposed that in the ancestral African wildcat, "palatability may play a role in its preference for prey, but only if that prey can be associated with its taste when caught. ... " Shrews, which are insectivores, might be less palatable than other small mammals, which does not stop house cats from catching and killing shrews, then leaving them uneaten (Bradshaw and Thorne 1992, von Borkenhagen 1979). To Bradshaw and Thorne (1992) and Bradshaw *et al.* (1996) this behavior indicates a weak association between taste and predation, perhaps a result of domestication, although it could simply be the discharging of a predatory urge.

When domestic animals other than the cat are fed diets deficient in a specific EAA, food intake diminishes and growth is retarded, the signs of an amino acid imbalance response (Harper *et al.* 1970). According to Morris (2001), in this situation *growth retardation results from diminished consumption caused by insufficiency of an EAA. What we see when any of the EAAs is deleted from a cat's diet and not replaced is reduced food intake accompanied by slow, steady weight loss, but no other clinical signs.* This reinforces two facts: (1) the functions of EAAs are to promote and regulate growth and tissue maintenance, and (2) clinical signs of EAA deficiencies are growth-related and non-specific (i.e. signs of deficiency are the same for all of them). The exception is arginine, deletion of which causes death. Remediation requires adding more of the missing EAA with any increases in dietary protein. In other words, simply raising the protein content is inadequate if the deficient EAA does not rise in tandem (Hammer *et al.* 1996). Stated in a slightly different context, the correlation between a given EEA and the concentration of protein in domestic omnivores is not just positive, but strongly so. When a specific EAA is deficient, growth actually increases as protein concentration declines. Cats, in contrast, display little or no amino acid imbalance response: food intake increases with a rise in dietary protein regardless of EAA status, and neither the feeding response nor growth is necessarily tied to a concomitant rise or fall in the deficient EAA. As an example, kittens continued to grow when protein went from 200 to 300 g/kg of food at all levels of threonine tested (4, 5, 6, 7, 9, and 12 g/kg of food). Interpreting their findings, Hammer *et al.* (1996: 1502) wrote: "These results suggest that cats do not have tight control on the efficiency of utilization of protein, which is probably secondary to nonadaptive protein catabolic enzymes in this species. ... " In practical terms, cats do not suffer amino acid imbalances if the protein they ingest is animal-based and consumed in sufficient amounts. This statement holds true for arginine too: the chances of experiencing hyperammonemia are nonexistent if the sole food is animal tissue.

Cooking has no significant effect on the protein composition of raw meat (Kerr *et al.* 2012). However, heat treatment of canned cat foods during pasteurization lowers the digestibility of all amino acids, and this is true especially at high temperatures when amino acids become cross-linked within and between proteins, possibly causing interrupted or diminished enzyme penetration (Hendriks *et al.*1999).

Feces of laboratory cats fed raw and cooked beef contained a higher proportion of propionate and a lower proportion of butyrate compared with others fed a dry diet, indicative of altered carbohydrate fermentation (Kerr *et al.* 2012). The dry-diet group also excreted higher concentrations of ammonia, isobutyrate, valerate, isovalerate, and total branched-chained fatty acids (Table 7.5). Overall, raw meat diets are more completely digestible on a DM basis, and lower volumes are consumed (Table 7.5) by both domestic cats and zoo specimens of their close relatives the African wildcat (*Felis libyca*) and sand cat (*Felis margarita*). The body masses of laboratory cats were

Table 7.5 Fecal quality and the concentrations (μmol/g DM) of ammonia, short-chained fatty acids (SCFAs), and branched-chained fatty acids (BCFAs) excreted by domestic cats ($n = 9$), African wildcats ($n = 6$), and sand cats ($n = 8$) fed a high-protein extruded diets (DRY), raw beef (RB), cooked beef (CB), or raw meat ("RM"). See sources for levels of significance.

Item	Dry[a]	RB	CB	±SEM	Dry[b]	"RM"[d]	±SEM	Dry[c]	"RM"[e]	±SEM
		Domestic laboratory cats				African wildcats			Sand cats	
Water					5.2	61.8	–	6	68	
Dry matter					94.8	38.2	–	94	32	
Organic matter					89.2	91.8	–	–	–	
Protein					52.9	44.9	–	40.2	57.2	
Fat					23.5	36.9	–			
Carbohydrate					–	–	–			
Fiber					2.0	4.8	–			
Energy (kJ/g)					23.03	26.79	–	23.24	24.62	
Fecal score[f]	3.3	2.9	3.8	0.2						
Ammonia	190.4	69.4	72.0	17.9	190.3	137.2	36.4			
Acetate	214.6	178.2	275.3	48.9	173.5	195.9	22.0			
Propionate	50.9	65.3	102.7	16.6	42.7	67.7	9.1			
Butyrate	38.2	21.2	25.5	3.2	30.0	18.4	4.4			
Total SCFA[g]	305.1	266.3	404.7	66.9	246.2	282.0	32.75			
Isobutyrate	10.1	4.9	5.1	0.9	8.5	5.8	1.3			
Valerate	18.3	6.0	5.3	1.9	11.9	8.4	2.5			
Isovalerate	15.3	6.7	6.4	1.3	10.7	7.0	2.05			
Total BCFA[h]	43.7	17.6	16.8	3.5	31.1	21.2	3.7			

[a]Dry – Natura Pet Products Inc., Fremont, NE.
[b]Dry – Natura Manufacturing Inc., Fremont, NE.
[c]Dry – Mazuri Diet 5M54, Purina Mills, St. Louis, MO.
[d]"RM" – Nebraska Brand® Special Beef Feline, Nebraska Packing Inc., North Platte, NE): Beef, meat by-products, fish meal, soybean meal, dried beet pulp, calcium carbonate, dried egg, dried brewer's yeast, Nebraska Brand feline vitamin premix, salt, Nebraska Brand trace element premix.
[e]"RM" – Nebraska Canine Diet (Animal Spectrum, Inc., North Platte, NE): Horsemeat, meat by-products, dried beet pulp, vitamin and salt additions.
[f]Scale of fecal scores: 1 = hard, dry pellets; 2 = dry, well-formed; 3 = soft, moist; 4 = soft, unformed; 5 = watery and containing liquid that can be poured.
[g]Total SCFAs, acetate + propionate + butyrate.
[h]Total BCFAs, isobutyrate + valerate + isovalerate.
Sources: Data from Kerr *et al.* 2012 for Dry[a], RB, CB (domestic cats); Vester *et al.* 2010 for Dry[b], "RM"[d] (African wildcats); Crissey *et al.* 1997 for Dry[c], "RM"[e] (sand cats).

unaffected by any of their diets (raw beef, cooked beef, or high-protein kibbles), but fecal output was twice as great in the kibbles group (36.1 vs. 17.5 g/d), evidence of its reduced digestibility and the need to eat more of it to maintain body mass (Kerr *et al.* 2012). Feces of these cats were also looser than cats fed only beef, either raw or cooked. Ammonia and branch-chain fatty acids are products of putrefaction from unused amino acids, and both were far higher in the dry diet, pointing to lower digestibility of the crude protein fraction (Kerr *et al.* 2012: 518, Table 2). Albumin levels were also elevated in these cats, which could be simply an effect of dehydration from eating dry food and not drinking enough water.

7.4 Fats

Fat's three functions are to (1) provide concentrated energy, (2) supply essential fatty acids (EFAs), and (3) serve as carriers of fat-soluble vitamins (MacDonald *et al.* 1984a, Rivers and Frankel 1980). What the "fatty acid requirement" actually means is that every mammal evidently requires all the *cis*-polyunsaturated fatty acids, which are divided into two "families" labeled n6 and n3 (McLean and Monger 1989). Most fatty acids in cats are synthesized in the adipose (i.e. fatty) tissues, not the liver, and acetate is the primary carbon source of their *de novo* formation instead of glucose (Richard *et al.* 1989). Cats can digest high concentrations of fat (MacDonald *et al.* 1984a). Adding them to commercial diets for house and laboratory cats can increase palatability. Fats, or lipids, occur in animal and plant tissues. These and processed animal-based fats are easily digested (Kane *et al.* 1981a). Cats, like other mammals, require dietary linoleate (18:2*n*6) and *n*3 fatty acids (Morris 2002). Their capacity to synthesize arachidonic acid (20:4*n*6) from linoleate is evidently limited, as is the capacity to form eicosapentaenoate (EPA, 20:5*n*3) and docosahexaenoate (DHA, 22:6*n*3) from α-linolenate (18:3*n*3), as shown by Bauer (1997) and MacDonald *et al.* (1983).

McLean and Monger (1989) wrote that the classic fatty acid deficiency is an *n*6 shortage, arising in most mammals from lack of dietary linoleic acid. However, *n*3 deficiencies occur too, and these result from inadequate intake of α-linolenic acid. The signs are comparatively vague, their experimental induction more difficult. Dogs – but not cats – convert linoleic acid to arachidonic acid and meet their arachidonate requirement by eating plants. Arachidonate, as mentioned above, is a polyunsaturated *n*6 fatty acid. In general, mammals convert linoleate to arachidonic acid sequentially with $\Delta 6$ desaturation and chain elongation preceding $\Delta 5$ desaturation (Sinclair *et al.* 1979). However, chain elongation from γ-linolenic acid (GLA, 18:3*n*6) to dihomo-γ-linolenic acid (DGLA, 20:3*n*6) and then to arachidonate can occur via the $\Delta 5$ desaturase pathway (McLean and Monger 1989, Sinclair *et al.* 1979). The $\Delta 6$ and $\Delta 8$ desaturase activities are low in cats, marginalizing desaturation (MacDonald *et al.* 1984a, McLean and Monger 1989, Pawlosky *et al.* 1994, Rivers and Frankel 1980), although feeding high concentrations of γ-linoleate appears to bypass the $\Delta 6$ desaturase step and permit arachidonate to form directly via $\Delta 5$ desaturation (Trevizan *et al.* 2012). With the capacity to synthesize 20:4*n*6 from 18:2*n*6 compromised, dietary sources of arachidonic acid are necessary, especially for pregnant females to proceed through gestation avoiding stillbirths and kittens with deformities (Pawlosky and Salem 1996).

Feeding cats adequate γ-linoleate while withholding arachidonate causes other maladies including listlessness, dry coats, heavy dandruff, poor growth, fatty infiltration of the liver, underdeveloped testes, and failure to complete estrus (Rivers *et al.* 1975). Animal fats are assured sources of arachidonate, and it seems sensible that any cat would fare best on diets containing preformed long-chain PUFAs derived from them (McLean and Monger 1989). Although all EFAs are PUFAs, not all PUFAs are essential (Rivers and Frankel 1980). What constitutes a PUFA is its molecular structure, which is a chemical designation. Labeling it an EFA refers to its biological function, and this varies by species (Rivers and Frankel 1980).

What scavenging cats eat – or fail to eat – obviously affects their health and survival. Excessive quantities of fish oils raise the plasma and skin concentrations of eicosapentaenoic and docosahexaenoic acids, two PUFAs, potentially inducing compromised inflammatory responses and immune suppression (Park *et al.* 2011, but see Jaso-Friedmann *et al.* 2008 and Mazaki-Tovi *et al.* 2011). Cats scavenging at fish-waste dumps where fish offal is the only source of food might ingest large amounts of PUFAs and acquire a vitamin E deficiency (MacDonald *et al.* 1984a). There is also older evidence that fish oils low in vitamin E cause reduced tissue concentrations of arachidonate in cats ingesting them (Stephan and Hayes 1978, but see Rivers and Frankel 1980). The extent to which large quantities of PUFAs harm or benefit free-ranging cats is therefore unknown. As for house and laboratory cats, Plantinga *et al.* (2011) made the point that commercial diets contain fats from captive domestic animals, and these yield fatty acid profiles different from the wild animals preyed on by feral cats.

7.5 Carbohydrates

Carbohydrates make up a large proportion of the human garbage ingested by scavenging cats, and house and laboratory cats fed commercial diets also consume carbohydrates in unnatural quantities. Dietary carbohydrate determined by analysis is often expressed differentially as nitrogen-free extract (NFE), a misnomer because it is neither an "extract" nor does it have any relationship with nitrogen. Nitrogen-free extract ordinarily refers to soluble, or "digestible," carbohydrates (i.e. starches and sugars); the remaining insoluble portion is fiber. Cats are described as "obligate carnivores," implying that meat is a dietary requirement. The reasoning goes that as a result of this irreplaceable need, and because carbohydrates are scarce in animal tissues, carbohydrates are neither required (MacDonald et al. 1984a) nor especially useful to cats. What cats actually require is a high amount of dietary protein containing the requisite amino acids – not necessarily meat – and carbohydrates, although not substitutes for amino acids, are digestible if delivered in proper forms, and capable of supplying energy. Were this not true then no cat could survive on today's commercial cat foods, which contain 5–20 times the carbohydrate of natural prey on an energy basis (Eisert 2011), or more in a single small meal than a feral cat ingests over several days of small meals.

Carbohydrates, or saccharides, are described chemically by the general formula $C_m(H_2O)_n$ and include sugars and stored plant and animal compounds (e.g. starches, glycogen). Animal tissues contain small quantities of glucose, glycogen, glycoproteins, glycolipid, and pentose, but not starch (Plantinga et al. 2011), which only plants produce. The inclusion of carbohydrates in commercial cat foods – or diets of humans, for that matter – is unnecessary given adequate amounts of protein and fat, and as feral cats are pure carnivores they often have trouble processing them. Adult cats do not need carbohydrate in any form, and kittens might not either if provided with triglycerides and sufficient amounts of glucogenic amino acids (MacDonald et al. 1984a).

Fats are a natural ingredient of animal tissues, but carbohydrate is not. Commercial cat foods typically include >30% carbohydrate by DM. The question is, can cats use it effectively? As mentioned (Section 7.2), carbohydrates make up <2% by mass of the food intake of feral cats living on live prey and carrion (Plantinga et al. 2011). Animal flesh contains almost none, and what gets eaten consists principally of glycogen and plant-based carbohydrates in the digestive tracts of omnivorous and herbivorous prey (De Wilde and Jansen 1989), which turns out to be very little. Strays scavenging on garbage (e.g. dining hall waste) are likely to ingest items containing a high percentage composition of cooked starches (Table 7.6), and confined cats are often fed diets of up to 60% carbohydrate by DM, more than half of which is starch (De Wilde and Jansen 1989, Funaba et al. 2001).

Cats have what Morris (2002: 164) termed "an abridged pattern" of enzymes standing ready to metabolize carbohydrates, which is not a burden if the diet consists solely of animal tissues. Several biochemical factors account for the cat's limited capacity to use most forms of carbohydrate. Cats lack salivary amylase activity (Buddington et al. 1991, Plantinga et al. 2011) and display a low activity of pancreatic and intestinal amylases (Kienzle 1993a, 1993b), making them unresponsive to changes in dietary

Table 7.6 Proximate composition of different food sources shown as percentage wet mass (e.g. g/100 g); energy in kJ/g DM.

Food	Water	Dry matter	Protein	Fat	Carbohydrate	Fiber	Ash	Energy	Source
Atka mackerel	78.3	21.7	15.67	2.96	–	–	2.71	–	Van Pelt et al. (1997)
Beef trimmings	71.3	28.7	18.91	5.54	–	2.01	–	25.1	Kerr et al. (2011)
Bison trimmings	64.0	36.0	13.98	13.68	–	2.41	–	28.5	Kerr et al. (2011)
Canned cat food[a]	25	75	9.4 (37.6)	5.5 (22.0)	7.1 (28.4)	1.2 (4.8)	1.8 (7.2)	4.6	Hill's Science Diet® Adult Gourmet Beef Entrée
Canned cat food[a]	22	78	10.0 (45.4)	5.0 (22.7)	1.9 (8.6)	1.0 (4.5)	3.0 (13.6)	7.9	Friskies Classic Paté Salmon Dinner
Chicken breast	74.0	26.0	22.8	11.3	–	–	–	–	López et al. (2011)
Cockerel (skinned)	77.8	22.2	13.14	5.97	–	–	1.80	20.0	Lavigne et al. (1994)
Cockerel (whole)	75.0	25.0	15.0	7.02	–	–	1.77	25.8	Lavigne et al. (1994)
Dining hall waste[b]	65.3	34.7	5.16	6.06	22.39	–	1.12	20.2	Ferris et al. (1995)
Dining hall waste[c]	70.8	29.2	5.12	6.33	16.58	–	1.16	20.6	Ferris et al. 1995
Dry cat food	5.6	94.4	33.4	22.2	37.8	1.0	5.6	17.0	Hill's Science Diet® Adult Optimal Care
Dry cat food	12	88	34	13	34.6	4.5	7.0	7.6	Purina Cat Show Complete Formula
Elk muscle	71.3	28.7	22.61	1.55	–	2.64	–	22.6	Kerr et al. (2011)
Goat (cooked composite)	57.4	42.6	11.25	6.01	–	–	0.51	9.96	Johnson et al. (1995)
Goat (raw composite)	68.5	31.5	6.11	3.81	–	–	–	–	Johnson et al. (1995)
Grasshopper	65.9	34.1	21.4	8.5	10.0	–	1.4	–	Ademolu et al. (2010)
Grasshopper (adult)	65.9	34.1	21.38	0.85	10.02	1.23	1.40	–	Ademolu et al. (2010)
Green anole	70.6	29.4	19.83	–	–	–	–	–	Cosgrove et al. (2002)
Guinea pig (whole)	69.3	30.7	17.8	13.8	–	–	0.03	–	Clum et al. (1996)
Horse trimmings	65.4	34.6	20.62	9.03	–	2.46	–	25.9	Kerr et al. (2011)
Japanese quail (whole)	65.4	34.6	23.3	10.3	–	–	0.032	–	Clum et al. (1996)
King pigeon (breast)	66.5	33.5	7.90	2.37	–	–	0.37	–	Pomianowski et al. (2008)
Kitchen garbage	–	–	21.1	7.0	30.1	–	2.0	–	Sehgal and Thomas (1987)

(continued overleaf)

Table 7.6 (continued)

Food	Water	Dry matter	Protein	Fat	Carbohydrate	Fiber	Ash	Energy	Source
Lab mouse (skinned)	67.7	32.3	14.47	15.02	–	–	3.33	29.2	Lavigne et al. (1994)
Lab mouse (whole)	64.4	35.6	15.20	16.55	–	–	2.70	29.2	Lavigne et al. (1994)
Lab rat (whole)	66.2	33.8	26.6	11.8	–	–	0.029	–	Clum et al. (1996)
Municipal garbage[d]	7.9[c]	92.1	24.9	13.0	13.0	38.8	10.3	–	Johnson et al. (1975)
Pacific cod	77.8	22.2	18.62	1.80	–	–	–	16.8	Anthony et al. (2000)
Rabbit (domestic, age 45 d)	70.8	29.2	19.4	6.4	–	–	3.4	–	Xiccato et al. (1999)
Rabbit (wild, adult)	75	25	23.7	0.2	–	–	1.2	–	González-Redondo et al. (2010)
Turkey breast	69.8	30.2	29.1	5.4	–	–	1.9	–	Smith et al. (2012)
Turkey breast	72.6	27.4	19.7	3.83	–	–	2.9	–	Fanatico et al. (2005)

[a]Values taken or recalculated from website information. Parenthetical values for canned cat food represent recalculated percentage composition based on dry matter for comparison with dry cat foods.
[b]University Dining Center, Kansas State University, Manhattan, KS. Plate leftovers (service plate scrapings) and production food wastes combined ($\bar{x}$ = 3276 meals/d served).
[c]Noncommissioned Officers Academy Dining Facility combined with 134 Armor Dining Facility, Fort Riley, KS. Plate leftovers (service plate scrapings) and production food wastes combined ($\bar{x}$ = 911 meals/d served).
[d]After aerobic digestion.

carbohydrate concentrations. They can nonetheless digest cooked starches efficiently (Kienzle 1993b, Morris *et al.* 1977). Glucokinase, the predominant hexokinase in animals, is absent from cat livers, which is consistent with carnivory (Morris 2002). Cats find sugars either unpalatable or are unable to taste them (Beauchamp *et al.* 1977, White and Boudreau 1975). A case has been made for the second possibility based on the existence of an unexpressed pseudogene labeled *Tas1* (Li *et al.* 2005). Cats, however, show a preference for sucrose and lactose in milk, perhaps because of the change in texture and increased density (Beauchamp *et al.* 1977). Lactose is present in cat milk, and nursing kittens obtain some of their energy from it (MacDonald *et al.* 1984a). Habit formation can be a strong influence early in life (Chapter 6), and becoming habituated to the specific properties of lactose prior to weaning is a possibility for its acceptance in adulthood, but this hypothesis has not been examined. During their evolution, tigers and cheetahs also underwent pseudogenization of the sweetness reception gene, an event that perhaps occurred in all cats as they evolved into pure carnivores (Li *et al.* 2005).

The carbohydrate component of fresh animal tissue is often barely detectable, but in commercial cat foods, and extruded dry kibbles especially, its percentage might equal or exceed that of protein. A preferable commercial food would approximate the nutrient profile of the cat's evolutionary diet (Plantinga *et al.* 2011). Converting the findings of Plantinga and colleagues to "as fed" reveals the "average" feral cat's intake to comprise 69.5% water, 19.1% protein, 6.9% fat, 0.85% NFE, and 3.65% ash. Starch is a small portion the NFE. Of the fatty acids, $n6/n3$ exist at a ratio of 2, which is substantially lower than the ratio found in commercial cat foods (range 5–17). By my estimation, a suitable basic commercial cat food might be a canned (wet) product offered *ad libitum* and having an "as fed" proximate composition of 67% water. The remainder would contain 22% animal protein (based either on nitrogen or EAA content), 8% animal fat, 3.0% ash, no digestible carbohydrate, and no fiber. Such a diet ought to be nutritious and maintain adequate water balance (Chapter 8).

Good growth and maintenance have been seen with diets based on vegetable protein (e.g. Morris 1999: 904, Table 1), and it could be argued that animal protein is expensive while offering no tangible benefit. However, animal proteins like fish meal might be superior in other ways, such as helping to prevent urolithiasis and constipation (Funaba *et al.* 2001), but these and other possible advantages need considerably more evaluation. The natural sources of derived EFAs for cats are animal products, and their highest concentrations occur in organ meats (e.g. kidney, heart, liver, brain). The internal organs are important dietary components, and feral cats feeding on small mammals and consuming most of their carcasses are unlikely to become deficient. Even so, as Rivers and Frankel (1980: 94) wrote, "Since triglycerides provide only low levels of [derived] EFA and most of the variations in fat content of meat is due to variation in triglyceride content, there is a poor correlation between the fat content of the meat and its [derived] EFA content."

Cats have strict protein requirements, and raising the amount of dietary carbohydrate fails to compensate for a shortage (Jones 1989). For example, pregnant cats fed diets deficient in protein are more likely to abort (Gallo *et al.* 1984). As mentioned, whereas raw plant materials are indigestible to cats, starches added to commercial cat foods become digestible after having been milled and cooked (see later). Still, adding certain ingredients – especially corn – induces a postprandial glucose spike, and corn and other plant-based starches cause blood insulin to rise (de-Oliveira *et al.* 2008). The postprandial effect is less marked in cats than in omnivores like humans and dogs, perhaps because cats digest and absorb starch less efficiently – that is, slowly – and the effects detected by blood sampling are delayed (de-Oliveira *et al.* 2008, Hewson-Hughes *et al.* 2011b). This probably reflects the generally inferior capacity of cats to process carbohydrates mentioned earlier (e.g. lack of salivary amylase, low pancreatic amylase activity) as a result of evolution toward strict carnivory.

Commercial cat foods are formulated and manufactured under cost restraints, and animal protein is an expensive major ingredient. Filling out part of the protein composition with plant protein and adding a large measure of carbohydrate contributes to bulk and potential energy without compromising digestibility. Some essential compounds (e.g. taurine) are absent from plant proteins and must be added separately if animal proteins are omitted. Many forms of carbohydrate can be rendered digestible to cats, cellulose being an exception. The sugars glucose, sucrose, lactose, and dextrin are significantly more digestible than starch, although inclusion of lactose at 200 g/kg of food reduced the digestibility of crude protein (Morris *et al.* 1977). Starches (e.g. wheat, maize) are digestible by cats if milled to a fine powder and cooked (de-Oliveira *et al.* 2008, Morris *et al.* 1977). Starch fed raw or improperly milled or cooked is digested poorly, resulting in gut fermentation and interfering with the digestion of protein. Raw starch thus reduces protein digestibility (Kienzle 1994), but starch prepared correctly then added to commercial diets exerts no such effect (De Wilde and Jansen 1989).

Three dry diets containing low, medium, and high starch content (Hewson-Hughes *et al.* 2011b: S106, Table 1) induced graded effects when fed to cats (Fig. 7.3). Protein and fat in all the diets had been raised or lowered to keep the water content comparable at 7.3–8.0%. Blood glucose levels from the medium-starch diet (31.0%) did not change from pre-meal values through post-feeding, although the concentration fell significantly 3–7 h after ingestion of the low-starch diet (12.3%). Compared with the pre-feeding concentration, cats fed the high-starch diet (41.4%) had plasma glucose levels significantly higher starting 11 h post-feeding, and they stayed significantly higher thereafter.

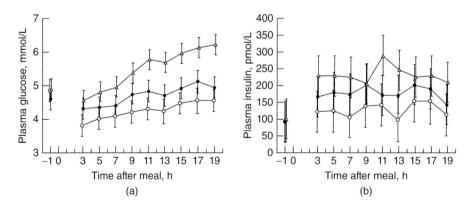

Fig. 7.3 Postprandial responses of plasma (a) glucose and (b) insulin of laboratory cats after ingesting low-starch (o, 12.3%), medium-starch (•, 31.0%) and high-starch (Δ, 41.4%) dry diets. Data shown are $\bar{x} \pm 95\%$ least significant difference intervals (vertical bars). Source: Hewson-Hughes *et al.* 2011b. Reproduced with permission of Cambridge University Press.

Variation in the digestibility of starches and the insulin response to them depends on their sources (e.g. corn gluten, rice, sorghum), fiber content and fiber structure (de-Oliveira et al. 2008, Sunvold et al. 1995), and probably particle size after processing, with smaller particles being used more efficiently. Feral cats, as emphasized, eat little or no carbohydrate. Stray cats undoubtedly ingest some while foraging on garbage. The high carbohydrate component of commercial cat foods is comparable to garbage: both contain starch that has been cooked and rendered more digestible, in the latter case originally for humans (Table 7.6).

Whatever its origin, cooked starch has been gelatinized, after which water acts as a plasticizer to heighten both digestibility and the subsequent glucose response. Gelatinization was ≥92.7% in all three diets used by Hewson-Hughes et al. (2011b), indicating adequate processing to enhance digestibility, and this was evident in the medium- and high-starch diets (Fig. 7.3). Their experiments, in concert with those of de-Oliveira et al. (2008) and Hoenig et al. (2007), demonstrated that high- and medium-starch diets are not beneficial to house and laboratory cats and potentially harmful if fed exclusively over long periods. Even if cooked and processed until rendered digestible the high carbohydrate content of today's commercial cat diets serves no useful purpose other than to reduce manufacturing costs and "spare" protein and fat (i.e. limit their concentrations used as energy sources), which do not require "sparing" if present in sufficient amounts. In the case of feral cats, animal tissue is sufficiently high in fat that amino acids are "spared" from supplying energy and reserved instead for protein synthesis and gluconeogenesis (Beliveau and Freeland 1982). Hewson-Hughes et al. (2011a) showed that adult cats preferentially select a daily intake of ~420 kJ from protein (52%), 280 kJ from fat (36%), and ~100 kJ (12%) from carbohydrate. According to Plantinga et al. (2011: S41), the choice of these ratios by cats appears to represent "sensitive metabolic regulation mechanisms to consume an overall dietary macronutrient profile close to their evolutionary diet."

Cat food manufacturers would prefer that any protein added be "spared" from also supplying energy if that requirement can be met more cheaply by fats and carbohydrates. Consequently, cat foods contain high levels of both fats and starches (Table 7.6). Such diets are obviously "unnatural." Plantinga et al. (2011) wrote, "Almost all the metabolic adaptations [of cats] related to the carbohydrate component of the diet indicate the lack of this nutrient in the evolutionary diet." That cats survive on them is proof of their flexibility. Commercial cat foods have an overall composition remarkably close to that of kitchen and dining-hall garbage, which is also high in carbohydrates and fats (Table 7.6) and relatively low in protein, but humans are omnivores, not carnivores. The qualities of our diets are largely unsuitable for cats, even an unsophisticated stray picking through a trash bin. Despite the low regard some have for scavenging cats the food we throw away seldom meets their standards. If we consider as ideal a diet adapted to an animal's evolutionary needs then feral cats subsisting on other animals have diets superior to the commercial foods eaten by a pampered house cat or stray relegated to scavenging human leftovers.

7.6 Fiber

Fiber is a common additive to commercial cat foods, although the reason is unclear. That dietary fiber might aid omnivorous humans (references in Sunvold et al. 1995) is not necessarily relevant to the carnivorous cat. Digestibility of DM, and of organic matter generally, is higher in cat diets to which no fiber has been added (Sunvold et al.

1995). Some sources of fiber included in cat foods lower the digestibility of other nutrients, besides increasing wet fecal output (Sunvold *et al.* 1995).

7.7 Vitamins

Cats are typical of other mammals in their vitamin requirements except for A, D, and niacin (MacDonald *et al.* 1984a). They are unable to synthesize retinol (vitamin A) because the enzymes needed to cleave carotene have been deleted (Morris 2001). This prevents them from satisfying vitamin A requirements indirectly by eating plants, which omnivores like dogs can do. Preformed dietary sources are therefore necessary, and in consuming typical prey (rodents and birds) this need can be met. An obligation to also consume vitamin D and niacin is traceable to the high activities of the enzymes involved in catabolizing the precursors of these vitamins to different substances (Morris 2001).

Most animals, but not domestic cats, are able to generate vitamin D in the presence of sunlight (Morris 1999). Despite the propensity to "sun" themselves, doing so does nothing to maintain physiologically necessary levels of vitamin D. Synthesis is prevented by enzymatic activity that limits the availability of a precursor (Morris 2002), although adequate amounts of vitamin D are available in prey tissues (Morris 1999), and as with vitamin A, cats are unusually tolerant of high dietary concentrations (Morris 2001, 2002; Sih *et al.* 2002). Changes in plasma 25-hydroxyvitamin-D (25-OHD, or cholecalciferol), can be used to monitor the vitamin D status of cats (Morris 1999). Plasma 25-OHD declined in kittens fed a diet deficient in vitamin D whether they were exposed to sunlight or kept indoors, or if exposed to ultraviolet (UV) lamps, or not. When plasma 25-OHD fell to <5 nmol/L the hair on the backs of some was shaved to expose more skin directly to UV light. Some were also administered an inhibitor of the enzyme 7-dehydrocholesterol, or 7-DHC-Δ^7-reductase (EC 1.3.1.21), in their food. Only those receiving it showed a steady increase in 25-OHD (Fig. 7.4), and their skin contained 5 times more 7-DHC than skin samples of controls. The inhibitor – BM 15.766 (4-[2-[1-(4-chlorocinnamyl)piperizin-4-yl]ethyl]benzoic acid) – reduced the conversion of 7-DHC to cholesterol and allowed vitamin D synthesis to proceed. The enzyme 7-DHC therefore interrupts vitamin D formation by cats (Morris 1999). Commercial cat food manufacturers spike their diets with cholecalciferol. In the presence of sufficient calcium, a concentration of 6.25 μg (250 IU) cholecalciferol/kg of diet is adequate to prevent clinical signs of vitamin deficiency (Morris *et al.* 1999). As mentioned, feral cats ordinarily eat most of the bodies of their prey including the bones, and in doing so ingest sufficient vitamin D and calcium over the course of a day.

Cats lack the capacity to synthesize niacin from tryptophan via nicotinic acid, (Mercer and Silva 1989). This pathway is blocked by the unusually high activity of picolinic carboxylase, which prevents nicotinic acid from forming at physiologically useful concentrations. Ultimately, niacin formation is precluded too (Morris 2002). Animal tissues contain adequate niacin, eliminating any need for its synthesis from tryptophan. This is another example of the synthesis of an important vitamin being limited by a precursor. Feeding nicotinamide at 2 mg/d is adequate (Mercer and Silva 1989).

The cat's requirement for choline in the B-complex vitamins is unusually high. The NRC (2006: 365, Table 15–9; 367, Table 15–11; 369, Tables 15–13) gave the minimum daily requirement for all cats – weaned kittens, adult cats, and pregnant and lactating females – as 2040 mg. Deficiency of thiamine, another B vitamin, causes Chastek paralysis in carnivorans, which is fatal if not remedied quickly. The

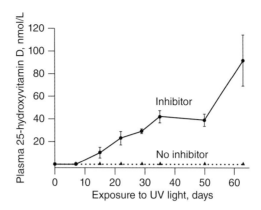

Fig. 7.4 Changes in plasma 25-hydroxyvitamin D (25-OHD) in two pairs of cats depleted of vitamin D and exposed to ultraviolet (UV) light 2 h/d. Both pairs were fed diets containing no vitamin D, and the hair on their backs was clipped weekly. The pair designated "inhibitor" received 250 mg of an inhibitor of 7-dehydrocholesterol-Δ^7-reductase/kg added to the purified diet. The inhibiting compound is BM 15.766 (4-[2-[1-(4-chlorocinnamyl)piperizin-4-yl]ethyl]benzoic acid). The group designated "no inhibitor" received just the vitamin D-free diet. Values are $\bar{x} \pm$ SEM ($n = 4$). Source: Morris 1999. Reproduced with permission of American Society for Nutrition.

disease is similar to Wernicke–Korsakoff syndrome of humans, also caused by lack of vitamin B_1. Chastek paralysis was first described in 1932 in red foxes at a fur farm owned by J. S. Chastek near Glencoe, Minnesota, United States; Everett (1944) reported identical effects shortly thereafter in cats. Neurological signs are preceded by inappetence and subnormal body temperature (Evans *et al.* 1942). The disease is characterized by liver degeneration, abnormal gait, ataxia, spasms, moaning, convulsions, and death, usually within 24 h of onset of ataxia and spasms. Brain lesions are apparent at necropsy. Chastek paralysis occurs in many animals, including rats and pigeons, but its onset in captive carnivorans most commonly follows the addition of certain fishes, most of them freshwater (e.g. European carp, *Cyprinus carpio*), to the diet, the tissues of which are thiaminase-positive. The result is destruction of dietary vitamin B_1 in the carnivoran that eats them (Green *et al.* 1941). Changing the diet of foxes to 20% carp induced the disease within 3 weeks, but foxes administered 25 mg/d of thiamine stayed healthy (Evans *et al.* 1942). Spiking the diet with thiamine or a natural product high in thiamine (e.g. fish-liver oil) is an effective preventative.

8 Water balance and energy

8.1 Introduction

The perception of energy consumption used in the following sections requires explanation. I accept that not all protein, fat, or carbohydrate is digestible, but energy, unlike these other factors, is intangible. I reject the equivalent terms "digestible energy" and "metabolizable energy" on logical grounds: energy remains in the potential state until released by oxidation, which imposes the dilemma of what could constitute "indigestible energy." Assuming "digestible energy" is always – and can only be – 100% of the energy released during metabolism, the term is tautologous. Maintenance energy traditionally includes basal (fasting) heat production, in addition to level of activity and heat increment (MacDonald *et al.* 1984a) and represents the amount required to sustain physiological functions. Metabolizable energy is then the percentage of gross (overall) – but still *potential* – energy ingested that is retained and not lost in feces, urine, or gases. Here I combine these and other categories and refer to the heat produced by any process simply as "energy" and give it the symbol E regardless of subsequent function (e.g. "maintenance," "metabolizable"). This simplifies the discussion and avoids the antithetical by considering E to be a measurable derivative instead of a fractional component from some larger hypothetical pool.

8.2 Water balance

A cat loses water continuously through respiration and intermittently during defecation, urination, and evaporation off its body surfaces. These processes result in a deficit that must be made up from three sources: drinking, water contained in food, and water produced metabolically by oxidation of nutrients (i.e. proteins, fats, and carbohydrates). Metabolic water supplies ~10% (Anderson 1982). A cat fed canned food obtains ~91% of its water needs from food alone (depending on the food's water content), but a diet of kibbles spiked with the same amount of table salt (1.3% DM) required 96% of water intake to derive from drinking, and this was insufficient to compensate for the diet's water deficit (Burger *et al.* 1980). Ordinarily, house and laboratory cats drink little or no water when fed canned foods (Kane *et al.* 1981b), but if fed kibbles their water intake exceeds that contained in the food. Seefeldt and Chapman (1979) fed laboratory cats kibbles or canned cat food. The same cats fed kibbles ($\bar{x}$ = 8.8%water) drank 6 times more water than when consuming the canned diet ($\bar{x}$ = 76.8%water). Cats on either diet maintained the same body masses, and water turnover did not vary, but neither were they subjected to a variable environment. Drinking water provided ~14% of body water turned over on the canned diet with the balance taken in with food; on the kibble diet, drinking water accounted for >90% of water turned over.

Free-ranging Cats: Behavior, Ecology, Management, First Edition. Stephen Spotte.
© 2014 John Wiley & Sons, Ltd. Published 2014 by John Wiley & Sons, Ltd.
Companion Website: www.wiley.com/go/spotte/cats

It has been proposed (Jackson and Tovey 1977, Seefeldt and Chapman 1979) that in assessing water balance, substituting water turnover-to-DM intake for water intake-to-DM-intake might be a more accurate measure because it includes metabolic water. Seefeldt and Chapman (1979: 184, Tables 1 and 2) showed metabolic water (i.e. water turned over minus total water consumed) to be 65% greater when feeding their cats canned food (12.7 vs.7.7 mL/d on kibbles). In still another study, Thrall and Miller (1976) measured metabolic water loss at 39% in adult cats fed kibbles, with another 25% excreted in the feces. Of total water turned over ($\bar{x}$ = 59.9 g/kg M/d, including metabolic water), 25% accompanied the feces ($\bar{x}$ = 12.7 mL/kg M/d), 39% ($\bar{x}$ = 29.0 mL/kg M/d) was lost "insensibly" (i.e. through respiration, surface evaporation, salivation; "insensible" loss ordinarily includes the feces), and 36% was urinary water ($\bar{x}$ = 18.2 mL/kg M/d).

Cats can survive an extraordinarily long time without food and water (Fig. 8.1). They adeptly resist hydropenia, which has allowed them colonize hot, arid places where collected rainwater is intermittent, seasonal, or even nonexistent (Fig. 8.2). Without water to drink, they adapt by relying heavily on tissue fluids from prey, from producing metabolic water, and by concentrating their urine. Given food containing its normal complement of moisture (e.g. cod fillets, ground fresh-frozen salmon, ground beef), laboratory cats can survive months without drinking. During this time they maintain body mass, behave no differently than other cats, and sustain normal plasma osmolarity (Prentiss et al. 1959). If the food is partially dehydrated, signs of hydropenia appear, and plasma osmolarity is not sustained. When both water and food are withheld and cats are then given an opportunity at rehydration, recovery is slow. Cats have a low thirst drive (MacDonald et al. 1984a) and are lax to replenish a water debt, even a slight one (Adolph 1947). A cat drinks like it eats, ingesting small volumes often (Kane et al. 1981b). Plasma osmolarity declines, as expected, once drinking commences. A cat that lost 51.3% of its initial body mass over the course of an experiment drank 0.173 kg of water in 80 min, the equivalent of 10.5% of its current body

Fig. 8.1 Ni Hao ("Hello" in Mandarin Chinese), an adolescent stray shortly after his rescue in July 2012. Ni Hao had stowed away inside a cargo container at Shanghai and was discovered a month later when the ship docked in Los Angeles, California. He survived the entire voyage without food or water. Source: County of Los Angeles Department of Animal Care and Control.

Fig. 8.2 Urban cats drink from a curbside gutter in Aghios Nikolaos, Crete, shortly after shop owners have opened their doors and hosed off the sidewalks. Source: Stephen Spotte.

mass. Plasma osmolarity fell, although not to normal values, and concentrations of potassium, nonprotein nitrogen, and hematocrit actually rose (Prentiss et al. 1959). The implication of these findings is that feral cats in arid climates can survive without external water by capturing and eating fresh food, but not by scavenging desiccated animal carcasses.

Cats can delay hydropenia or even overcome a previous water deficit by mariposia (i.e. drinking seawater). In the laboratory they thrive on a diet of partly desiccated salmon mixed with seawater in volumes of 45–180 mL/d (Wolf et al. 1959). When given artificial seawater to drink *ad libitum* along with just partly desiccated salmon, laboratory cats drank an average of 73 mL/d. When partly desiccated salmon was mixed with artificial seawater at 9–38 mL/d they became hypersalemic. When fed undesiccated salmon mixed with artificial seawater in amounts from 0 to 80 mL/d the cats did well, but became hypersalemic when undesiccated salmon was mixed with 165 mL/d. Attempts to maintain or restore euhydration were erratic and varied by individual, some cats ingesting too much and others too little artificial seawater when it was available *ad libitum* with undesiccated or partly desiccated food.

Cats are relatively resistant to high temperature, withstanding dehydration to 17–23% of original body mass before revealing signs of hyperthermia (Adolph 1947). Laboratory tests showed the limit to be ~58°C at 23% relative humidity (RH) in ≤2.0 h, and a cat can withstand a temperature 20°C higher at 20% RH than at 100% RH (Adolph 1947). Other factors affect water balance and heat tolerance. Body size matters because small individuals retain and accumulate heat faster, and the extent of evaporative cooling is also important. Cats open the mouth and salivate, and panting is an adequate substitute for sweating. Most of a cat's sweat glands are located on its back, opening like the sebaceous glands into the funnels of hair follicles (Schummer et al. 1981: 491). Sweat glands also occur on the toe pads, sole pads, and near the teats, but are rudimentary elsewhere. Cats do not visibly sweat. They also lick their fur when overheated. The deposition of saliva promotes surface evaporation.

These behaviors are effective when the ambient air is dry, but not when water-saturated. Under severe heat stress a cat loses body mass at 1–2%/h, and neither this rate nor the rate of panting diminishes with increasing dehydration (Adolph 1947). Under conditions of severe heat stress a dog's respiration rate rises by 12–20 times, a cat's only 4–5 times (Anderson 1982). Panting in dogs is effective, resulting in dispelling up to 57% of body heat. Cats are comparatively inferior, losing ~9%. Finally, adequate dietary linoleic acid is necessary for regulation of skin permeability to water, which increases during times of EFA deficiencies accompanied by reduced *BMR* (McLean and Monger 1989), and can be traced ultimately to the cat's limited $\delta 6$ desaturase activity (Rivers and Frankel 1980, Trevizan et al. 2012). Evaporative water loss could be compromised further in cats under heat stress if food and not water is unavailable.

If evaporative cooling stops suddenly while a cat is losing body water at 1%/h, its deep-core temperature starts to rise at ~7°C/h. Deep-core temperature also cools at this rate if a heat-stressed cat is transferred to a place where the evaporation rate does not decrease. As Adolph (1947: 569) wrote: "The initial stages of recovery from hyperthermia are made rapid by the same means that hyperthermia itself is resisted, namely, evaporative cooling." *The majority of heat dissipated by a hyperthermic cat is not metabolic but originates in the environment.* The actual internal heat generated is slight, <24.28 kJ/kg/h.

Hyperthermia can be lessened or avoided if a cat is allowed to drink, especially if water is available from the beginning (Adolph 1947). When exposed to temperatures of 48–57°C and given water *ad libitum*, 5 laboratory cats were able to hold dehydration at 6.6–7.9% of starting body mass. Cats therefore drink at rates sufficient to withstand large deficits of tissue water, but in Adolph's experiments they did less well when facing moderate deficiencies (Fig. 8.3), seeming insensitive to their situation. The resulting disparity was labeled "terminal voluntary dehydration" (Adolph 1947: 573), making the cat's thirst response to dehydration appear incomplete compared with that of the dog. A dog can replenish water deficits up to 8% of M in minutes (Adolph 1947). Cats ordinarily quit drinking before making up even 4%, thus prolonging the return to normal water balance. Even if sufficiently hydrated a house or laboratory cat is likely to reduce its water intake if offered only low-moisture foods such as kibbles (Anderson 1982).

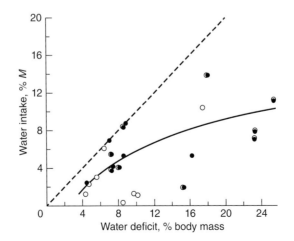

Fig. 8.3 *Ad libitum* intake of water in 23 tests on cats ($n = 5$) that had been dehydrated to different extents. Dashed line shows data on cats not exposed to heat. Source: Adolph 1947. Reproduced with permission of American Physiological Society.

A cat's state of hydration and ambient air temperature act in concert to influence evaporative heat loss. Any substantial rise in plasma osmolality might then lower the rate of evaporative cooling enough to be life-threatening. Normally hydrated laboratory cats that were cannulated and infused with 30% saline (1.5 mL/kg M) at 38°C demonstrated reduced evaporative heat loss by $\bar{x} = -0.21$ W/kg M while body temperature rose simultaneously by $\bar{x} = +0.43$°C (Baker and Doris 1982). Neither factor differed statistically from pre-infusion values when tested at 25°C. A reverse effect was seen by perfusing distilled water (15 mL/kg M) into both hydrated and dehydrated cats at 25°C. This stimulated a significant increase in evaporative heat loss in dehydrated individuals ($\bar{x} = +0.35$ W/kg M) while body temperature fell by $\bar{x} = -0.45$°C. The effect, as expected, was less marked in animals already hydrated ($\bar{x} = +0.13$ W/kg M and $\bar{x} = -0.03$°C). Osmotic interaction during regulation of body temperature alters responses of the hypothalamus when a cat encounters rising ambient temperature, and it influences thermoregulation once dehydration has set in (Baker and Doris 1982). This has important consequences for cats inhabiting arid regions where water is scarce, and availability of game – often the only source of water – is unpredictable.

A feral cat can survive without free water if food is adequate (Jones 1989). The evolved physiological mechanisms are augmented by life-style modifications for water conservation. Feral cats in arid regions might spend much of the day inside rabbit burrows or other enclosed spaces to retain moisture and hunt at night when the air is cooler. These and other adjustments in behavior and time budgeting have enabled cats to colonize central Australia and similar desert regions. The capacity to conserve water leaves cats disinclined to drink even when dehydrated. In house and laboratory cats this induces common kidney and urinary tract diseases, urethral obstructions, and related maladies accounting for ~8% of all morbidity of cats examined at veterinary teaching hospitals in the United States and Canada (Lekcharoensuk et al. 2001). Some of these problems are more prevalent in animals confined indoors and those fed only dry commercial cat foods (MacDonald et al. 1984a, Segev et al. 2011), but they are also related to the existing state of hydration. Canned cat foods, the natural prey of feral cats, and probably garbage eaten by strays, contain ~70–80% water, which increases urine frequency and volume and helps prevent urolithiasis (Kane et al. 1981b). Extruded dry foods, in contrast, are typically <10% water and ~84% digestible (Kerr et al. 2012, Kuhlman et al. 1993). Cats fed kibbles exclusively produce less urine and more fecal water than cats consuming wet diets (Jackson and Tovey 1977). The urine is more concentrated, having a higher osmolality and specific gravity, and contains struvite microcrystals, all indications that the resulting dehydration is not relieved by drinking. This led Jackson and Tovey (1977) to recommend that the ratio of water to DM (presumably by mass) should be >3.

8.3 Energy

That animals eat to obtain energy, once an axiom chiseled in granite by nutritionists, is true, but there are other reasons as well, one being to juggle multiple nutritional requirements. The proportionate composition of protein, fat, and carbohydrate – not total energy – actually determines both diet selection and regulation of food intake by cats (Hewson-Hughes et al. 2011a). As discussed previously (Chapter 7), in cats this includes the extreme need for nitrogen caused by the low capacity of the hepatic catabolic enzymes to adapt (Morris 2002, Rogers and Morris 1982). An adult laboratory cat's total daily energy use varies from 83.74 to 418.68 kJ/kg M of which 60–75% is attributable to resting energy expenditure (Center et al. 2011). Adult cats need ~65%

more dietary protein than adult rats, dogs, and humans (Rogers and Morris 1982), but they also have lower E requirements (Kendall *et al.* 1983). The NRC (2006: 354) recommended a daily energy intake of 1046.7 kJ for a 4-kg adult neither pregnant nor lactating, a value that appears to have been overestimated for house and laboratory adults.

In a meta-analysis, Bermingham *et al.* (2010) collected published data on energy needs of nonpregnant adult cats ($n = 42$ reports, 141 treatment groups comprising 1933 cats). Not all the information surveyed was complete in every respect, but results yielded a mean daily value of 929.89 kJ/d (± 22.19 SEM, range 512.88–1678.91, $n = 115$) to supply the E requirement of an "average" domestic cat; that is, an animal of mixed breed ranging in age from 0.5 to >7 y and in body mass from < 3 to 5.5 kg, whether neutered or intact, male or female, provided the females are neither pregnant nor lactating. This value represents the energy a free-ranging cat must acquire daily to maintain a steady body mass (Fig. 8.4).

Because there is no such thing as an "average" cat the value of E varies by individual and depends on age, health status, how starting body mass has been classified (e.g. underweight, "normal" weight, overweight), reproductive status, sex, and other variables. Bermingham *et al.* (2010) used the standard allometric equation

$$E = aM^k \tag{8.1}$$

in which $E = y$ on the ordinate (dependent variable) in kJ, a = the coefficient constant (kJ/kg), M = body mass (kg), and k = the allometric exponent. The range in E given in the preceding paragraph is large and likely derived from inconsistencies caused by failure to properly categorize subjects used in the different experiments (Bermingham *et al.* 2010). Combining all categories of cats yielded the allometric equation

$$E = 324.89 \, M^{0.711} \tag{8.2}$$

with $R = 0.67$ ($n = 115$ treatment groups), indicating to me that in this configuration body mass accounts for ~45% of a cat's daily maintenance energy requirement.

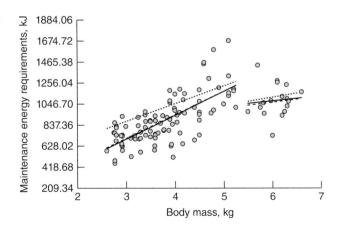

Fig. 8.4 Effect of body mass (*M*) on the energy requirements (*E*) of domestic cats (●, solid lines) compared with predicted requirements from a meta-analysis ($n = 42$ reports) of Bermingham *et al.* (2010, dashed line) and the NRC (2006, dotted lines). Allometric equations predicted from Bermingham *et al.* (2010) are 235.30 kJ/kg $M^{0.966}$ for cats designated light- and normal-weight combined, and 550.98 kJ/kg $M^{0.366}$ for cats designated as heavy. Corresponding predictive equations from the National Research Council (2006) are 418.68 kJ/kg $M^{0.667}$ and 544.28 kJ/kg $M^{0.40}$. Source: Bermingham *et al.* 2010. Reproduced with permission of Cambridge University Press.

Bermingham et al. (2010) classified cats as light ($\bar{x} = 2.8$ kg ± 0.02 SEM), normal ($\bar{x} = 4.0$ kg ± 0.07 SEM), or heavy ($\bar{x} = 6.0$ kg ± 0.07 SEM). When shown as a proportion of body mass, E values of light cats ($\bar{x} = 262.93$ kJ/kg $M \pm 11.72$ SEM) were significantly higher than those for normal ($\bar{x} = 235.72$ kJ/kg $M \pm 5.02$ SEM) and heavy ($\bar{x} = 183.80$ kJ/kg $M \pm 7.12$ SEM) cats. Feeding rates of animals increase with increasing body mass, but in less than one-to-one fashion so that a large animal consumes less food daily than predicted from its size; in other words, allometric slopes are usually <1.0 (Nagy 2001), and a 5-kg cat does not eat 5 times more than a cat of 1 kg. Heavy cats had lower maintenance energy requirements than light cats or those of normal weight. When light and normal cats were combined for comparison with values presented by NRC (2006) the allometric equation was 235.30 kJ/kg $M^{0.966}$ with k near unity ($R = 0.69, n = 98$ treatment groups), or lower than the NRC's recommendation of 418.68 kJ/kg $M^{0.67}$ for cats this size.

Age also exerts an effect (Bermingham et al. 2010). Expressed as a proportion of M, values of E were higher in young cats ($\bar{x} = 248.60$ kJ/kg ± 8.79 SEM) than adults ($\bar{x} = 202.64$ kJ/kg ± 7.54 SEM). However, data for senior cats (>7 years old) assessed separately did not differ significantly from those of young cats or the other adults ($\bar{x} = 202.64$ kJ/kg ± 17.58 SEM):

$$E = 276.75\, M^{0.878} \quad \text{young cats } (0.5-2.0\,\text{y}) \tag{8.3}$$

$$E = 303.12\, M^{0.703} \quad \text{adult cats } (2.0-7.0\,\text{y}) \tag{8.4}$$

$$E = 283.03\, M^{0.781} \quad \text{senior cats } (> 7.0\text{y}) \tag{8.5}$$

Younger cats have higher energy needs than older ones. At $R = 0.77$ for young cats in the meta-analysis, M explained ~59% of variation in the data for E. Explanations were less well conserved for adults with even lower regression coefficients ($R =\sim 0.67$) but weak in the case of senior cats ($R = 0.42$), explaining only ~18%. The reports surveyed vary considerably on how age relates to E requirements (references in Bermingham et al. 2010).

Sex and sexual status (intact vs. neutered) also affect the requirement for E. When expressed simply as energy intake per day, values of intact males ($\bar{x} = 1173.56$ kJ ± 76.62 SEM) exceed those of intact females ($\bar{x} = 900.16$ kJ ± 51.92 SEM). When expressed as a proportion of M, there appears to be no significant difference. The value of E was reduced 11.7% after neutering (males and females combined) to 236.97 kJ/kg M (± 8.37 SEM) compared with their intact counterparts combined ($\bar{x} = 264.606$ kJ/kg $M \pm 9.21$ SEM).

Neutering cats – whether confined or free-ranging – affects their metabolism and capacity to control food consumption, causing them to gain weight, much of it as fat (Alexander et al. 2011, Fettman et al. 1997, Flynn et al. 1996, Kanchuk et al. 2003, Nguyen et al. 2004, Scott et al. 2002). This occurs even when activity levels are the same as intact controls (Flynn et al. 1996). Neutered cats require less energy (Alexander et al. 2011, Hoenig and Ferguson 2002), and their fasting metabolic rate declines (Fettman et al. 1997). Because neutered laboratory cats seemed incapable of regulating food intake, they left little food uneaten (Flynn et al. 1996), and their increase in body mass following surgery was attributable to eating more, not to reduced energy expenditure (Kanchuk et al. 2003). Nguyen et al. (2004) showed that neutered cats on a high-fat diet gained more weight than a comparable group fed a diet low in fat.

Increased protein intake stimulates fat loss while also reducing the loss of lean body mass during weight reduction (Laflamme and Hannah 2005), bringing up the possibility that strays scavenging on garbage likely ingest a lower percentage of protein than feral cats subsisting by predation alone (Table 7.6).

Changes in body mass and insulin sensitivity after neutering can be caused by excessive dietary carbohydrate (Section 8.6), but dietary fat intake is also an influential factor. Backus *et al.* (2007) fed cats purified diets of constant protein/E ratios in which the nonprotein component was either carbohydrate or fat at 9, 25, 44, and 64%. The diets were administered over 13 weeks. The cats (12 males, 12 females) were then sterilized and the program continued another 17 weeks. All individuals gained in body mass after neutering, especially the females. Plasma glucose, triacylglycerol, and leptin were unaffected by percentage of dietary fat, neutering, or weight gain, but plasma insulin rose with fat-induced gain in body mass. Backus *et al.* (2007: 641) interpreted their results as evidence that "high dietary fat, but not carbohydrate, induces weight gain and a congruent increase in insulin, [but neutering] increases sensitivity to weight gain induced by dietary fat." Hoenig and Ferguson (2002) reported no weight change in females after neutering, but significant increases in males 8–16 weeks afterward.

Given choices, cats attempt to regulate and thereby control their energy intake, evidence that carnivores do not differ in this respect from herbivores or omnivores (Hewson-Hughes *et al.* 2011a, Mayntz *et al.* 2009); in other words, they are not insensitive to energy intake and eat simply to maintain consistent bulk as earlier investigators speculated (e.g. Kanarek 1975, Kaufman *et al.* 1980, Mugford 1977). Because animals in these other two categories ingest foods over a range of variety and quality it was once believed that this simple flexibility would balance everything in the end. Carnivores, some thought, had no need to control their intake of different nutrients. The foods they consumed were more nutritious but relatively rare (Chapter 2), and unlike browsers, scavengers, and grazers the main limitation was locating enough to eat. Feral cats with access to live prey seldom find this troublesome because the composition and quality of fresh animal tissues vary little compared with grasses and browse. Confined house and laboratory cats, and strays forced to subsist on garbage, are more likely to experience unbalanced diets than a feral cat living on mice and chipmunks.

Strays might have to consume more bulk if sufficient energy, protein, and fat are unavailable in scavenged food. Laboratory cats in some experiments have compensated for dilution of their diets, maintaining constant energy intake by eating more (Castonguay 1981). Results of other work (e.g. Hirsch *et al.* 1978, Kanarek 1975) show the opposite, that cats ingest about the same volume regardless of the dilution factor and ignore the caloric density of what they eat. The latter experiments have been criticized because the diets used were low in energy (Castonguay 1981, Kane *et al.* 1987). In any case, few experiments have used the same diluent, some investigators using water, others inert DM (e.g. kaolin, celluflour). Later work (Kane *et al.* 1987) indicates that cats indeed regulate gross energy intake, consuming less high-fat food if a low-fat alternative is available.

Apparent from the discussion so far is the fact that energy requirements vary substantially across estimates even for house and laboratory cats. In the case of free-ranging cats, all are models to some extent, having incorporated assumptions where data are missing. The assumptions have seldom been uniform. Reports differ on estimations of the energy content of prey, frequency of feeding, quantity ingested at each meal, effect of weather on foraging and energy requirements, the limitations imposed by competitive interactions, and so forth. Consequently, none can claim reproducible accuracy and precision except within the experimental restrictions imposed by the investigators. According to MacDonald et al. (1984a), quantitative differences between E and basal metabolic rate in the cat are small, and they concluded that E was 1.5–2.0 times BMR. After a literature review they felt that $E = 334.9$ kJ was reasonable, which included a BMR multiplier of 1.5. The exponents subsequently discussed were the traditional 0.67 and 0.75 (Chapter 2). Expressing the second in an allometric formula gives $334.9M^{0.75}$. Following through, this is the equivalent of $E = 902.5$ kJ/d for a 3.75-kg cat, in line with findings of Kane et al. (1981b) who measured the daily intake of inactive laboratory cats fed wet and dry commercial diets at ~297.3 kJ/kg M/d (1114.9 kJ/d for a cat of 3.75 kg).

To make this information easier to visualize, think in terms of a novel energy expression, the "mouse unit," which is the energy contained in one "standard" mouse or its equivalent in other animal matter. The energy content of a mouse is ~25 kJ/g DM (Litvaitis and Mautz 1980). My prototypical "standard" mouse of 16 g is considered perfect prey size for a cat (Short et al. 1997). Assuming it contains 69% water the remaining 31% supplies 4.96 kJ/g, or ~125 kJ of total energy, a convenient number to use in computations. Were mice its only prey an adult cat of 3.75 kg would have to consume 7 of them (i.e. 7 mouse units) within each 24-h period to acquire the 902.5 kJ/d necessary to survive. Using $303.12M^{0.703}$ (Bermingham et al. 2010, Equation 8.4 above) yields $E = 767.6$ kJ/d for a cat of 3.75 kg, the equivalent of 6.5 mouse units, or close enough. These values match other findings. Laboratory cats monitored by Kane et al. (1981b) consumed 96.3 kJ/meal, which these investigators considered the equivalent of a 20-g mouse.

8.4 Energy needs of free-ranging cats

In their meta-analysis of the energy intake of feral cats, Plantinga et al. (2011), using the method of Fitzgerald and Karl (1979), calculated that the mean daily E requirement of an adult feral cat was 1258 kJ. This makes a convenient starting place to examine the energy needs of a cat living on its own.

Conditions under which data are obtained can affect the results. MacDonald et al. (1984a) used values calculated from caged laboratory cats allowed limited activity, fed and watered adequately, and held at room temperature. Feral cats living in temperate and subantarctic climates apparently use energy faster (Table 8.1), which is not surprising if conditions are persistently windy, wet, and cold, food quality uncertain, and food availability unreliable. At Marion Island, for example, it rains an average of 300 d/y (van Aarde 1978). A 3.75-kg adult on the Falkland Islands, not pregnant or lactating, takes in an estimated 2131 kJ/d (Matias and Catry 2008) as derived from $1670M^{0.869}$, a general allometric formula applied to carnivorans for "field metabolic rate" (Nagy et al. 1999: 260, Equation 8.5). The result is equivalent to 17 mouse units when divided by 125 kJ, a large daily consumption. Pregnant females about the same size were thought to consume 2508 kJ/d, or 20 mouse units. Fortunately for these cats, New Island also has rats, rabbits, and birds, which are bigger than mice. Applying a multiplier for freshly dead mammals (6.24 kJ/g) to a typical Falkland Islands rabbit of 317 g (Matias and Catry 2008: 613, Table 2) yields an energy value of 1978 kJ, equivalent to 15.8 mouse units, which still seems high. This assumes most of the rabbit is edible. Konecny (1987b: 24) estimated the daily energy intake of Galápagos cats at 711.76 kJ, which he considered "the caloric break even point for non-pregnant females, slightly below that for adult males and pregnant females and well below that for lactating females." The Galápagos, being equatorial, are much warmer than the Falklands.

Table 8.1 Mass constant values of mammals and birds used to calculate relative contributions of each in terms of fresh body mass and energy to the diets of free-ranging cats at New Island, Falkland Islands, over summers 2004/2005 and 2005/2006 (data combined, sample sizes for body mass determinations in parentheses). Energy equivalents used were 10.9 kJ/g fresh mass for birds, 6.24 kJ/g of fresh mass for mammals, and 29 kJ/g dry mass for eggs. Other assumptions are described in the source.

Species or prey item	Body mass, g	Estimated maintenance energy (E), %
Mammals		
Black rat (*Rattus rattus*)	124 (1)	14.60
House mouse (*Mus musculus*)	16 (2)	3.20
Rabbit (*Sylvilagus* sp.)	317 (3)	45.77
Birds		
Anatidae (unidentified)	409 (7)	3.80
Austral thrush (*Turdus falklandii*)	82.5 (8)	0.89
Kelp goose (*Chloephaga hybrida*)	409 (7)	0.63
Long-tailed meadowlark (*Sturnella loyca*)	110 (9)	0.51
Rockhopper penguin (*Eudyptes chrysocome*)	120 (6)	0.56
Thin-billed prion (*Pachyptila belcheri*)	154 (2)	20.02
Upland goose (*Chloephaga picta*)	409 (7)	0.63
Unidentified passerines	96 (10)	2.38
Unidentified birds	115.5 (11)	5.18
Bird egg	31 (12)	0.85

Source: Matias and Catry 2008. Reproduced with permission of Springer Science+Business Media.

Liberg (1982) considered only part of a prey animal to be digestible and added a "correction factor" to account for the indigestible or inedible material (e.g. hair, large bones and feathers, teeth) in the feces of Swedish feral cats. Of an adult European rabbit (*Oryctolagus cuniculus*), the upper jaws, ears, dorsal skin with tail, part of the spine, and lower legs were usually left uneaten by the feral cats and captive controls he studied, and some of the material ingested was passed unchanged in the feces. Combined, they could account for perhaps 25% of the original carcass mass, or ~250 g for a 1-kg rabbit. From results of feeding experiments in the laboratory and comparing scats from captive and feral cats, Liberg estimated that feral animals consume ~282 g/d of animal matter. He assumed the energy in a mouse to be 6.11 kJ/g fresh M and from literature values set an adult cat's daily requirement at 334.94 kJ, exactly matching the estimate of MacDonald *et al.* (1984a). Cats living on their own typically defecate once daily (Apps 1983, Konecny 1987b, Liberg 1982), although Jackson (1951) observed laboratory cats taking ~2 d to pass rodent remains, and considered the 500 scats he collected from urban Baltimore strays the equivalent of 220 d of defecation for a single cat. The average scat picked up by Konecny (1987b) in the Galápagos Islands was 7.42 g ($n = 640$). This information, combined with a defecation rate of once daily as observed by Liberg (1982b) in his captive cats, meant that a 4.3-kg feral cat would need ~236 g/d of digestible matter (~239 g/d by my recalculation of Liberg's data), an approximate 18% difference from the estimated values (projected and actual).

Evidence presented so far paints the cat as highly carnivorous, in the words of some a "hypercarnivore," a creature that evolved to exist on animal tissues exclusively. Its evolution in this direction has been accompanied by other nutritional divergences, including an inability to synthesize vitamin D, niacin, and taurine in adequate concentrations, and convert β-carotene to vitamin A (Chapter 7). Perhaps most interesting is the cat's restricted capacity to regulate protein oxidation during times when dietary protein is low (Morris 2002), and its subsequent failure to conserve nitrogen (Chapter 7). Experiments to describe nitrogen turnover, notably those of James G. Morris and Quentin R. Rogers starting in the 1970s, placed the situation in clear perspective, and Regina

Fig. 8.5 Linear regression plot of protein intake vs. protein oxidation in laboratory cats. Each data point represents a cat that ate 1 of 4 diets containing a different protein concentration: 7.5% (o), 14.2% (●), 27.1% (□), 49.6% (■). Regression equation: $y = 0.93x \pm 0.04 + 1.17 \pm 0.77$ ($R^2 = 0.946$, $p < 0.0001$). Source: Green et al. 2008. Reproduced with permission of American Society for Nutrition.

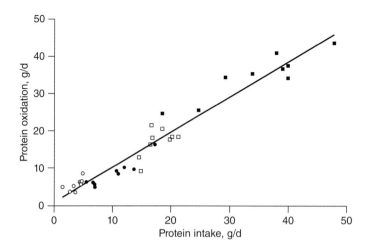

Eisert offered a thorough and compelling hypothesis giving it a mechanistic foundation. I retrace the path of her reasoning in the following discussion.

Eisert (2011: 2) proposed that "gluconeogenesis from amino acids in cats is constitutive and represents a significant metabolic sink for amino acids that increases the minimum protein requirements … above that of non-hypercarnivorous mammals." Evidence laid out by Morris, Rogers, and other of her predecessors illuminates the way. In practical terms, carbohydrates and other dietary "fillers" are not suitable substitutes for protein despite the potential energy they contain. Cats can modulate protein oxidation to accommodate a range of concentrations taken in, so long as protein comprises at least 14–20% of dietary energy (e.g. Green et al. 2008; Russell 2002; Russell et al. 2003), which is considerably more than the 6–8% required by omnivores (e.g. Green et al. 2008, Morris 2002), and adult laboratory cats fed a purified diet containing 7.5% protein showed a net loss of nitrogen that was independent of energy balance (Green et al. 2008), in line with findings of Burger et al. (1984) that 10% of energy derived from dietary protein is near the threshold of nitrogen balance. At higher concentrations protein oxidation closely approximated protein intake, and cats could modulate protein oxidation to sustain nitrogen balance over a range of diets containing ~14–50% protein as E (Fig. 8.5).

According to general thinking the lower threshold to catabolize amino acids depends on an animal's rate of whole-body protein turnover and its accompanying obligation to lose nitrogen. Cats are puzzling by turning over nitrogen more slowly than other species (Russell et al. 2000, 2003). Why are they trapped on a catabolic treadmill that forces the breakdown of amino acids so relentlessly and results in such large losses of nitrogen? The answer, Eisert posited, starts with the cat's high brain-to-body mass ratio, which has led to the brain's strong glucose demand (Fig. 8.6). According to Karbowski (2007), the scaling exponent of the mammalian brain for glucose consumption, relative to its volume, is 0.86 (±0.03 SEM), substantially exceeding 0.75 and 0.67, values often used for whole-body *BMR* predicated on body mass (Fig. 8.7). Feral cats rely on gluconeogenesis using amino acids in the proteinaceous tissues of their prey, and they manage this without falling back on hyperketonemia. The result meets tissue energy demand through obligatory gluconeogenesis; in other words, *while proceeding independently of carbohydrate consumption*. Obligatory gluconeogenesis, which is

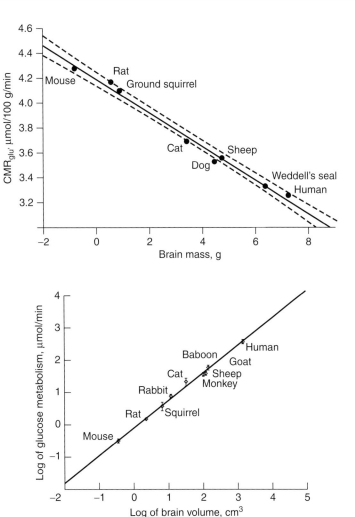

Fig. 8.6 Log-normal regression plot of brain mass vs. glucose cerebral metabolic requirement (CMR_{glu}) for different mammals from the literature (see source for references). The ground squirrel is the thirteen-lined ground squirrel (*Spermophilus tridecemlineatus*) Solid line, Deming linear regression model; dashed lines, 95% confidence intervals: $y = -0.135x$ (± 0.006) + 4.19 (± 0.03), $R = -0.995$. Source: Eisert 2011. Reproduced with kind permission from Springer Science+Business Media.

Fig. 8.7 Log-normal regression plot of whole-brain volume vs. rate of glucose metabolism for different mammals. Scaling exponent = 0.86 (± 0.03). Regression equation: $y = 0.86x - 0.09$ ($R^2 = 0.989$, $p < 0.0001$, $n = 7$). Source: Karbowski 2007. Open Access/CC-BY-4.0.

always operating at maximal rates in cats (Silva and Mercer 1985), raises their minimum protein requirement compared with omnivores of similar size because the lower limit is regulated by loss of endogenous nitrogen. Therefore, although cats do not need dietary carbohydrate they need physiological glucose, as demonstrated by Rowland (1981) when he induced feeding behavior by injecting cats with either insulin or 2-deoxyglucose. *In practical terms the requirement for glucose is not solved by adding carbohydrate to the diet, and what should be added instead is protein.*

Stores of glycogen are limited (references in Eisert 2011), so how is the high demand for it met by a carnivorous diet? Not from eating plant material in the digestive tracts of herbivorous and granivorous prey, which even if cats did so would be inadequate (Eisert 2011: 12–13, appendix). Cats seldom consume the digestive tracts of rodents and lagomorphs, and ordinarily not before removing the contents when they do (Leyhausen 1979: 63). Leyhausen (1979: 62) wrote that cats "squeeze out the contents of the stomach and the small intestine, and still more often of the lower intestine." He continued (Leyhausen 1979: 62–63): "The tongue, slightly rolled under, squeezes a section of the intestine against the palate and presses it outward toward the incisors, then the intestine is snapped [off] a little way and the next section pressed out."

Authors who have seen feeding cats ignore the digestive tract's contents made their assessments based on the prey being terrestrial mammals, not carnivorous birds. Results could be different on oceanic islands where cats feed heavily on piscivorous seabirds and might actually eat the gut contents preferentially. Ashmole (1963: 329) implied this happening at Ascension Island where the feral cats, having become sated after thousands of sooty terns (*Onychoprion fuscatus*) had settled to court and reproduce, began to eat only certain body parts. Most birds killed were adults, many captured after having returned from feeding at sea. As the breeding season progressed "the cats normally only removed part of the contents of the abdomen, and sometimes also the breast." Whether "contents of the abdomen" includes the digestive tract and what it held needs to be investigated at another location now that Ascension Island has been cleared of feral cats.

A cat feeding on fresh prey could take up some glycogen along with liver tissue, but not much from an animal the size of a mouse. A cat that ate 6–12 small rodents (~1000 kJ/d) would assimilate 0.6–1.1 g of dietary carbohydrate (Eisert 2011). Using an "as fed" fat content of 12% would add another 1.8 g of glycerol, yielding a hypothetical amount of 1.8 g of glucose, assuming all associated fatty acids have been oxidized completely. However, this contingency alone restricts the usefulness of glycerol and renders it a minor source. Eisert (2011) pointed out that the amount of carbohydrate ordinarily present in animal tissues of a prey-based diet (range 2.4–2.9 g/d) is insufficient to supply the net glucose demand of the brain (3.1 g/d). And this is just the brain exclusive of other tissue demands. The cat's eye perfused *in vitro* shows impressive sensitivity to changes in glucose concentration as small as ±1-3 mmol/L, and whole-body circulating reductions of ≤2 mmol/L produce inflections in the central nervous system that cause dramatic electrophysiological disruptions accompanied by behavioral abnormalities (Macaluso *et al*. 1992).

Plantinga *et al*. (2011) presented an exercise similar to Eisert's using digestive-tract contents of domestic rabbits, pointing out that the data extrapolate weakly to wild rabbits, which feed on low-starch, high-fiber grasses and leafy weeds, materials of little use to a cat. The remaining major nutrient in the cat's diet is fat, and fatty acids, although sources of energy, are not demonstrated precursors of glucose. *To meet its continuous need of glucose, amino acids are a feral or stray cat's only reliable option.*

The brain functions poorly when only partly energized, and glucose is its "default fuel" (Eisert 2011). Furthermore, brain metabolic rate varies allometrically with brain-to-body mass ratio (Karbowski 2007). A case has been presented for gluconeogenesis as a main pathway of amino acid disposal, and for cats needing to ingest a high-protein diet to complete glucose synthesis effectively. To Eisert (2011: 6-7) the remaining question was "whether gluconeogenesis in the cat is simply a means of disposing of excess amino acids … or whether gluconeogenesis from amino acids continues despite relative protein deficiency and the threat of a negative nitrogen balance." In omnivores (e.g. humans, pigs, dogs, rats), "fasting urinary nitrogen loss" (FUNL), which could also be called amino acid oxidation, greatly exceeds "endogenous urinary nitrogen loss" (EUNL), another name for obligatory amino acid oxidation, in the fed (i.e. not fasting) state (Eisert 2011: 2, Table 1). A helpful aspect arising from the cat's loose enzyme regulation is its capacity to rapidly catabolize amino acids for both energy and gluconeogenesis, which allows a starving cat to maintain its blood glucose levels more efficiently than a food-deprived omnivore or herbivore. During starvation the catabolism of additional amino acids drives gluconeogenesis. Thus glucose production overrides the body's conservation of nitrogen, suggesting that in cats *the protein requirement is higher than in omnivores as a result of obligatory gluconeogenesis*. A nonadjustable urea cycle augmented by a high-protein diet seems to reinforce the idea that cats "waste" amino acids at times of reduced protein intake when the reason is actually an unremitting need for glucose. In addition, although cats demand large amounts of amino nitrogen, their requirement for EAAs is not inordinately great (Rogers and Morris 1979).

Cats have growth limits, and a positive nitrogen balance must be relieved in some way other than through the usual channels of shed hair and skin, urine, feces, and skin-gland secretions. In fed animals, adjustment of the "obligatory nitrogen loss" (ONL) occurs at a certain point, and nitrogen intake must exceed ONL. During starvation this is not the case for omnivores, and FUNL then exceeds ONL and EUNL. The measured EUNL of cats fed a protein-free diet (360 mg N/kg $M^{0.75}$/d) reported by Hendriks *et al.* (1996) exceeded the mean FUNL that Biourge *et al.* (1994) observed in cats starved for 1–6 weeks (283 mg N/kg $M^{0.75}$/d), or the reverse of how omnivores respond. Thus gluconeogenesis in fed cats equals or exceeds that of starved cats (Kettelhut *et al.* 1980, Rogers *et al.* 1977), further illustrating the pressing need cats experience for gluconeogenesis driven by amino acid oxidation.

According to Eisert's hypothesis the protein requirement of cats is not great *per se*; instead, *they require protein in large amounts so they can respond to the brain's high glucose demand*. Gluconeogenesis is carried out in feral cats on a diet naturally devoid of carbohydrates but loaded with amino acids. Stray and house cats, in contrast, are often fed diets high in carbohydrates, predisposing them to hyperglycemia, dulled insulin resistance, and the metabolic consequences of both (Section 8.6).

This discussion has provided a partial answer to whether carbohydrates are necessary in the diets of cats and their ingestion harmful or benign. Data accumulated over many years point to carbohydrates being unnecessary dietary components for cats. *Some published work shows evidence of high digestibility after processing, which is not confirmation of a useful function. Nor do such findings indicate benign status, especially when accounting for carbohydrate's low target ceiling*. The capacity to metabolize excess dietary carbohydrates has limitations, and the reaction products have nowhere to go. Storage capacity for glycogen is limited, and feline adipocytes favor acetate over glucose as a carbon source during *de novo* lipogenesis, and glucose does not effect conversion to fatty acids in the cat's hepatocytes (Chapter 7). Glucokinase, which operates at high capacity, is one of the rate-limiting enzymes participating in glycolysis. Cats, however, do not demonstrate glucokinase activity (Ballard 1965), relying instead on the low-capacity hexokinase to clear circulating glucose and sustain a concentration gradient steep enough to permit further uptake (Tanaka *et al.* 2005). As a result, glucose clearance is subjected to slowing as the gradient levels out. This could create a bottleneck to rapid postprandial oxidation of carbohydrates if their dietary concentrations are high. Furthermore, undigested carbohydrate, especially if poorly processed, might be harmful by providing a substrate for microbial fermentation if it passes through the digestive tract and into the colon (Hewson-Hughes *et al.* 2011a, Kienzle 1994), a problem potentially worsened by the cat colon's general dryness (Kienzle 1993b). Fermentation has been validated during *in vitro* experiments using inoculums from cat feces (Sunvold *et al.* 1995). Altering a cat's diet for several weeks (e.g. switching from wet food to kibbles), changes the composition of its fecal bacterial populations (Bermingham et al. 2011), but this is probably harmless over time.

Dietary carbohydrates have been implicated in other metabolic dysfunctions (Section 8.6) and their use even at very low concentrations would appear to offer no tangible benefit. Hewson-Hughes et al. (2011a) defined "target intake" as the quantity and composition of a diet that supports optimal fitness. Consequently, laboratory cats regulate dietary intake toward a target composition closer to commercial wet foods than to kibbles, but these are not ideal either. Hewson-Hughes and colleagues concluded that the intake target approximates 26 g/d of protein, 9 g/d of fat, and 8 g/d of carbohydrate, or an energy composition derived from 52% protein, 36% fat, and 12% carbohydrate. These values are similar to those of Plantinga et al. (2011), whose meta-analysis of free-ranging cat diets showed a daily energy intake based on 52% protein, 46% fat, and 2% carbohydrate (Chapter 7). It seems reasonable that the best commercial cat diets would be optimally prey-based; that is, have contents similar in quality and quantity to those of the cat's natural prey. Of commercial products, wet foods containing high amounts of protein, medium levels of fat, and little or no carbohydrate approach this ideal, which in my opinion most nearly matches the whole-body proximate composition of a well-nourished, freshly killed mouse minus the contents of its digestive tract.

8.5 Energy costs of pregnancy and lactation

Pregnancy is costly in terms of energy expenditure, but less so than lactation. Pregnant cats gain weight, the percentage in excess of premating body mass correlating directly with the number of developing fetuses (Table 8.2). Nursing kittens drain a mother cat's physiological resources: their energy gained to some extent mirrors her loss. The mother's mean body mass tends to drop over the first 5 weeks postpartum and then recover, in one group changing from a mean value of 3.24 kg on day 7 to 3.12 kg on day 35 (Martin 1986), or a difference of ~4%. In most cases the disparity is greater, and considerable loss can occur during the first week of lactation (Martin 1986), reflecting the sizeable shift in energy requirements as lactation commences.

The amount of food consumed by nursing cats fed *ad libitum* increases with litter size. Mean daily intake of energy in week 1 of lactation for a cat with 1 kitten was 883.41 kJ compared with 1490.50 kJ for a cat with a litter of 4, a difference of 69% (Loveridge 1986). During week 3, these respective values rose to 1126.25 and 2022.22 kJ (an 80% difference). By week 7 the combined energy intake of the cat and her single kitten reached 1339.78 kJ, or 52% more than in week 1. The other cat and her 4 kittens consumed 4299.84 kJ, a 188% increase over week 1 and 221% more than the cat with the single kitten.

Mothers with large litters lose weight faster (Loveridge and Rivers 1989). The number of kittens still alive after every successive week obviously affects the mother's condition, drawing down her energy reserves a little more as each offspring survives into the following week prior to weaning. Females in one study lost an average of 5.7 g/d over the 8 weeks spent with their kittens, a mean daily loss in body mass of 0.16% (Deag et al. 1987). A female ending gestation with more than adequate body fat loses proportionately less weight while lactating and is able to support a larger biomass of kittens, a trend evident from the start: small mothers tend to have smaller kittens, large mothers larger kittens (Deag et al. 1987, 1988). However, this disparity tapers off during weeks 6–7 and disappears in week 8, indicating that accelerated growth of the

Table 8.2 Cumulative gain in body mass of female laboratory cats during gestation and subsequent litter size. Gains are percentage of body mass at mating.

Week of gestation	Number of kittens in litter				
	1	2	3	4	5
1	−0.3	−0.4	0.2	−0.2	−0.5
2	3.1	2.7	3.4	2.8	2.3
3	6.7	7.4	7.9	7.1	6.7
4	11.1	10.4	12.0	11.2	11.4
5	17.2	15.7	17.0	16.4	18.5
6	21.9	21.7	25.1	24.2	26.0
7	27.4	28.6	30.8	30.6	33.9
8	29.7	34.3	35.9	34.1	40.3
9	32.4	37.5	40.2	39.4	45.7

Source: Loveridge and Rivers 1989. Reproduced with permission of Cambridge University Press.

kittens, typically starting about week 5, might signal increasing nutritional independence from the mother (Bateson and Young 1981; Deag et al. 1987, 1988), although the effect can be concealed by individual variation (Fig. 5.6). No proportional difference is apparent by sex of the kittens (i.e. number of females in a litter vs. number of males), but kittens from big litters showed significantly greater discontinuity in growth around this time, and kittens of lighter-weight mothers experienced the interregnum at an earlier age and weighed less when it occurred. Also, kittens of large litters were more apt to display growth discontinuity at lower body mass and to have mothers suffering the signs of nutritional stress described later (Deag et al. 1987).

Martin (1986) estimated the energy intake per family for 7 families of cats. Each consisted of a mother cat fed *ad libitum* and her 2 kittens. Energy intake rose from 2140 kJ/d for the first week postpartum to >3000 kJ/d in week 10. Energy intake by the mothers represented the total intake during the first 3 weeks because the kittens were not eating solid food. Females averaged 2131 kJ/d, or 670 kJ/kg M/d. If a kitten needs ~19 kJ/g M for normal growth (Martin 1984c), energy requirement to weaning is ~475 kJ/d (Martin 1986). Assuming domestic cat milk to contain 6 kJ/g (reference in Martin 1986), growth alone requires 79 g/d.

If growth represents one-third of energy needs (Martin 1984c), then $79 \times 3 = 237$ g/d are necessary to supply energy for growth and maintenance. In other words, a 3.2-kg cat must produce 200–300 g of milk each day during lactation. Even though the cats Martin (1986) studied doubled their daily food intake and had food available at all times they still lost >120 g each through lactation until weaning despite rearing only 2 kittens. In another experiment (Wichert et al. 2011), individual cats varied during lactation from +0.04 kg (cat with a single kitten) to −1.24 kg (cat with 4 kittens), and this trend appears to be general.

The composition of cat milk is dynamic, changing over the course of lactation and depending partly on the percentage fat of the mother's diet (Jacobsen et al. 2004). Crude protein started to rise ~3 weeks postpartum and continued in a steady trend to the end of lactation. Fat decreased slightly at ~3 weeks postpartum then increased, ending slightly higher than its starting concentration by week 6. Kittens of these laboratory cats gained weight at comparable rates whether their mothers were fed a low-fat (LF) or high-fat (HF) dry diet (Jacobsen et al. 2004). Total solids (TS) rose or fell depending on the concentrations of protein and fat, making them higher in the HF group. Lactose peaked at week 2 postpartum, increased up to week 4, and then diminished over the final 2 weeks. Ash varied little throughout. The mothers fed HF food lost less body mass during lactation. Over 6 weeks the LF group ($n = 7$) lost an average of 660 g, the HF group ($n = 4$) lost 252 g. Small sample sizes reduce the credibility of this information in quantitative terms, but the pattern is distinct. Nipple preferences (Chapter 6) might give some kittens an advantage: tentative evidence indicates that rear teats might provide slightly more lactose and bigger volumes of milk (Ewer 1959, Jacobsen et al. 2004).

Loveridge (1986) monitored changes in body mass of 160 females from gestation through weaning and estimated their energy consumption. Litter size ranged from 1 to 5. Mean energy intake of cats producing only single kittens rose from 883.42 kJ/d in week 1 to 1339.78 kJ/d in week 7, an increase of 52%. A cat rearing 5 kittens consumed 8 times this amount ($\bar{x} = 3747.19$ kJ/d), or 196% more. Mean loss of body mass was 1% between mating and weaning (3030 vs. 2994 g). Body masses of females increased during pregnancy, the relative amount depending on ultimate litter size, ranging from 989 g (litter of 1, a 32% increase) to 1474 g (litter of 5, an increase of 46%). This difference represents a mean increase in mass during gestation of 121 g for each additional kitten. Wichert et al. (2011) obtained comparable results: females gained an average of 1.4 kg (range 0.6–2.2 kg) during gestation, equivalent to 39.2 % above pre-mating body mass, the cat producing a single kitten gaining the least (14.9%).

Cats lose weight at parturition (Loveridge 1986). For a cat with 1 kitten the loss at 24 h postpartum ($\bar{x} = 254$ g) was 6% of pre-parturition body mass. Comparable loss for a cat producing 5 kittens ($\bar{x} = 638$ g) represented 14%. In experiments of Wichert et al. (2011), median loss of body mass at parturition was 0.4 kg (0.1–0.8 kg), the cat with only 1 kitten losing 2%, the cat with 5 kittens losing 15%.

Compared with lactation, gestation is relatively undemanding of mother cats (references in Deag et al. 1987, Loveridge 1986), as seen by comparing changes in the body mass of females during both conditions (Fig. 8.8). Body mass rises steadily in a near-linear fashion between conception and parturition (Loveridge 1986; Wichert et al. 2009, 2011) then falls between parturition and weaning, most of the loss originating from fat reserves gained during pregnancy (Loveridge 1986, Wichert et al. 2011). However, lean body mass also accumulates over the gestation period, and enough is lost during lactation to induce a negative protein balance in the second week postpartum

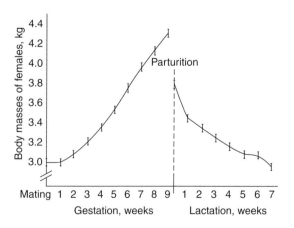

Fig. 8.8 Mean body masses (± SEM) of females ($n = 160$) during gestation and lactation. Source: Loveridge 1986, figure 1.

(Wichert *et al.* 2009), suggesting that the daily protein recommendations (not the minimum requirements) of the NRC (2006: 368, Table 15–13) of pregnant (21.3% DM) and lactating (30.0% DM) cats based on a diet containing $E = 16.75\,\text{kJ/g}$, are too low.

In Loveridge's 160 cats, mean body mass of 4319 g immediately prior to giving birth exceeded that at mating by ~40%, exactly the percentage lost ($\bar{x} = 512\,\text{g}$) within 24 h of parturition. Mean loss of body mass during lactation, measured as the difference at parturition vs. at weaning ($\bar{x} = 717\,\text{g}$) was 19% for cats producing 1 kitten and 21% ($\bar{x} = 864\,\text{g}$) for mothers with litters of 5 (Loveridge 1986). The duration of pregnancy through weaning resulted in a mean increase of 18 g (0.6%) for cats with 1 kitten and a mean loss of 29 g (0.8%) for those producing 5 kittens. Thus the mean mass of 2994 g at mating was within 1% of that at weaning, showing that the balance of the 60% pre-parturition gain was lost during lactation. In contrast, 4 females in the experiments of Wichert *et al.* (2011) were 5–33% heavier at weaning than at mating, and gain in body mass during gestation more than offset the losses incurred while lactating. Cats with >3 kittens lost 7–18% of body mass, and a starting and ending body mass within the range of ~10% is probably normal. Female cats vary widely in the final change in body mass at weaning.

The physiological demand imposed by lactation exceeds energy intake during this time, requiring the female to consume energy stored during pregnancy (Loveridge 1986, Loveridge and Rivers 1989) and raising the possibility that cats kept as laboratory or companion animals could become energy- or protein-deficient during pregnancy and lactation. However, not all cats lose weight. In one experiment, laboratory cats fed commercial dry food *ad libitum* ($E = 19.24\,\text{kJ/g}$) weighed more after 8 weeks of lactation than prior to mating (Wichert *et al.* 2009). This led Wichert *et al.* (2011) to consider whether commercial canned cat food would perform similarly. Canned cat food, with its higher water content, contains less protein and energy by wet mass compared with dry food (Chapter 7). In practical terms, even though both components might be comparable as assessed on a dry-mass basis, larger amounts of wet food must be ingested to achieve equal benefit, at which point bulk could be a limiting factor. As it turned out, results from dry and canned food were comparable. Although such experiments deepen our understanding of important nutritional issues, they seem trivial in the lives of feral and stray cats where opportunism dictates the rules, and the high species requirement for protein still allows cats to be successful in nearly all habitats. Were this not true they would have become extinct long before the need to invent cat food.

A protein-deficient diet late in gestation appears not to affect the number of kittens born, but a rapid drop in protein during gestation's early stages can cause abortion (Gallo *et al.* 1980, Smith and Jansen 1977a). Lactation, in contrast, is extremely energy-demanding, often the most expensive component of a female's investment in her offspring (Martin 1984a). Lactating cats sometimes suffer "nutritional stress," recognized by unsteadiness, weight loss, depressed appetite, and a general scroungy appearance (Deag *et al.* 1987). The condition is more likely in mothers with large litters ($\bar{x} = 4.80$ kittens vs. $\bar{x} = 3.63$ kittens). When the number of kittens is controlled statistically, these stressed mothers tend to be small. Predictably, sex of the kittens is irrelevant, nor does their survival to week 8 depend greatly on the mother's nutritional status.

The time postpartum when nutritional stress was first diagnosed varied among mother cats but averaged 37 d, and lighter mothers tended to become stressed sooner

(Deag et al. 1987). Stressed mothers leaned toward being smaller with larger litters. These data were obtained from laboratory cats receiving routine veterinary care and nutritional supplements when diagnosed; the effects would likely be more severe in feral and stray cats.

Pregnancy and lactation are accompanied by metabolic costs. Loveridge and Rivers (1989) showed that food consumption during pregnancy rises from the start in parallel with the increase in body mass. A subgroup of 10 of the 75 cats they studied took in a mean of 1030 kJ/d prior to mating. During pregnancy this rose 70%, peaking at 1737.5 kJ/d before falling to 1553 kJ/d during the week before giving birth. Overall, average gain per cat of 1.5 kg came at an energy cost of 28.9 kJ/g, attributable both to the female's heightened metabolism and her increase in size. The mean cost of growth determined by regressing gain in body mass against energy intake per unit of mass (Fig. 8.9) was 18.8 kJ/g, the estimated E was 280.5 kJ/kg M/d.

Weight gain during pregnancy rose in linear fashion with litter size, shown by

$$y = 106x + 888.9 \tag{8.6}$$

where x = number in the litter. Weight loss at parturition was described by the regression equation

$$y = 108.1x + 159.1 \tag{8.7}$$

where y = loss of body mass at parturition (g) and x = number of kittens in a litter. It mirrors Equation 8.6, fitting well with the mean M of 106.2 g for kittens at birth in this group.

As shown by comparing Equations 8.6 and 8.7, 729.8 g of body mass gained was retained by the mothers after giving birth (888.9 − 159.1), and this was not associated with litter size. Loveridge and Rivers (1989: 128) wrote: "Although the fact that heavier cats had larger litters meant that the postpartum weight as a percentage of the pre-mating weight declined slightly with increasing litter size, the range was small, animals being 119% to 126% of pre-mating weight."

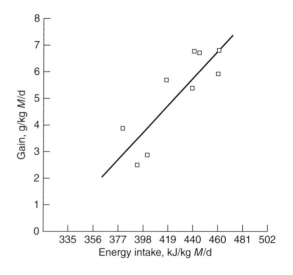

Fig. 8.9 The energy cost of gain in body mass of pregnant cats ($R = 0.845$, $p < 0.001$). Data points are means for each week of gestation. Equation for the line of best fit: $\Delta M/M = 0.194(1/M) - 14.68$; M = body mass, ΔM = gain in body mass (g/kg M/d), $1/M$ = energy intake (kJ/kg M/d). The authors stated that $n = 10$, but only 9 data points are shown. Source: Loveridge and Rivers 1989. Reproduced with permission of Cambridge University Press.

The weight gained in pregnancy was lost in lactation until at weaning the females were within 2% of pre-mating body mass. The regression equation describing the effect of lactation on body mass is:

$$y = 33.1x + 669.5 \tag{8.8}$$

where y = body mass lost in lactation and x = litter size. The correlation coefficient was moderately strong ($R = 0.70$), explaining approximately half the variation.

Energy intake of lactating females (Table 8.3) shows clearly that litter size affects energy intake and its rate, and the effect is greatest in the first 3 weeks. That of a female feeding 1 kitten rises from 242.8 to 406.1 kJ/kg M/d (68%) between weeks 1 and 6 and triples for a cat rearing 5 kittens. Loveridge and Rivers (1989: 130) referred to this situation as "relative anorexia," pointing out that "even if no demands were being made for milk production for the young … then food intake is inadequate for maintenance." They proposed regression equations modeling the energy yield from catabolism of a female's tissues, noting their similarities but different intercepts (i.e. the energy expenditure if no milk is produced). The intercepts represent E requirements for a lactating animal, which are sure to be higher than for a cat not lactating. They suggested that $E = 1214.2–1256.0$ kJ represented a reasonable value with mostly fatty tissue lost having an energy value of 25.1–37.7 kJ/g. Energy demand during the first 3 weeks of lactation for every extra kitten in a litter is ~355.9–393.6 kJ, or sufficient to meet the requirements. Wichert et al. (2009) confirmed that gain in body mass during gestation is linearly independent of the number of kittens and noted too that females lost weight while lactating, but were near pre-mating weight or heavier 2 weeks after lactation ended. Energy consumption during gestation was higher than reported previously (502 kJ/kg M/d), 1.8 times E in week 4 and twice E in week 7 with considerable variation among individuals. As mentioned, energy and protein intake during lactation exceeded the NRC's recommendations. Median gain in body mass in their 11 cats was 1.5 kg (range 1.0–2.1 kg). The cats at the two ends of the range had 4 and 3 kittens, respectively. At parturition the median loss was 520 g (320–940 g), or 12.8% (6.8–25.3%).

The experimental animals monitored by Wichert et al. (2009) were laboratory cats. Much of their gain was fat, resulting partly from low activity and a schedule of *ad libitum* feeding. The food contained 33% protein by DM. Nonetheless, Wichert and colleagues wondered if sufficient amounts could be supplied to lactating cats even at

Table 8.3 Weekly energy intake (kJ/kg of M/d) of lactating females ($n = 75$) by litter size.

Week of lactation	Number of kittens in litter				
	1	2	3	4	5
1	242.8	326.6	364.25	414.5	494.0
2	288.9	385.2	515.0	519.2	611.3
3	330.75	460.5	490.3	632.2	766.2
4	334.9	523.35	690.8	782.9	933.65
5	385.2	586.1	799.7	954.6	1235.1
6	406.1	657.3	963.0	1243.5	1482.1

Source: Loveridge and Rivers 1989. Reproduced with permission of Cambridge University Press.

higher concentrations. Feral and stray cats no doubt respond similarly when pregnant and lactating, although the range and limits of their responses are likely to be different.

The inability of females to obtain sufficient calories during gestation and lactation might culminate in higher sex-biased mortality on some oceanic islands. Jones (1977), for example, reported a heavily biased male/female ratio at Macquarie Island of 2.5. Jones and Coman (1982a) suggested that of cats inhabiting the arid regions of Australia, females were under greater nutritional stress than males as a result of reproduction. Feral cats in the Galápagos Islands were strongly biased toward males (Konecny 1983, in 1987b), and their mean body sizes were smaller than usually reported for adult feral cats (males = 2.7 kg, females = 2.1 kg).

Instead of gorging, adult laboratory cats eat 8–20 small meals spread randomly over the diel cycle, often consuming just a few grams each time if food is offered *ad libitum* (Bradshaw and Thorne 1992; Feldman 1993; Kanarek 1975; Kane 1989; Kane et al. 1981b, 1987; Kaufman et al. 1980; Martin 1986; Martin and Rand 1999; Mugford 1977). Cost–benefit analysis indicates that as the effort to obtain food increases, a cat's response is to increase meal size and decrease meal frequency, following what Kaufman et al. (1980: 137) called the "economics of abundance." Under these conditions loss of body mass can be expected, but cats appear to have considerable latitude before such fluctuations become an impairment to health (Kaufman et al. 1980). Gluconeogenesis in carnivorans peaks postprandially as opposed to peaking during the post-absorptive state or while fasting, as in omnivores (MacDonald et al. 1984a). This could explain why cats eat small amounts at frequent intervals (Eisert 2011).

As determined by Mugford (1977), each of a cat's many meals is ~125.6 kJ, the equivalent of a 16-g mouse (i.e. 1 mouse unit), adjusting volume to maintain a nearly constant intake of energy. If the food is diluted to lower its energy content a cat adjusts its intake to compensate (Castonguay 1981; Kane et al. 1981b, 1987). In such situations a cat eats just as frequently while ingesting smaller amounts at each meal. Kane et al. (1987) fed cats *ad libitum* diets containing 15% or 45% fat. When switched from the low-fat to the high-fat diet, the cats ate just as often but decreased their food consumption at each meal.

8.6 Obesity

Cats rely heavily on gluconeogenesis and obtain much of their basic energy needs from oxidation of fatty acids (Mercer and Silva 1989). The glucose entry rate (i.e. irreversible loss) in cats fed a high-protein diet and then fasted has been measured at 3.1 mg/kg/min, or ~46% that of fasted rats (Kettelhut et al. 1980). That gluconeogenesis proceeds regardless of how much carbohydrate appears in the diet is contrary to the situation of healthy rats and humans, which can shift from high-protein to omnivorous diets and back again while gluconeogenesis and glucose tolerance adjust in tandem with each oscillation. House cats fed high-carbohydrate diets are more prone to hyperglycemia, hyperinsulinemia, insulin resistance, and obesity (Eisert 2011), and effects vary depending on the source of carbohydrate (Appleton et al. 2004). Offering isocaloric high-protein, low-carbohydrate diets can increase insulin sensitivity while causing a loss of body fat (Hoenig et al. 2007, Rand et al. 2004). Insulin resistance and obesity are linked to the onset of noninsulin-dependent (type 2) diabetes mellitus (Appleton et al. 2001), and feeding house cats high-carbohydrate commercial diets is considered a risk factor (Rand et al. 2004). If high dietary protein improves the glucose homeostasis of overfed house cats, perhaps it reduces the insulin requirement. If so, then it affects insulin sensitivity (Hoenig et al. 2007) and vindicates current thinking that diets high in protein should be prescribed for diabetic cats (e.g. Rand *et al.* 2004). Certain cats are predisposed to being overweight (Wichert *et al.* 2012), and these individuals are perhaps more likely to become diabetic. At the least, high-protein diets increase heat production in cats, especially lean cats, indicating facilitation of fat metabolism (Hoenig *et al.* 2007).

For confined cats, protein aids weight loss in several ways (Laflamme and Hannah 2005). The inherently high rate of nitrogen metabolism helps, even on low-protein diets. Protein also stimulates postprandial thermogenesis, which raises protein turnover. The heat generated after a cat consumes a high-protein diet exceeds that of an isocaloric carbohydrate diet ~68% (caused by protein turnover). Increased synthesis of protein accounts for ~20% of energy expended after a high-protein meal compared with 12% after an isocaloric meal high in carbohydrates. Consequently, protein releases less net energy, meaning that adenosine triphosphate (ATP) yielded from oxidizing fat, carbohydrate, and protein are, respectively, 90%, 75%, and 55%. Stated differently, protein makes less dietary energy available than fat or carbohydrate, and the results are clear when laboratory cats are maintained on experimental diets controlled for these nutrients. In cats fed diets containing either a normal concentration of protein (30%) or a high concentration (40%) by DM the high-protein group lost significantly more fat and less lean body mass (Laflamme and Hannah 2005). Lean mass comprises mainly skeletal muscles (Millward 1995). Typical commercial cat foods contain enough fat and carbohydrate to "spare" protein. When energy is restricted, as in low-fat, low-carbohydrate diets, more protein is catabolized to supply energy. In the study of Laflamme and Hannah (2005), 45% of the energy from this type of diet originated from the high-protein component, enabling the cats to lose fat while conserving lean body mass.

A stimulus change, which could involve sensory contrasts indicating novelty, affects a diet's acceptance (Mugford 1977). Meal size and duration of feeding can be increased if cats are offered different foods in succession. It would appear that one way of inducing obese house cats to lose weight is simply by feeding them once daily instead of *ad libitum* and not varying their diet.

Accumulated adipose tissue helps sustain metabolic balance by providing reservoirs of energy, and by generating adipokines (e.g. adiponectin and leptin). A cat's body fat is related inversely to blood concentrations of adiponectin and directly with the amount of circulating leptin, making cats similar in this regard to humans and other mammals (Hoenig *et al.* 2007, Mazaki-Tovi *et al.* 2011). Adiponectin promotes an insulin-sensitizing effect by increasing the efficacy of fatty acid oxidation in skeletal muscle and liver tissues and lowering triglyceride concentrations. At the same time it promotes uptake of glucose by skeletal muscles while inhibiting glucose production in the liver, the result being a decline in blood glucose. Leptin modulates the mass of adipose tissue and increases insulin sensitivity. As in humans, obesity in cats reduces adiponectin and is associated with several metabolic maladies including insulin resistance and type 2 diabetes mellitus. Increased leptin is associated with insulin resistance.

Consumption of PUFAs has long been touted as beneficial to humans, partly by mitigating disorders associated with low levels of circulating adiponectin. Diets containing large amounts of long-chain *n*3 PUFAs (e.g. EPA, DHA) are considered helpful in delaying glucose intolerance, and obese cats have been shown to benefit in similar ways (Wilkins *et al.* 2004). Consequently, the concentrations of *n*3 PUFAs and adiponectin should correlate directly (Mazaki-Tovi *et al.* 2011). Serum EPA and leptin showed this, but EPA was not in step with levels of glucose or cholesterol, nor was a direct correlation found between DHA or α-linolenic acid and any of the other dependent variables (e.g. insulin, glucose, triglyceride). A rise in serum EPA of 1 mg/100 mg fatty acids in cats of normal body mass is accompanied by a 62% decrease in mean adiponectin and an increase in mean leptin of 43%. These and other findings of Mazaki-Tovi *et al.* (2011) demonstrated that in normal-weight cats the levels of adipokines (adiponectin and leptin), insulin, and triglyceride are associated with circulating levels of long-chain *n*3 PUFAs, but not with shorter-chained versions like α-linolenic acid.

Such results are difficult to interpret. In cats of normal body mass the relationship between EPA and adiponectin was negative, the concentration of the first explaining 18% of the concentration of the second. Whether this is enough to suggest a shift from positive but weak correlation into the restricted realm of cause is unknown, the question being whether fatty acids exert a causal effect on adiponectin production. The EPA concentration similarly explained 23% of variation in leptin levels of normal cats, but what this means exactly is uncertain too, conforming to some studies and at odds with others (references in Mazaki-Tovi *et al.* 2011). Feral and stray cats are seldom obese. This much is apparent: in overweight house and laboratory cats the concentrations of EPA correlate directly with adiponectin levels and inversely with insulin and triglycerides. A trend like this could benefit obese cats for the regulatory effect fish oils seem to exert on adiponectin and leptin production, insulin sensitivity, and fat metabolism in general. In contrast, we have no knowledge of what it could mean to a free-ranging cat of normal body mass or less.

9 Foraging

9.1 Introduction

As defined elsewhere (Spotte 2012: 119), "hunting is the act of finding, pursuing, catching, and killing prey, scavenging the act of feeding on carcasses or separating edible components from human refuse and other inedible materials." *Foraging* – the act of searching for food – encompasses both terms. Previous chapters covered the nutrients and vitamins cats require with special attention given to compounds they are unable to synthesize. Here the focus shifts to how feral and stray cats acquire energy and satisfy other dietary needs by foraging. Emphasis is on hunting because it requires greater skill and effort than scavenging, although Chapter 7 makes clear that what strays eat is far more dynamic and difficult to determine than diets of feral cats, obviously because the composition of garbage varies by location and even by garbage bin; that of a mouse or pigeon is similar everywhere.

The daily survival of a feral or stray cat is uncertain, its source of food often unreliable and barely adequate. Not surprisingly, life in alleys and scrublands is usually short and mean, and an individual that survives long enough to reproduce has fulfilled its genetic mandate. In contrast, a cherished pet is likely to be neutered and its need to forage offset by high-energy food provided in ample quantity, routine veterinary care, and guaranteed shelter, which combine to extend life-span into decades (Chapter 5) instead of compressing it into a few months or years.

In this chapter I dip tentatively into the deep pools of the senses, hoping to take the predatory behavior of cats beyond simple descriptions of motor patterns that dominate the literature. We know cats hunt mice, and we know why (to eat them). Among my objectives is to investigate how cats detect prey, but also to examine the situation from a mouse's position. Hunting and avoiding predation is a game of cat-and-mouse, and success does not always go to the hunter.

The interpretation of sensory functions is a darkened moonscape of craters awaiting the unwary, turning especially treacherous when we attempt to align sensory biology with the outward expressions of behavior. In a landmark 1979 article, Stephen J. Gould and Richard C. Lewontin emphasized the futility – and foolishness – of attributing an evolutionary adaptation to every tangible feature of nature. They called it the "Panglossian paradigm" after Dr. Pangloss, mentor of Candide in Voltaire's eponymous novella. Dr. Pangloss was an indefatigable optimist: everything happens for the best, and every incident or event, no matter how awful, is always for the greater good. Gould and Lewontin transposed this deranged philosophy to biology and called it the "adaptationist programme," an unrelenting march toward evolutionary perfection through

---

Free-ranging Cats: Behavior, Ecology, Management, First Edition. Stephen Spotte.
© 2014 John Wiley & Sons, Ltd. Published 2014 by John Wiley & Sons, Ltd.
Companion Website: www.wiley.com/go/spotte/cats

natural selection. Gould and Lewontin (1979: 581) wrote of the "adaptationist programme" that "It proceeds by breaking an organism into unitary 'traits' and proposing an adaptive story for each considered separately." They continued, "Trade-offs among competing selective demands exert the only break upon perfection; non-optimality is thereby rendered as a result of adaptation as well." Clearly, the adaptationists among us had created a world in which even halting flaws are only temporary detours to perfection.

Perhaps nowhere has this thinking veered more out of control than the meeting place of descriptive behavior and sensory reception. It seems like every snort and sniff, each piloerection and narrowed pupil, has been granted space in an "ethogram" and tagged with an attendant function, nearly always speculative but stated as fact. In trying to explain adaptations that make behavioral responses possible, glands have been invented that excrete unknown compounds into the environment to serve as signals not yet shown to exist. Pheromones are assigned roles even when their presence has not been demonstrated.

Developmental biology follows close behind the senses as an arena rife with conjecture. In a nod to Dr. Pangloss, Kitchener (1999: 237) wrote in reference to cats: "The longer development period required for the permanent dentition to develop … increases the period of dependence on the mother and, consequently, the opportunities for social learning. … " Thus prolonged tooth development is an adaptation for social learning, and "large felid species take advantage of an extended period of development to optimise prey size and minimise injuries/mortalities from the capture of large prey through social learning."

9.2 Cats as predators

The domestic cat's genome is conserved in many of its outward expressions (Murphy *et al.* 2000, O'Brien and Yuhki 1999), one being the retention of hunting skills (Bradshaw 2006). Hunger and the motivation to hunt are thought to be under separate neurological control (Adamec 1976, Polsky 1975). Others have reported structural behaviors that show an incomplete decoupling of hunger and predation. Robertson (1998) failed to demonstrate a positive correlation between predation and several independent variables (e.g. body condition, body mass, number of times fed daily, whether treats were given by its owner throughout the day). Turner and Meister (1988) found that fed cats hunted less, but did not relinquish hunting. Whether or not this is true, well-fed cats can be relentless predators (Davis 1957, Toner 1956). Sterilization does not reduce feeding activity (Fettman *et al.* 1997), nor does it release cats from the motivation to hunt. The effect of a cat's age on the numbers of animals killed does not usually vary by sex (Churcher and Lawton 1987).

Strong motivational ties bind killing and hunger, but association with the consummatory act is not so obvious. Hungry cats sometimes abandon kills, returning to feed after minutes or hours (Biben 1979), although I found just one report of free-ranging domestic cats caching their kills and covering them with leaf litter (Macdonald *et al.* 1987: 18). Moreover, satiated cats frequently kill when opportunity arises and still eat all or most of the carcass (Biben 1979). In general, hunger raises the probability of killing while satiation tends to decrease it, as would be expected. Consistent with this observation, killing latency correlates negatively with consumption; in other words, cats kill less when not hungry and take longer to eat the prey.

Laboratory experiments showed killing and hunger to correlate directly; killing and time correlated inversely (Biben 1979). This is because time since the last kill could not be less than the time since the last meal (i.e. hunger), although it could be longer if the cat did not kill for several trials or kill at all. When killing and hunger were controlled for time they retained a positive correlation while the negative correlation between killing and time increased, indicating that when hunger is controlled the probability of killing decreases strongly since the last kill (Biben 1979). An unknown portion of these effects can be attributed to individual differences among cats, some of which are more effective predators than others.

The daily expense of sustaining homeothermy places limits on both predator and prey. Proximate composition and energy content of mammals and birds thus fall within a narrow range, making all more or less equally useful to a cat and differing mainly in bulk. What generally holds true is that big animals contain more of everything, just as a large sack of corn has more of each constituent found in corn than a small sack. For example, a live adult laboratory mouse of 36 g contains 50 µg of α-tocopherol/g DM; a live pinkie mouse at 1.7 g has 35 µg/g DM (Douglas *et al.* 1994). The pinkie holds more vitamin E per unit of mass, but the adult, being 21 times bigger, contains more vitamin E overall. Much of a feral cat's dilemma is obtaining enough volume of prey to meet its dietary requirements, a daily chore every bit as important as what it catches.

The previous two chapters show why cats need to eat often. Maintaining a positive nitrogen balance means consuming a high-protein diet in adequate quantity (Chapter 7). Where energy is concerned, homeothermy is an expensive life-style (Chapter 8), and among vertebrates, birds are the most high-maintenance. On a mass basis a bird eats ~31% more food every day than a mammal, and birds and mammals combined need 8–11% more food each day than a reptile (Nagy 2001).

Free-ranging cats – even house cats – can be formidable predators, especially in wild and rural areas. Over 2 y, 2 house cats in Scotland brought home 444 prey animals (birds, mammals, and parts of a lizard) comprising at least 26 species (Carss 1995). Most birds were newly fledged or first-year, and 85% were passerines. A single well-fed farm cat owned by Bradt (1949) killed 1628 mammals in 18 months, 1200 of them meadow voles (*Microtus pennsylvanicus*). Another delivered 2–3 meadow voles to its owner daily (Toner 1956). Over its 17-y life a neutered female house cat in suburban New Zealand brought home 221 mice, 63 rats, 35 rabbits, 4 hares, 2 weasels, and 223 birds and was never seen >600 m from its residence (Flux 2007). It hunted mostly on the ground, but climbed trees to heights of 15 m in search of prey. After observing cats hunting mice in open fields, Pearson (1964: 183) wrote: "The consistent success of their vigils beside runways makes *Microtus*-catching seem absurdly easy." George (1974) reported that his 3 free-ranging cats in a southern Illinois agricultural region killed hundreds of small mammals (mainly rodents) and 47 eastern cottontail rabbits (*Sylvilagus floridanus*) over 3 y of observation. Their combined hunting range covered <10 ha. These numbers represent prey taken home; the number eaten or left in the field was unknown.

Predation can be intense even where garbage is plentiful and feeding supplemented by well-meaning humans. Page *et al.* (1992) observed stray cats at Avonmouth Docks preying regularly on birds, including pigeons and collared doves (*Streptopelia decaota*). Rodents were plentiful too. One cat was seen to capture and eat 2 rats within 40 min. In contrast, Calhoon and Haspel (1989) recorded just one predatory encounter during 180 h of watching urban strays in Brooklyn, New York. Garbage and food set out by residents were plentiful, but so were birds and rodents. At other locations, rats are killed and partly eaten or not consumed at all. Childs (1986) witnessed 24 predation events in urban Baltimore in which 11 (45.8%) of the carcasses were not eaten, 8 (33.3%) were eaten partially, and only 5 (20.8%) were consumed with the exception of the head, skin, tail, and feet. These cats had ample garbage available and were also fed by local residents. Of rats eaten the most common parts consumed were the brain, muscles of the neck and back, and the entrails, which were probably garbage-filled.

Why cats are size-selective when preying on rats is unknown, but an adult rat can be a dangerous opponent. Throughout development, young laboratory cats displayed arch-back more often toward rats than mice (Adamec et al. 1980b), indicating some level of unease or fear. Cats and adult rats observed by Beck (1971) and Childs (1986), coexisted peacefully in Baltimore's urban alleys. Cats are less likely to attack an animal that defends itself or reacts aggressively (Adamec et al. 1980a), and adult rats often qualify (Clevenger 1995, Leyhausen 1979: 76–77). Adamec et al. (1980a) screened 82 adult female laboratory cats for their killing prowess, and none would attack a rat. Although rodents are a principal prey of domestic cats, predation on them is evidently size-dependent with cats tending to avoid large specimens. Urban strays in Baltimore each killed an estimated 28 rats annually (Jackson 1951). Most were juveniles or young adults (<200 g). However, these were a minority component in trapped samples obtained later in urban Baltimore (10 of 109 rats captured, or 9.2%), mean size being 392.9 g (±124.9 SD) and the range 55.4–610.0 g (Childs 1986). Feral cats on the Mediterranean island of Cabrera preyed heavily on adult house mice, but most of the black rats (*Rattus rattus*) killed were juveniles (Clevenger 1995), and the same was true in Baltimore, where the brown rats (*R. norvegicus*) killed were typically <200 g (Childs 1986). Feral cats in Australia (Jones and Coman 1981) and New Zealand (Fitzgerald 1978) were heavy predators of black rats.

That meadow voles of any age are killed and eaten indiscriminately (Christian 1975) indicates that body size and not age determines prey selection (Childs 1986), especially if the prey is potentially dangerous. Because of their size, adult rabbits are also difficult to capture and kill, perhaps explaining why neonates, juveniles, and young adults are attacked preferentially (Bonnaud et al. 2007, Catling 1988, Eberhard 1954, George 1974, Jones 1977, Keitt et al. 2002, Liberg 1984b, Short et al. 1997). Nonetheless, feral cats in central New South Wales, Australia, increased their predation on adult rabbits as the rabbit population declined during a drought, at which time 82% of rabbits were >800 g (Molsher et al. 1999). Hungry cats at Macquarie Island were never seen to attack king penguin chicks (*Aptenodytes patagonicus*), even unguarded ones, probably because they stood taller than a cat (Jones 1977). Other birds that behave aggressively, such as skuas (*Stercorarius* spp.) and southern black-backed gulls (*Larus dominicanus*), are also known to repel stalking cats (Jones 1977, Lesel 1971).

Domestic cats are not extraordinarily strong for their size, although capable of killing animals larger than themselves so long as the prey remains reasonably docile. The strength of skeletal muscles varies according to cross-sectional area (Millward 1995). In humans, body volume and body mass vary with length, and this should hold true for mammals generally. If strength is to increase in proportion with M, a cross-sectional area of the muscles under momentary stress (i.e. during an act of predation) must vary with length of the bones being moved, as must muscle mass. The result of the cat's deficiency is sometimes apparent. Australian Eric C. Rolls (1969: 119) attempted to keep domestic rabbits (New Zealand Whites and Angoras) as pets for his children, but feral cats eventually killed them all, including "a big New Zealand buck that must have weighed at least four pounds heavier than the cat which took him." However, adult wild rabbits were not such easy prey, as evident from this scene witnessed by Rolls (1969: 119):

> I have watched bush cats stalking rabbits. The cat does not always win. Animals that fight by raking with the hind claws lie on their backs for support. It is the one on top that is losing. A scrawny ginger tom who leaped on a big buck from a wire-grass clump near a dung-hill one day had much the worse of it. The tom closed his teeth in the skin at the back of the rabbit's neck. It rolled over and fur flew in long strips from the cat's belly. Once or twice the cat rolled to the bottom and then the rabbit's fur flew. But finally the buck brought both his hind legs up near his opponent's chest and kicked him several feet away. The rabbit showed plenty of speed to its burrow but the cat limped away at a walk.

Eberhard (1954) reported that adult rabbit parts found in the stomachs of free-ranging domestic cats were probably scavenged from carrion. This is perhaps not surprising. The European wildcat (*Felis silvestris*) has a similar preference for young rabbits, which Gil-Sánchez *et al.* (1999) considered opportunism, reasoning that young rabbits are merely less experienced, which makes them easier to capture than adults.

9.3 Scavenging

Free-ranging cats the world over scavenge household scraps, carrion, and garbage (Coman and Brunner 1972), and sometimes the items have questionable value (Table 9.1). Those in urban areas often become adept at panhandling, frequenting outdoor cafés (Fig. 9.1) and docks (Fig. 9.2), and waiting expectantly underneath kitchen windows hoping to be thrown scraps (Fig. 9.3). Scavenging in garbage-strewn alleys and at waste-disposal sites is a common activity of stray cats, as is feeding on carrion by feral cats and strays. Carcass parts of eastern grey kangaroos (*Macropus giganteus*) scavenged by feral cats in central New South Wales included 17% of stomach contents by frequency and 8% by volume (Molsher *et al.* 1999). Where food is clumped, females tend to feed only at one site, but adult males might visit several, even though their daily nutritional requirements could be met by using just one (Say and Pontier 2004).

At Ainoshima Island, kittens followed their mothers to specific locations where garbage was available (Izawa and Ono 1986). It would seem that both the location and philopatry to a feeding site are fixed by habituation. Apps (1986b: 121) observed feral cats on Dassen Island scavenging bird carcasses in loose groups, doubtfully interpreting their behavior as "ready reversion to the more typical solitary foraging [indicative of] incomplete adaptation to the resource distribution. ... " Actually, this is the expected behavior when the distribution of food is dispersed instead of clumped, and if after feeding the cats went separate ways it simply illustrates their solitary natures. As emphasized earlier (Chapter 3), cats seen together in a garbage-strewn alley are likely to be there for the food, not for each other. Remove the food permanently and the predicted response is for them to disperse. Brickner-Braun *et al.*(2007: 136), after tracking house cats and strays in different parts of Israel, wrote: "A clear association between permanent food source (Dumpster or feeding site) and home range position is observed." They continued, "The nearest-neighbor distance between cat positions and the nearest food source was significantly clumped (i.e., the observed nearest-neighbor

Table 9.1 Recognizable nonplant scavenged materials found in the stomachs of free-ranging cats.

Material	Source
Apple (*Malus domestica*) core	McMurry and Sperry (1941)
Blackberry (*Rubus fruticosus*)	Molsher et al. (1999)
Bologna	McMurry and Sperry (1941)
Bread	Coman and Brunner (1972), McMurry and Sperry (1941), Parmalee (1953), Winslow (1944a)
Cellophane	McMurry and Sperry (1941)
Cereal	McMurry and Sperry (1941)
Charcoal	McMurry and Sperry (1941)
Charred bones	McMurry and Sperry (1941)
Cheesy material	Errington (1936)
Chicken foot	McMurry and Sperry (1941)
Chicken heads	McMurry and Sperry (1941)
Chicken skins	McMurry and Sperry (1941)
Cooked meat	Errington (1936)
Cooked meats	McMurry and Sperry (1941)
Cooked potatoes	McMurry and Sperry (1941)
Cooked rice	Errington (1936)
Cooked vegetables	McMurry and Sperry (1941)
Corn	Weber and Dailly (1998)
Corn husks	McMurry and Sperry (1941)
Custard pudding	Errington (1936)
Duck heads	McMurry and Sperry (1941)
Edible fig (*Ficus carica*)	Clevenger (1995)
Egg shells	McMurry and Sperry (1941)
Ensilage	Errington (1936)
Fatty offal	Errington (1936)
Fish bones	McMurry and Sperry (1941)
Leaves	Jackson (1951)
Paper	Jackson (1951), McMurry and Sperry (1941)
Pepper tree (*Schinus areira*)	Molsher et al. (1999)
Sanitary napkin	Korschgen (1957: 51, Table 11)
Spurge olive (*Cneorum tricoccon*)	Clevenger (1995)
String	Jackson (1951)
Sweet briar (*Rosa rubiginosa*) fruits	Molsher et al. (1999)
Table scraps	Korschgen (1957: 51, Table 11)

distance was significantly shorter than expected by the random distribution of food source … an indication that these food sources are an important predictor of … movement pattern." These authors emphasized that cats having a consistent feeding location – whether pets or strays – seldom wandered far.

Scavenging cats consume nearly anything organic, including vegetable matter (Table 9.1), and even laboratory cats will eat it if the alternative is nothing. One captive group was fed potatoes and condemned meat, "which were boiled and minced together; a sprinkling of salt and appropriate minerals were added." (Robinson and Cox 1970: 100). Winslow (1944a) reported that some of his laboratory cats preferred bread to prepared cat food, and scavenging cats eat bread too (Coman and Brunner 1972). For unknown reasons, cats sometimes eat grass (Jones 1989) despite its lack of nutritional value.

Feral cats can obtain all their dietary needs by hunting, assuming sufficient prey animals are available (Chapter 7). A cat scavenging on garbage, especially during hot

FORAGING 187

Fig. 9.1 A stray cat panhandles at a streetside café. Kotor, Montenegro. Source: Stephen Spotte.

Fig. 9.2 A stray cat waits expectantly near a fishing boat that has just arrived in port. Kuşadasi, Turkey. Source: Stephen Spotte.

Fig. 9.3 A stray cat waits outside an open window to be thrown table scraps. Kotor, Montenegro. Source: Stephen Spotte.

weather, is likely to encounter rancid fats. Foods containing PUFAs are especially susceptible to rancidity. Among the maladies potentially caused by consuming them are lowered growth rate, weight loss, reduced concentrations of plasma protein and arachidonic acid, intestinal inflammation, and interference with the absorption of nutrients (Chamberlin *et al.* 2011).

At a practical level the heterogeneity of municipal wastes – including human food wastes on which stray cats scavenge – leaves their compositions almost impossible to assess except as processed sludge. Even then the results are likely to be biased toward carbohydrate if paper has not been separated (Zaher *et al.* 2009). Nonetheless, analytical methods have been devised based on standard analyses used in wastewater treatment (e.g. chemical oxygen demand, total organic carbon, total ammonia-N). The results show high correlation with proximate analyses (Zaher *et al.* 2009), indicating that the proximate composition of organic matter at waste-disposal sites can be determined with excellent accuracy and precision even after reduction to sludge. This allows comparison with commercial cat foods and the animals on which feral cats prey. Limited data from dining halls and kitchens give even more direct insight into what stray cats are likely to encounter while Dumpster-diving (Table 7.6).

Waste-dump sites can offer foods that are both tasty and nutritious for scavenging cats. By-products of fisheries and aquaculture are examples. Cooked commercial shrimp waste is 94.6% protein and 4.2% fat by dry mass and contains 17 amino acids

(Mandeville *et al.* 1992), a composition making it an attractive ingredient for commercial cat foods. There seems to be no shortage. Crustacean waste from the fishing industry composes ~70% of the total landings, much of which ends up in landfills (Mandeville *et al.* 1992).

9.4 When cats hunt

Diel timing of foraging depends on several factors, including air temperature, food availability, and degree of risk. Farm cats might hunt nocturnally in summer because of cooler temperatures, the increased activity of nocturnal prey like mice, and perhaps to avoid humans and dogs (Chapter 2). Others hunt at midday in winter when temperatures are warmer, becoming crepuscular in other seasons (George 1974). If strays or house cats are fed during the day, they are likely to be active then. If any sort of pattern can be ascertained it seems to be bimodal activity during warm seasons at higher latitudes and in consistently warm climates, peaking at twilight, but unimodal in cold seasons at temperate latitudes (Goszczyński et al. 2009), peaking at midday.

Cats have been reported hunting at all hours. George (1974) monitored predatory activities of the free-ranging cats living at his farm house in southern Illinois and reported that diurnal hunting accounted for 49.8% of predation, crepuscular hunting for 20.1%, and nocturnal forays for 30.1%. George's cats hunted during the middle 6 h of the day in winter and longer on clear than on overcast days. They also reduced their hunting activities at night and on dark days when the temperature was $<-9.44°C$. In summer and autumn they hunted less during hot periods and became more active during crepuscular times and at night. At all times they seemed undeterred by precipitation. The hunting success of free-ranging house cats in an English village varied little regardless of temperature, but increased in calm, dry weather (Churcher and Lawton 1987). Feral cats in the Galápagos Islands hunted at most times except midday when few prey species were active, air temperature was highest, and evaporative water loss was greatest (Konecny 1987b). Feral cats on grasslands of the Otago Peninsula, South Island, New Zealand, were active both day and night, but moreso at night (Alterio and Moller 1997). Cats in the Mackenzie Basin became more diurnal in winter (Pierce 1987, in Alterio and Moller 1997).

Marion Island cats were principally nocturnal, emerging from rabbit burrows only on days when there was no rain (Derenne 1976). Feral cats at Stewart Island/Rakiura appeared to be mainly nocturnal too, an assessment based on most specimens having been live-trapped at night (Harper 2004), perhaps because rats, their main prey, are most active after dark. Feral cats and other predators of ground-nesting birds at the Upper Waitaki Basin, South Island, New Zealand, were recorded on video 90% of the time between sunset and sunrise (Sanders and Maloney 2002). At Réunion Island, southern Indian Ocean, cats prey heavily on endemic Barau's petrels (*Pterodroma baraui*), which breed in burrows and are easy to find and kill. The petrels are active at night, and so are the cats (Faulquier *et al.* 2009). Cats in the Kerguelen Islands were mainly nocturnal (Lesel 1971).

9.5 Food intake of feral cats

A hungry feral cat can consume up to 10% of its body mass at one sitting, the amount of fresh rabbit consumed per meal ranging from 184 to 470 g (Jones and Coman 1981). The food required by an adult feral or stray cat to meet its nutritional requirements has been estimated several times: 328 g/d at Isla Natividad, México (Keitt *et al.* 2002), and 294 g/d (absolute intake) in southern Sweden (Liberg 1984b). Keitt *et al.* (2002)

recalculated van Aarde's (1980) estimate for Marion Island cats using Nagy's formula (Nagy 1987), arriving at 260 g/d for females and 210 g/d for males. Both values are probably low. In the harsh subantarctic climate of Macquarie Island, Jones (1977) estimated that the daily intake of an adult cat was 300 g/d. Fitzgerald and Karl (1979) estimated the daily mean intake of feral cats living in a New Zealand forest at 170 g. Feral cats at Port-Cros Island in the Mediterranean Sea consumed an estimated daily mean biomass of 201.2 g (±17.7 SD), and consumption varied seasonally from 171.6 g (±118.6 SD) in spring to 227.3 g (±113.3 SD) in autumn (Bonnaud et al. 2007), but notice the large standard deviations. An adult male caged for 15 d consumed 2696 g of rodents and road-killed birds ($\bar{x}$ = 180 g/d), equivalent to 9.5% of its body mass (Howard 1957). The most it ate was 17 laboratory mice (413 g) in 26.5 h.

Feral cats in the montane wet forest of Mauna Kea, Hawai'i, consumed an estimated 1 rat and 1 mouse each day and 4 birds every 5 d (Smucker et al. 2000). Using conversions of 101.5 g (rat), 15.5 g (mouse), and 17.8 g (bird) gave an estimated daily intake of 135.6 g, of which rodents composed 90%, forest birds making up the remaining 10%. Among Galápagos Islands feral cats, rodents (36.1%) and birds (38.2%) contributed about equally to food intake with lizards (19.1%) covering most of the balance (Konecny 1987b). These cats also ate lots of invertebrates, especially insects. Vertebrate items made up 17.9% of scats by mass but contributed 93.4% of total daily energy intake.

Apps (1983) estimated 547 g/d at Dassen Island, South Africa. Feral cats there fed mostly on European wild rabbits (*Oryctolagus cuniculus*) and scavenged seabird carcasses (Apps 1986a, 1986b). Each cat consumed ~200 kg of mammals and birds annually (Apps 1983) comprising an estimated 134 European wild rabbits, 24 house mice (*Mus musculus*), 37 jackass penguins (*Spheniscus demersus*), 25 Cape cormorants (*Phalacrocorax capensis*), and 31 other birds of various species. Some of the birds appear to have been scavenged carcasses. When supplied with fresh whole carcasses of rabbits and birds, a caged young adult male ate an average of 540 g daily and a young adult female consumed 440 g daily.

Daily food consumption (FC) by mammalian predators in relation to body mass has been calculated by Harestad and Bunnell (1979) at

$$FC = 1.7 M^{0.68 \pm 0.02} \tag{9.1}$$

where M = body mass of the carnivoran (g).

9.6 How cats detect prey

In common with all mammals, free-ranging cats navigate the world using multiple sensory modalities and motivated by two factors: acquiring enough energy to live another day, and to reproduce. Cats obviously tune into the diel activity cycles of the animals they hunt, but ascertaining when their prey is active only starts a sequence that must then proceed sequentially through detection, location, stalking (or waiting), capture, killing, and ingestion. The relative importance of the different senses while foraging has been difficult to assess in the field. Laboratory experiments using blinded adult cats brings into question the utility of olfaction to a hunting cat. According to Crémieux et al. (1986: 231), "tests on the search for olfactory targets (smelling food) showed that the blind cats proceeded almost at random, except when the target was at a close distance (about 0.2 m)." On the contrary, the same cats "had strong orienting reactions and showed auditory pursuit," indicating that auditory cues might be more important.

Olfaction could be important if the prey gives off a strong, distinctive odor. Feral cats in the Turks and Caicos Islands, British West Indies, appeared to locate curly-tail lizards (*Leiocephalus* sp.) and juvenile rock iguanas (*Cyclura carinata*) from the distinctive odor of their shallow burrows in the sand. According to Iverson (1978: 70–71), when captive juvenile iguanas were placed in plastic bags and taken inside, cats were scratching at the door to gain entry within minutes. "If allowed to enter they would always find the lizards within 30 seconds, even when the plastic bags were hidden under cloth bags on top of an overhead shelf."

Sensitivity of the minimum auditory field (MAF) for cats and humans is nearly identical between 0.0625 and 0.50 kHz (Miller *et al.* 1963). From 0.5 to 4.0 kHz the cat is more sensitive by ~8 dB, and at frequencies >4 kHz the cat's advantage rises sharply. Humans hear to ~20 kHz, the cat to at least 60 kHz. Thus the cat is clearly superior at frequencies >0.5 kHz. That cats hear well into the ultrasonic ranges offers intriguing possibilities for how they might use this capability while hunting rodents.

Infant rodents of many species produce sounds at what to humans are ultrasonic frequencies (i.e. >20 kHz), including the house mouse (*Mus musculus*), brown rat, golden hamster (*Mesocricetus auratus*), bank vole (*Clethrionomys glareolus*), and species of other genera including *Acromys, Apodemus, Dicrostonyx, Meriones, Microtus, Peromyscus, Spermophilus,* and *Thamnomys* (Allin and Banks 1972, Hahn and Lavooy 2005, Sales and Smith 1978, Wilson and Hare 2004). Species of all these genera are actual or potential prey of cats. During agonistic interactions, dominant Norway rats emitted sound bursts of ~50 kHz lasting 3–65 ms (Sales 1972a). Submissive rats responded with synchronous sounds extending to 3400 ms at ~25 kHz. Depending on age and situation, infant, juvenile, and adult rodents vocalize in ultrasonic ranges under many conditions including agonism (Griffiths *et al.* 2010, Sales 1972a, Takahashi *et al.* 2010), while feeding (Takahashi *et al.* 2010), during courtship and mating (Griffiths *et al.* 2010, Hanson and Hurley 2012, Holy and Guo 2005, Musolf *et al.* 2010), aggression and social investigation (Griffiths *et al.* 2010, Scattoni *et al.* 2011), in response to pain (Williams *et al.* 2008), when distressed (Allin and Banks 1972, Hahn and Lavooy 2005, Sales 1972b, Sales and Smith 1978), in anticipation of play (Knutson *et al.* 1998), when signaling alarm (Randall *et al.* 2005), and when simply moving about (Takahashi *et al.* 2010).

Bats and other predators with keen hearing detect and locate prey by listening in on their intraspecific vocalizations (e.g. Akre *et al.* 2011, Page *et al.* 2012, Siemers *et al.* 2012), but whether cats eavesdrop on cryptic rodents based on the high-frequency sounds they make is unknown. Small rodents are the principal prey of free-ranging cats. They demonstrate not just impressive diversity but a highly conserved capacity to vocalize in the ultrasonic ranges. In the family Muridae, deer mice (*Peromyscus* spp.) alone comprise >50 species, and all evaluated so far can produce high-frequency sounds of apparent intraspecific relevance (Kalcounis-Rueppell *et al.* 2010). Many members of the family Cricetidae (hamsters and gerbils) possess this ability too (Griffiths *et al.* 2010).

As used here, *acuity* in the context of sensory modalities (e.g. audition, vision) means the capacity for distinguishing sharpness of detail; *resolution*, or *contrast sensitivity*, refers to the smallest separation that can be made between images against a highly contrasted background (e.g. in visual resolution the capacity to distinguish two thin black lines against a white background). From what we know, cats seem to be mainly visual and auditory hunters, approaching prey with head up, eyes and ears directed forward, then moving directly toward it (Sanders and Maloney 2002), not that this behavior excludes other sensory modalities. At such times the nose is also directed forward. Because vision and audition are joined by tight neurological links, I consider them as composing a single hunting modality (see later).

Cats have dichromatic vision under photopic conditions, rendering them unable to distinguish between red and green hues (Huberman and Niell 2011). According to Kang et al. (2009), their mesopic vision permits excellent contrast sensitivity in the dim illumination of twilight. Rod densities are especially suited for seeing in low-light conditions (Jarvis and Wathes 2012), making the cat's scotopic acuity superior (Kang et al. 2009). Technically, retinal illuminance in failing light increases with dilation of the pupil, lowering photon noise and altering spatial contrast sensitivity function (CSF) and visual acuity (Jarvis and Wathes 2012). What it means comparatively in a functional context is this. Humans see best in photopic, or cone-mediated, light conditions. Humans have superior visual acuity compared with cats. Our optical resolution is 3–5 times better than theirs (Bonds 1974) and shows bigger changes in resolution (Pasternak and Merigan 1981). Over a luminance range of 6 logarithmic units a cat's contrast sensitivity decreases ~1 log unit, a human's ~1.6 log units, with resolution of the two species converging at the lowest luminance. The brightness threshold of humans in photopic conditions is much lower than the cat's, but the two are similar in very low light, and in actual darkness the cat's surpasses ours. As Pasternak and Merigan (1981) noted, this disparity would be even greater were comparative results expressed as retinal illumination instead of luminance. This is because the cat's larger pupil, combined with shorter focal length and the presence of a tapetum lucidum, provide enhanced retinal illumination, lowering the cat's absolute threshold to 6 times below a human's (Gunther 1951).

As luminance diminishes from daylight through twilight into night, a human's accelerating loss of visual acuity remains superior until at very low luminance the cat's acuity starts to approach ours (Jarvis and Wathes 2012, Pasternak and Merigan 1981) and eventually surpasses it. Mechanistically, the shift from cone to rod vision affects quantum efficiency of humans and cats differently, our acuity and perception worsening, theirs changing much less, thus explaining the species disparities detectable in CSF under a steadily darkening sky (Fig. 9.4). It stands to reason that cats can hunt effectively throughout the diel period, sometimes being at a disadvantage when stalking diurnal prey in bright light, and effective although not necessarily superior at night when some nocturnal prey have excellent scotopic vision too.

Mice, like cats, are dichromats (Huberman and Niell 2011), but they see at much lower resolution. Were we to take a mouse to an optometrist for an eye examination it would test out the equivalent of 20/2000 (Huberman and Niell 2011), an order of magnitude worse than a legally blind human. Such poor vision is partly a result of having small eyes and relatively few photoreceptors.

In addition, the mouse eye lacks a fovea, a part of the retina that substantially enhances the acuity and resolving power of central vision (i.e. the central 1° of vision), representing the locus of sharpest eyesight. Diurnal primates have clear, crisp sight in photopic conditions because most cones occur in the fovea where they enhance acuity and resolution, providing excellent distance vision. This gain is offset by rapidly attenuating vision in peripheral fields. Consequently, primates can see large objects far

Fig. 9.4 Comparison of cat and human contrast sensitivity (resolution) over a 6-log unit range of luminance plotted against spatial frequency: cat functions, heavy lines; human functions, thin lines. Data are from one cat and one human, both tested with their pupils in the natural position. Source: Pasternak and Merigan 1981, figure 4, p. 1336. Reproduced with permission of Elsevier.

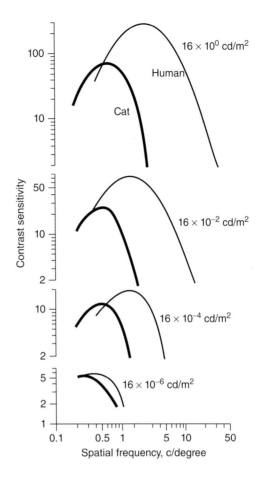

away and small objects up close, especially moving ones, if they look directly at them. *Visual eccentricity* is the angle relative to the fovea, and it increases with increasing distance away from the fovea. The more eccentric an object's position inside the visual field the more difficult it becomes to distinguish. In cats and humans, acuity and resolution thus diminish with increasing eccentricity (Pasternak and Horn 1991), as expected. The cat's oculomotor range is $\sim \pm 25°$ from center (Guitton *et al.* 1984). Without a fovea the mouse's whole retina is never more effective than a primate's peripheral retina; that is, it sees an object with approximately the same clarity that you see in the lateral visual field while looking straight ahead. The mouse eye is only 12.5% the size of a rat's, the axial length of the rat's nearly twice as long (Remtulla and Hallett 1985). The rat is emmetropic and perhaps slightly myopic, and the mouse probably is too.

Neurons in deep layers of the brain's superior colliculus initiate behaviors associated with locating external stimuli and orientation toward them. Within this context sight and hearing are linked (Fig. 9.5). In the cat, 55–57% of neurons deep in the superior colliculus are multisensory, of which 86% in the sensory reception fields overlap, heightening the response to an external event when both senses are stimulated (Meredith and Stein 1996, Meredith *et al.* 1991). Functionally, a cat's eye movements

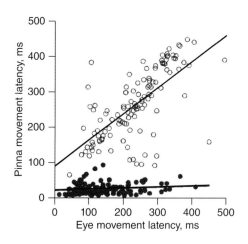

Fig. 9.5 Pinna movement latency to auditory (●) and visual (○) targets by cats ($n = 5$). Measurements taken from visual and auditory standard saccade trials to targets ipsilateral to the pinna at ± 18° and 0° are plotted against eye movement latency. Data ($n = 249$) originate from 126 auditory trials and 123 visual trials with 7 trials of latencies >500 ms excluded. Note that pinna movements toward the auditory source are tied directly to onset of the target and not to eye movement. Mean latency of pinna movements to auditory sources are significantly briefer than they are to visual targets. Slope of the regression for auditory data (0.03, $R = 0.16$) indicates no association between onset of pinna movements ($\bar{x} = 26.5\,\text{ms} \pm 15.2$ SD) vs. eye movements ($\bar{x} = 265.6\,\text{ms} \pm 77.8$ SD) toward the same targets. Slope of the regression for visual data (0.75, $R = 0.68$) suggests a direct association between eye ($\bar{x} = 233.8\,\text{ms} \pm 87.2$ SD) and pinna ($\bar{x} = 262.8\,\text{ms} \pm 96.2$ SD) movements. Source: Populin and Yin 1998b. Reproduced with permission of the Society for Neuroscience.

are accompanied by electrophysiological discharges from muscles controlling movements of the *pinnae* (external ears) so that orienting saccades and ear movements occur in tandem (Joseph and Boussaoud 1985); in the reverse case, cats direct their pinnae toward a target they are trying to locate visually (Populin and Yin 1998b). When cats turn their heads toward a visual stimulus, eye movement lags behind, and eye–head movement terminates with the head aligned to the target and eyes focused at its center (Guitton *et al.* 1984). Average latency in laboratory cats was ~40 ms, and few movements start when the eyes are >2° off-center (Fig. 9.6).

Reliance on multi-functional neurons requires incoming signals to converge and align before subsequent relay and activation of linked motor responses, which a split-second later switch on and then release predictable behaviors. In other words, the complementary sense comes into alignment when an auditory or visual signal is received – visual if the signal is auditory, auditory if the signal is visual – followed by appropriate behaviors.

The positions of a cat's ears and eyes affect sensory reception and consequently spatial location within its field of reception. A cat hearing a rustle in the dry grass close by looks in that direction and awaits another occurrence. Meanwhile, distance to the source is estimated. Eyes shift to the near-field and refocus. The body tenses as if in auscultation, but interest is entirely external. The pinnae show strong directional properties, and for most frequencies a single location of ipsilateral space exists in the frontal field where a tonal stimulus of consistent intensity is evoked (Phillips *et al.* 1982). Now they rotate like antennae, poised to modulate frequency and amplification while the

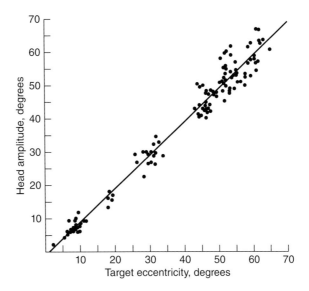

Fig. 9.6 Linear regression plot of target eccentricity vs. head amplitude for movements triggered visually. Eyes began saccade at an angle ≤1° from center. Slope of the regression shows perfect correlation (45°, $\beta = 1.0$); α (intercept) = -1.2 is near zero, $R = 0.99$, $n = 133$ data points. Source: Guitton et al. 1984. Reproduced with permission of American Physiological Society.

brain stands ready to filter signal from noise. Audition and vision actually work in sequence, although so quickly that the combined act seems instantaneous.

On hearing that rustle and following a brief delay the pinnae orient toward the signal, at which time amplitude of the movement can range to ~20° (Populin and Yin 1998b). The shift in position of the pinnae is followed by another latency as eyes are directed toward the target (Populin and Yin 1998b), often without a change in head position. *Simply shifting the eyes alters reception in the auditory field.* The reason for the first interval is unknown. The external ears of primates are essentially immobile, keeping pinnae and head in consistent alignment (i.e. the visual–auditory correspondence is one-to-one). The ability of cats to move their pinnae offers no apparent benefit when localizing sound in the lateral fields where binaural audition becomes less effective (Heffner and Heffner 1988). Pinna mobility ought to complicate sound localization by intensifying the "computational load" (Populin and Yin 1998b: 4233). Conversely, any capacity to alter how the pinnae are positioned raises three possible advantages. These are to (1) allow multiple sampling of the auditory field, (2) partition broadband signals for identifying unique sound spectra, or (3) adjust the pinnae for enhancing the signal-to-noise ratio (Populin and Yin 1998b). The first appears unlikely. Heffner and Heffner (1988) reasoned that the origin of long-duration sounds should be found quickly because of increased "scanning" time allotted. However, laboratory cats were unable to locate the source of a continuous sound any faster than one lasting 10 ms.

Sensory integration is most effective when external stimuli are weak. A faint rustle in the grass is more likely than a gunshot to elicit visual attention; not surprisingly, weak stimuli are also most apt to induce subtle behavioral adjustments. Slight rustles arouse attention, vision focused toward the sound, vision and hearing alert to any nuance. A loud noise, in contrast, elicits a startle response and rapid exit from the scene.

Cats consistently "undershoot" (i.e. underestimate the distance) when orienting toward eccentric angles (May and Huang 1996, Populin and Yin 1998a). The most accurate orientation responses occur when a sound originates directly in front of the subject, at azimuths and elevations of 0°. *Azimuth* is the horizontal angle measured clockwise; *elevation* (or altitude) the angle up or down from the horizontal. Both can have positive and negative values. Because orientation responses are plotted in double-pole coordinates, the center of mass (CM) for an elicited response is calculated by averaging head movements immediately after the sound stimulus has been applied.

May and Huang (1996) plotted deviation from speaker locations. In the horizontal plane, laboratory cats increasingly undershot the azimuth as it varied from 0°, and the same phenomenon (but to a lesser extent) was apparent as elevation shifted higher. Cats obviously have more difficulty locating sounds as the source shifts laterally or up, and in their attempts to locate it they undershoot the place of origin. The error, however, is typically <5° for targets within ±15° of either 0° azimuth or elevation, but increases with increasing eccentricity. At 90° azimuth, the extreme in the horizontal location, undershoot was >42°, indicative of poor orientation to sounds originating in the lateral field and illustrating that the most accurate orientation occurs near the median plane in both the horizontal and vertical and is best of all in terms of single-dimension response elevation in the interaural horizontal plane.

How are these deficiencies overcome and shifted to the cat's advantage while hunting? The most convenient way is by waiting for the sound to be emitted again. As explained above, orientation is not different whether sounds are received in single bursts of 10 ms or continuously. Repetitious sounds, in contrast, permit the rapid correction of orientation undershoots. Not only is a correction made, but a cat turning its head toward a sound's origin presents the frontal field where reception of both sound and vision is most effective. In the 3 cats tested by May and Huang (1996), orientation responses to single sound bursts within 15° of 0° did not vary during bursts of 40, 100, and 200 ms. However, target acquisition was achieved within four 100-ms intervals after the second burst when bursts were paired.

The cat's situation is complicated by its moveable pinnae, which can confound maximization of sound reception with head movement. In other words, it should not be assumed that a shift in head position alone serves to direct reception into the frontal field, and a cat waiting for a second sound has no doubt directed its head, ears, and eyes toward the anticipated site of origin. Pinnae-shifting probably happens quickly: latency of head movement has been timed at 300 ms (May and Huang 1996), and rotation of the eyes then typically follows (Guitton *et al.* 1984). Visual target acquisition therefore occurs after a combination of head and eye movements in which the head aligns with the target's angle of eccentricity, the eyes following and becoming centered when the response ends abruptly. The entire event will have taken <300 ms even for targets of 50° eccentricity. Reception of mid-frequency auditory cues is simultaneously enhanced by also being received frontally (Rice *et al.* 1992).

To a cat hoping to kill a mouse, it all means this. The mouse's hearing is acute, at least a match for the cat's, but the cat sees better. Comparison of olfactory capabilities are difficult to assess, but the mouse's is likely superior, perhaps by a sizeable margin. Vision and hearing are distance senses (Meredith *et al.* 2000); the sense of touch is proximate, making haptic senses effective only at close range. As Crémieux *et al.* (1986: 233) pointed out, "Vibrissae cannot participate in remote anticipation, especially during fast locomotion (a cat with vibrissae 5 cm long running at a speed of

1 m/s would have only 50 ms within which to process the information provided by its vibrissae before knocking into a barrier)."

Somatosensory functions of the vibrissae and body hairs represent a standoff, the cat having no reason to use haptic senses until after the mouse has been caught, and any value of the mouse's vibrissae by touching a cat and realizing its presence has come too late. Assume a mouse can smell a cat or hear its soft footsteps. Assume too that a cat can hear a mouse's movements and its ultrasonic vocalizations. Even humans can smell where mice have been by detecting the odor of their urine or feces, and no doubt so can a cat. However, smelling where a mouse *has been* is not the same as detecting where it is *right now*. Can a cat smell the odor of a living mouse and place it in context? I could find no data.

A motionless cat waiting beside a mouse run is probably invisible to the mouse considering its poor eyesight, although if the cat shifts position the mouse will notice movement. So will the cat when the mouse moves, although motion is not always a releaser for predation. Although the prey's movement can trigger pursuit, a cat's overall predatory behavior is not necessarily influenced by whether the prey is moving or still, and in captive cats most of the time was spent in inattentive structural categories

Table 9.2 Structural categories of behavior involved in predation and sometimes associated with play.

Structural category	Description
Group I	
Bat	Uses forepaws (one or both) to make the prey move or vocalize.
Carry	Grips prey firmly in its mouth but without chewing or mauling it.
Clutch	Grabs prey with forepaws and pulls prey toward its belly then holds prey in forepaws while kicking at it with the hindpaws.
Mouth	Muzzle makes contact with prey.
Recline	Alert cat lies on its side.
Toss	Flings pray some distance away, usually without injuring it.
Kill-bite	Normally a single bite to the head, body, or neck.
Group II	
Bat	Uses forepaws (one or both) to make the prey move or vocalize.
Tap	Gently taps at prey with a forepaw, claws retracted.
Herd	Forces prey to move and paces itself so that prey is not overtaken but its direction is controlled.
Spring	Belly is close to the ground, head stretched forward, tail straight out and hind legs flexed; springs toward prey.
Crouch	Belly flat against ground, forelegs flexed with elbows elevated over shoulder blades and forepaws directly underneath; tail might twitch gently at tip.
Rear	Rises on hind legs, extending forelegs out and up while keeping prey in front.
Attentive	Gazes at prey and follows its movements but without touching it.
Group III	
Cry	Vocalizes
Groom	Licks its fur or scratches itself.
Ignore	Prey is ignored.
Shut-eye	Closes its eyes
Jump	Jumps upward, perhaps onto a shelf
Avoid	Backs away or walks away from prey and seeks to actively avoid it.
Other	Visits water bowl, litter pan, or performs some other behavior not mentioned above.

Source: Data from Biben 1979.

(Table 9.2). Biben (1979) sorted the behavioral categories associated with predation into three groups. In the first, 5 categories (bat, carry, mouth, tap, and toss) increased in frequency up to the kill-bite; other categories diminished in frequency or did not change.

Nonetheless, cats see far better than mice under all conditions, easily resolving and recognizing a mouse's running form, and it can do this in daylight, twilight, or darkness. If unable to actually see its quarry the cat can hear faint rustling and scratching as it moves through grasses and ground litter, and the more often the mouse generates a sound from near the same angle of eccentricity, the more accurate the cat's detection of its exact location using audition and vision together.

Somatosensory neurons that modulate the vibrissae also overlap with those controlling other functions in the superior colliculus, but evidently this depends on which portion is stimulated (Meredith et al. 1991). Stein and Clamann (1981) were able to stimulate a few vibrissae and sometimes the entire mystacial pad, but Meredith et al. (1992) received no response from 182 neurons tested in the tecto-reticulo-spinal (TRSN) tract, 84% of which proved multisensory. Rodent TRSNs, in contrast, generate signals almost exclusively from the vibrissae (references in Meredith et al. 1992).

In lagomorphs and many rodents the somatosensory cortex contains discretely packaged neural arrays called "barrels" (references in Rice and Munger 1986, Rice et al. 1986). In these animals, and in cats too, the mystacial pad containing the vibrissae, or sinus hairs, are surrounded by pelage made up of guard and vellus hairs, the former being larger and coarser. The vibrissae, of course, are much larger than both (Fig. 2.5). Although cats lack barrel fields, the guard and vellus hairs around and among their vibrissae are far more intensively innervated (Rice and Munger 1986) than those of rodents and lagomorphs. The general pattern seems to be an inverse correlation between the absence (or degree of conspicuousness) of barrels and the density of intervibrissal innervation (Rice and Munger 1986). Just 25% of multisensory neurons in deep layers of the superior colliculus of the cat respond to somatosensory stimuli, but of these the piloneurons account for 78% of collicular somatosensory reception (Meredith and Stein 1996). How between-species disparities in somatosensory function translate is uncertain. Although evoking a barrel-field response raises glucose metabolism (references in Rice and Munger 1986), the question remains of what barrels actually do. Stimulating the intervibrissal pelage of rats and mice leaves barrels largely unresponsive, suggestive of vibrissal stimulation exclusively. This is perhaps not surprising in rodents living close to the ground, relying heavily on haptic senses for detection and orientation (Meredith et al. 2000).

The cat's vibrissae and associated innervation are well developed. A mouse has more pelage hairs per unit area than a cat, but proportionately fewer guard hairs overall and therefore fewer piloneural receptors. Being larger, cats have more of these receptors simply because the spaces between their vibrissae are much larger (3.5–4.0 mm^2). Rabbits are intermediate: they possess inconspicuous barrels and intervibrissal innervation that are less developed than the cat's.

Clipping off the vibrissae of blinded and sighted cats did not affect locomotor performance of sighted cats in the light, but it caused their performance to decline sharply in darkness (Crémieux et al. 1986: 233). Removal of the vibrissae had no effect on how blinded cats performed, and "this invalidates the hypothesis that vibrissae play an important rôle in the buildup of locomotor space in these cats. ... " Crémieux and colleagues speculated that learning to use vision and feedback from vibrissae might need to develop in tandem for proper attainment of spatial behaviors. Leyhausen (1979: 31) described how a cat's vibrissae change position during and immediately following a predatory act, writing that "at the time the cat springs on its prey its whiskers protrude as far forward as possible ... and obviously help to keep a check on the prey's movements after it has been seized." He noted that mice and similar small animals, once in the grip of a cat's jaws, "are practically enveloped by them. ..."

If a mouse's vision is substandard, its olfaction is sensational. Karlson and Lüscher (1959: 55) invented the term *pheromones* and defined them as "substances which are secreted [sic] to the outside by an individual and received by a second individual of the same species, in which they release a specific reaction, for example, a definite behaviour or a developmental process." They are hormone-like compounds, but unlike hormones, which are *secreted* into the circulatory system, pheromones are *excreted* into the external environment. The mammalian vomeronasal organ (VNO) takes up pheromones, their processing mediated by vomeronasal receptors 1 and 2 (V1Rs and V2Rs). Salazar and Sánchez-Quinteiro (2011) and Salazar et al. (1996) described the domestic cat's VNO.

Canids evidently have limited VNO function, and their capacity to produce and subsequently sense pheromones is doubtful (Spotte 2012: 65). Whether cats or other carnivorans are similar is unknown, although a specific felid pheromone has yet to be isolated and its function confirmed. Rodents and a few marsupials are the only mammals known with certainty to produce and respond to compounds that could properly serve pheromonal functions (Roberts et al. 2012, Spotte 2012: 59–66, Tirindelli et al. 2009: 923, Table 1). Major urinary proteins (MUPs) of mice operate as pheromones, pheromone carrier proteins, and environmental pheromone stabilizers (Beynon and Hurst 2003, references in Papes et al. 2010). Their molecular structures render them stable, refractory to degradation, and easily transmitted between individuals (Miyazaki et al. 2008, Papes et al. 2010). Major urinary proteins serve many functions, including involvement in individual recognition (Hurst et al. 2001, references in Miyazaki et al. 2008).

At this point we can say that the proteinuria often evident in healthy domestic cats is not evidence of renal pathology (Miyazaki et al. 2006a, 2008), as would be the diagnosis in most other mammals, but rather of high nitrogen turnover (Chapter 7). Nor do the excess proteins excreted by cats necessarily serve as carriers of volatile substances as they do in mice, perhaps acting as pheromones or facilitators of other, nonreproductive, interactions (e.g. individual or group recognition). Proteinuria in the domestic cat yields cauxin, a MUP that functions as an enzyme expressed in the kidney and regulates the synthesis of felinine (Chapter 2), a putative pheromone precursor in domestic cats (Miyazaki et al. 2006a, 2008).

Kairomones, such as those eliciting fear and defensive responses in mice subjected to the odor of a cat, function as olfactory cues transmitted between species that confer a selective *disadvantage* on the signaler and a simultaneous selective benefit to the receiver. A component in the high-protein diet of carnivorans might trigger these responses in rodents (Fendt 2006), although cat odor can originate from sources other than urine and feces, including saliva transferred to its own fur during grooming (Papes et al. 2010). For example, the cat's salivary glands generate large quantities of fel d 4, a cat homolog of rodent MUPs (Smith et al. 2004). As a potent allergen, fel d 4 is obviously transmissible between species.

Mouse kairomones are induced by a species-specific chemical belonging to the MUPs, homologs of mouse pheromones that promote aggression (Chamero et al. 2007, Papes et al. 2010). Kairomones, like pheromones, are detected and processed by vomeronasal receptors (VRs), or sensory neurons, in the VNO. Unlike pheromones, which are intraspecific, they can function between species. A pheromone is context-dependent, a kairomone independent of context. In other words, although a mouse's aggressive response is triggered by a pheromone it occurs only in the presence of another mouse. A mouse's response to a cat kairomone is released regardless of whether the odorant is presented by the cat itself or simply the cat's scent transferred to cotton gauze (Papes et al. 2010).

When recombinate MUP proteins activate pertinent VRs they process co-opting existing sensory mechanisms, explaining the mystery of how molecular detection can be both species-specific and interspecific by doubling as odorant receptors, thus conserving variety in fewer detectors (Papes et al. 2010, Tirindelli et al. 2009). The mouse VNO consequently has a role in pheromone detection, but also processes chemical signals from other species. In mice, these are compounds released by predators that function as kairomones. Mice with defective VNOs are unable to decipher them. As a result, they fail to sense the olfactory cues necessary to trigger defensive behavior and display fear of cats (Papes et al. 2010). The selective disadvantage to the cat (olfactory announcement of its presence) is then erased along with benefits ordinarily conferred on the mice (warning of danger and subsequent fear response).

Laboratory experiments demonstrate that odors can influence choices of foods, which in turn initiate and sustain feeding. For example, blowing "attractive" food odors in a continuous stream of air through a laboratory cat's maintenance diet raises its scalar quality, or "hedonic tone" (Mugford 1977: 43), rendering it more palatable. The result is to consistently increase both duration of feeding and extend food intake even after satiation. How this effect translates to the wild is less evident. Cats hunting dispersed mice would need to distinguish a telltale mousy odor from a smorgasbord of other olfactory stimuli carried on the wind, then follow a trail of increasing odorant concentration to the source. It seems doubtful that insular cats preying on seabirds rely much on olfaction. Having spent time in bird rookeries on subarctic oceanic islands during breeding season I can attest to the overpowering odors of excrement, of decaying birds and their broken eggs mingled with those of rotting kelp and dead seals and sea lions wafting up the cliff faces from below. These places are maelstroms of motion, cacophonous confusion, and general sensory overload. Cats hunting birds in such locations might count heavily on vision, although olfaction and audition could be important at close range.

9.7 How cats hunt

Cats obviously rely on a combination of stealth, athletic prowess, and physical attributes to capture prey. These are obvious and as isolated descriptions reveal little about the mechanistic processes. A hunt begins with a stimulus that sets the hunting sequence (Section 9.6) in motion: hunger (although not always), a perception of movement, a nearby sound. Perhaps a distinctive odor. If prey can be found in the immediate vicinity the cat hunts it. The species of prey killed by free-ranging house cats correlates directly with proximity of the prey within a cat's locus of activity (Churcher and Lawton 1987). The activity level of farm and house cats, in other words, declines with distance from the barn or house (Krauze-Gryz et al. 2012, Romanowski 1988). Because cats evolved as lone hunters, the presence of other cats

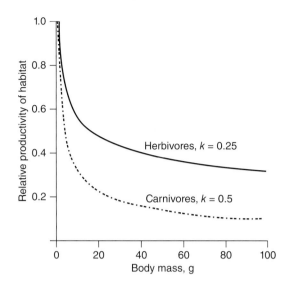

Fig. 9.7 Hypothetical relationships between density of energy production in the habitat and body mass of herbivores and carnivores. Because the larger predator must ignore smaller prey, the productivity of its habitat declines steeply with increasing body mass and more steeply than that of its prey. Source: Harestad and Bunnell 1979. Reproduced with permission of the Ecological Society of America.

probably lowers individual efficiency and puts all at a disadvantage. Hunting success has been shown to vary inversely with the density of free-ranging cats, indicative of some kind of interference (Churcher and Lawton 1987), although how this works has not been established.

In some ways a domestic cat is the ideal size for a carnivore. Food qualifying as both acceptable and accessible declines in availability with increasing body mass for any animal whether herbivore, omnivore, or carnivore, a relationship that holds more tightly in patchy habitats (Harestad and Bunnell 1979). Therefore, the likelihood of small or mid-sized species having unproductive patches inside the home range is less than it would be for a larger species. The decline in potential energy inside a given home range is steepest for carnivorans because predation is a high-energy endeavor compared with grazing or browsing, and because large predators need more energy per "package" of prey than smaller ones, resulting in smaller prey often being ignored (Fig. 9.7).

Home ranges of cats commonly overlap (Chapter 2), which requires that hunting spaces be shared too (George 1974), but domestic cats are lone hunters. A free-ranging cat could see its food intake reduced only if choosing to hunt as part of a group. The truly social lion gains little by cooperative hunting, much of which is apparently a consequence – not a cause – of group living (Elgar and Harvey 1987, Packer 1986, Packer and Ruttan 1988), and the same holds true for the considerably more social gray wolf (Spotte 2012: 141–142). In fact, group foraging is disadvantageous to lions, the ideal hunting "group" being a single individual (Packer 1986).

Gittleman (1989: 198) wrote that "felids rely … on an elaborate sequence of stalking, use of retractile claws for pulling down quarry, and truncated jaws used for a precisely oriented killing bite. … " Hunting cats either stalk prey or use the sit and wait method (Genovesi et al. 1995). Stealth hunting, which is used by most felids, appears to be an adaptation to solitary living (Kleiman and Eisenberg 1973). Solitary hunting, unlike group hunting, requires cryptic methods, and lone hunters usually prefer dense habitats rather than open landscapes (Gittleman 1989). Even among lions, females opting to live alone or forced to, eat as well as those in prides (Packer 1986).

Sometimes the sexes hunt differently. Male domestic cats have been seen moving continuously and stopping if sensing potential prey while females might sit and wait for a prey animal to approach (Macdonald and Apps 1978). A waiting cat uses less energy. The value of P (Equation 2.3) is expected to decline less rapidly for these lazier hunters (Harestad and Bunnell 1979).

Large prey animals are killed by a bite to the neck or base of the skull; small prey like mice are dispatched quickly with a bite to the head or body and swallowed whole (Jones 1989). Cats typically pin victims to the ground using one paw (Adamec et al. 1980a) before administering the killing bite. The use of both paws simultaneously is less common. Sometimes prey is grasped in both paws and the cat falls onto its side, kicking and clawing with its hind feet. Cats also tap the prey or strike it hard with one paw.

The digital organ of carnivorans is the claw (unguicula). The cat's claws are smooth, sharp-tipped, compressed laterally, scythe-shaped with borders "cut like a pruning knife" (Schummer et al. 1981: 495), and aptly termed "cutting claws." They can be retracted into pockets of skin and extended by contraction of the digital flexor muscles. To effect this the cat stretches out its forelegs. Muscles on dorsal and ventral sides of the paws contract, overriding the pull of ligaments that have kept the claws retracted and causing them to pop out. Claws of the forepaws, which are used to capture prey, are bigger, stronger, and more important than those on the hindpaws.

After making a kill a cat typically leaves the prey momentarily. It looks around and perhaps explores the area for a short distance if the location is not entirely familiar. Leyhausen (1979: 31) called this behavior "taking a walk" and considered it a mechanism for discharging excitement. Prey is seldom consumed where killed, but taken instead to a familiar, secluded location (George 1974, van Aarde 1980). This is contrary to Veitch (1985: 130), who stated that "Cats prefer to carry their catch to an open area." The head and brain is usually eaten first (Veitch 1985).

In some situations the prey animal requires little effort to find, subdue, and kill. Burrowing petrels on subantarctic islands lack behaviors that might protect them from predation by rats and cats. Adults, along with their eggs and chicks, are all vulnerable. Cats methodically search the burrows until a bird or egg is discovered, which is then dragged out with a forepaw and taken to a secure location to be eaten (van Aarde 1980). When actual hunting is necessary, not every attempt to make a kill is successful. Konecny (1987b) calculated the overall capture efficiency of Galápagos Islands cats pursuing a variety of vertebrates and invertebrates at 32%.

Predation in cats can be a stronger motivation than hunger, and they kill animals without eating them (Sanders and Maloney 2002). A cat will stop eating to kill a rat, immediately leave the kill, and return to its original meal (Adamec 1976). Although *surplus killing* (killing several animals without consuming them) is relatively common in certain canids (e.g. Spotte 2012: 148–149), this behavior is seldom seen in domestic cats. Peck et al. (2008) reported that 22% of sooty tern (*Onychoprion fuscatus*) carcasses on Juan de Nova Island, Mozambique Channel, represented surplus killings and considered it the first quantified observation of such an event.

The teeth of cats evolved for killing other animals and slicing tissues. Cats have long canines, small incisors, and no teeth with crowns that are large, flat, and adapted to chewing. According to Leyhausen (1979:58), "Cats in general do not masticate but, using their carnassials as 'shears,' they cut the prey up into small pieces or fairly long strips which they then swallow whole." The *carnassials*, or shearing teeth, comprise the third upper premolar and lower molar. The dental formula is I = 3/3, C = 1/1, PM = 3/2, M = 1/1 (Jones 1989). In effect, cats are unable to "chew" their food, making kibbles especially difficult to eat.

The variety of prey diminishes as feral cats age, acquire infirmities including conjunctivitis (Kalz et al. 2000) and generally slow down, and tooth condition directly affects longevity. The teeth are obviously important for both killing and sectioning prey. Feral and stray cats are prone to gingivitis and periodontal diseases, which probably shorten their lives (Kalz et al. 2000). Older cats are especially susceptible to tooth loss, a condition brought on by external resorption of the teeth by odontoclasts. The origin of this process is unknown but diagnosed by lesions at the enamel–cementum junction and at the surface of the tooth roots. The resulting loss of enamel, dentine, and cementum (DeLaurier et al. 2006) poses a risk especially to free-ranging cats deprived of veterinary care. Tooth loss thus contributes to a narrowing of prey choices, possibly leading to malnourishment and eventual starvation.

Cats sometimes skin birds, using their teeth to peel away the skin intact with feathers attached (Cuthbert 2003). They also start feeding on the larger pectoral muscles. Small lizards and birds are totally consumed (Veitch 1985), the latter sometimes after being partially plucked (Konecny 1987b). Veitch (1985: 130) wrote: "Cats do not deliberately pluck feathers before eating a carcass." This appears to be incorrect, unless Veitch meant that the entire carcass is not plucked prior to being eaten. Larger birds might be eaten entirely except for the wings. In the Galápagos Islands, grasshoppers were usually eaten entirely, but some cats ate only the abdomens (Konecny 1987b). Rats in the Galápagos were consumed more often in the wet season, birds, arachnids, crabs, and insects more often in the dry season. According to Leyhausen (1979: 44), birds smaller than the common European blackbird (*Turdus merula*), which is ~100 g, are seldom plucked but eaten whole (Fig. 9.8), the coarser feathers spat out as encountered. Big birds are plucked before being eaten (Fig. 9.9). According to Jones and Coman (1981), if the prey is a small rabbit (<400 g) nearly everything will be eaten except some of the viscera (e.g. the caecum). If the rabbit is an adult the head might be severed from the body before skin, muscle, bone, and organs are eaten (Jones and Coman 1981).

Prey abundance affects the rate at which prey is encountered. How much does the act of hunting affect the drive to hunt if hunger has been satisfied? Do cats hunt because they evolved to obtain energy and nutrients by being predators or are they satisfying an as yet unrecognized drive to acquire specific nutrients and other components of food? I have no answers to these questions, but the information available points out some intriguing possibilities. Obviously, feral and stray cats must hunt or scavenge to survive. Many free-ranging house cats, in contrast, are fed adequately, some *ad libitum*, outwardly giving them little motivation to kill and eat other animals, yet they do this anyway. Perhaps house cats hunt to satisfy some dietary component lacking in commercial foods, but if so this substance – assuming it exists – is not obvious. Or perhaps they miss a vague ancestral taste or texture acquired while stalking rodents through rain and snow, a gustatory memory that still lingers while dozing on the couch. Both are doubtful based on what we know about cat nutrition, although I came across one piece of intriguing information. Robertson (1998) reported that house cats fed commercial diets but not meat were more likely to kill wildlife, which suggests that a palatability factor (e.g. texture, odor, taste) might be missing or deficient.

Seawater contains every known element, and cats can sustain normal water balance by drinking it even at full strength (Chapter 8). In experiments using laboratory cats,

Fig. 9.8 Small birds (<100 g) are usually eaten whole, feathers included. Source: © Senkaya | Dreamstime.com.

Fig. 9.9 Large birds are plucked before being eaten; the mouthfuls of feathers are spat out. Source: © Luckynick | Dreamstime.com.

Wolf *et al.* (1959) noticed that intake of artificial seawater was sometimes sufficient to maintain or restore euhydration; at other times cats drank too much or too little. The tests extended over 295 d. Considering that conditions were similar for all cats and no extraneous sources of elements were available except in the food and water, this failure to regulate ionic intake indicates a certain insensitivity even to the osmolality of the fluid and making it unlikely that one purpose of mariposia – especially by a cat – is to replace a deficient inorganic element.

9.8 What cats hunt

Cats are well adapted for physiological water conservation, enabling them to inhabit arid regions where water is scarce or even absent (Chapter 8). This adaptive capacity makes location almost irrelevant to survival and offers a nearly limitless variety of potential prey.

Cats have historically been labeled bird killers (e.g. Forbush 1916), although they are far more adept at killing lagomorphs, rodents, and other small mammals. Mammals are consumed more often than birds or any other animal group (e.g. Bradt 1949, Dilks 1979, Eberhard 1954, Errington 1936, Fitzgerald 1978, Hubbs 1951, Jones and Coman 1981, Karl and Best 1982, McMurry and Sperry 1941, Molsher *et al.* 1999) except in certain circumstances: colony-nesting birds can become the dominant prey on islands (e.g. Harper 2010, van Aarde 1980), and reptiles can be important in arid regions (Dickman 1996b, Konecny 1987b). According to van Rensburg (1985), petrels composed nearly 93% of the total energy required by feral cats at Marion Island with house mice contributing just 4%. Cats are also egg predators, breaking the shells and consuming the contents both on oceanic islands and elsewhere (e.g. Amarasekare 1994, Ashmole 1963, Bloomer and Bester 1990, Matias and Catry 2008, Snetsinger *et al.* 1994).

Free-ranging cats are opportunists (Bonnaud *et al.* 2007, Clevenger 1995, Coman and Brunner 1972, Fitzgerald 1988, Konecny 1987b, Pearre and Maas 1998, Peck *et al.* 2008, Pontier *et al.* 2002), and many birds are difficult to capture unless unfledged or newly fledged and learning to fly (Toner 1956). Small mammals (mainly European wild rabbits) composed 88% of stomach contents by volume of 80 feral cats at mainland Australia (Victoria, New South Wales) with birds, cold-blooded vertebrates (e.g. lizards, frogs), and invertebrates combined accounting for only 5.2% (Coman and Brunner 1972). Similar findings have been reported from other locations where mammals are prevalent dietary items and birds, cold-blooded vertebrates, and invertebrates are of secondary importance (e.g. Bonnaud *et al.* 2007, Bradt 1949, Eberhard 1954, Errington 1936, Matias and Catry 2008, McMurry and Sperry 1941, Nogales *et al.* 1992, Parmalee 1953, Pearre and Maas 1998, Pontier *et al.* 2002, Woods *et al.* 2003). However, this is a trend and not a rigid pattern. What cats eat depends on what prey animals are available, their sizes and aggressiveness, and how difficult they are to catch.

On continents and large islands (e.g. New Zealand), mammals are far more common prey than birds by a ratio of ~6 (Pearre and Maas 1998), and small mammals far outnumber birds of similar size worldwide (Pearre and Maas 1998). Again, depending on circumstances, birds can be major items of prey. According to Anderson and Condy (1974) the feral cats on Marion Island killed mice and adult macaroni penguins (*Eudyptes chrysolophus*), but burrowing petrels comprised their principal prey: Salvin's petrel, or Salvin's prion (*Pachyptila salvini*), and the soft-plumaged petrel (*Pterodroma mollis*). They fed on young chicks of different species of albatrosses and also scavenged, primarily seal carcasses and the carcasses of penguins and petrels killed by great skuas (*Stercorarius skua*) and giant petrels (I assume the Antarctic, or southern, giant petrel, *Macronectes giganteus*).

Churcher and Lawton (1987) asked residents of Felmersham Village, Bedfordshire, United Kingdom, to keep prey items delivered by their free-ranging house cats ($n = \sim 70$) so they could be examined. The toll for 1 y was 1090 animals or their remains: 535 mammals, 297 birds, and 258 unidentified parts, an average of $\sim$14 carcasses per cat. There were 22 species of birds and 15 species of mammals. The bulk of the total comprised 17% wood mice (*Apodemus sylvaticus*), 16% house sparrows (*Passer domesticus*), and 14% short-tailed, or field, voles (*Microtus agrestis*). Birds made up 35% of the catch. This value, Churcher and Lawton (1987: 452) remarked, "is lower than folklore would suggest but higher than the results obtained in most research programmes."

All these data are no doubt conservative because a cat is unlikely to return home with every animal it captures (Churcher and Lawton 1987, George 1974). Some escape to die later, others are eaten near the capture site, and still others are simply killed and abandoned.

Opportunism and prey availability ultimately determine diet. Insular prey tends to be limited to a few species. At Port-Cros Island in the northwestern Mediterranean Sea, for example, yelkouan shearwaters (*Puffinus yelkouan*), black rats, and European rabbits constituted 93% of the prey, although evidence of one other small rodent and at least 19 additional species of birds were recovered from scats (Bonnaud et al. 2007). On San Clemente Island, California, the diet of feral cats overlapped with that of diminutive Island foxes (*Urocyon littoralis*), which at maturity are actually smaller than the cats. The cat diet was equal parts arthropods (47.9%) and vertebrates (44.2%) with mice predominating in the second category and lizards and black rats making up the balance (Phillips et al. 2007). Arthropods consumed were mainly grasshoppers, beetles, and ants with beetles and ants predominating.

At a multi-use land area in southeastern Brazil the diets of feral and stray cats and stray dogs were found to overlap almost completely (Campos et al. 2007). Cats preyed mainly on opossums (Didelphidae, 19.2% of total items) and Brazilian guinea pigs (*Cavia aperea*, 15.4%). Principal prey species in winter comprised the black-footed colilargo (*Olygoryzomys nigripes*, 21.4%); the nine-banded armadillo (*Dasypus novemcinctus*) and white-eared opossum (*Didelphis albiventris*) together made up 14.3%. In summer Didelphidae predominated (33.3%) with *C. aperea*, the threatened lesser grison (*Galictis cuja*), and Leporidae (rabbits and hares) each contributing 16.7%.

Cats are obviously generalist carnivores, feeding on everything from insects, centipedes, and spiders to mammals and birds up to $\sim$500 g (Bonnaud et al. 2007, Brickner-Braun et al. 2007, Nogales and Medina 1996, Panaman 1981, Tidemann et al. 1994, Turner and Bateson 2000). Other introduced mammals – especially rodents and rabbits – are often the principal prey, and nowhere is this more striking than oceanic islands (e.g. Apps 1983, 1986a; Fitzgerald et al. 1991; Furet 1989; Jones 1977; Medina et al. 2006; Pontier et al. 2002).

The frequency of insects in the diets of feral cats has probably been underestimated (Konecny 1987b, Pearre and Maas 1998), moreso in terms of numbers consumed than as percentage volume. Feral cats eat a variety of insects, but the total consumed and their dietary contribution overall is generally minor (e.g. Karl and Best 1982, Kirkpatrick and Rauzon 1986, Nogales and Medina 1996). Medina and García (2007) recovered 127 invertebrate prey items of 28 species from 500 scats at La Palma, Canary Islands. Parts of invertebrates occurred in 18% of scats but composed just 0.05% of the total biomass ingested. Of total invertebrates identified, 90.6% were insects.

Size of the prey is a factor in its selection, the incidence of predation decreasing with increasing prey size (Biben 1979). Perhaps not surprisingly, hungry adult cats that were experienced hunters are more likely to attack mice than adult rats in a laboratory setting, although both the probability of killing large prey and the speed with which it is accomplished are associated directly with state of hunger (Biben 1979). In general, adult cats prefer young rats (Furet 1989) and avoid attacking large prey unless very hungry (Adamec et al. 1980a). Biben (1979) could predict 95% of variability in the killing response if just state of hunger and prey size were known. In addition, hunting prowess varies among cats (Churcher and Lawton 1987), and an individual's inclination to hunt factors into its decision, not just prey size but whether to hunt at all, especially if other food is available. Adult white-chinned petrels (*Procellaria aequinoctialis*) are large and aggressive, and on Marion Island the smaller, weaker chicks were far more likely to be killed by feral cats (Bloomer and Bester 1990). Penguins can be difficult to kill too, and most remains found in the stomachs of feral cats were probably from scavenged penguins. Cats generally avoid attacking large aggressive birds (Apps 1986b, van Aarde 1980).

Mean prey size increases with latitude (Pearre and Maas 1998), not a surprising trend as seabirds become increasingly important in the northern and southern high latitudes, but this general statement does not hold true universally. In the equatorial Galápagos Islands, Konecny (1987b) saw feral cats attacking frigate birds (*Fregata* spp.), pelicans (*Pelecanus* spp), and flightless cormorants (*Phalacrocorax* spp.), all larger than the cats trying to prey on them.

9.9 Prey selection

The selection of prey depends on its size and availability. However, the structure of a prey organism's population and not simply its abundance affects which individuals are most susceptible (Jones and Coman 1982b). If selection is by size and individuals of smaller size are preferred, this introduces a seasonal component. For example, cats have trouble killing adult European rabbits, and 81% of rabbits eaten at Macquarie Island were <600 g (Jones 1977), indicating a strong preference for younger – and thus smaller – rabbits.

A predator's own population is food-limited, forcing it to rise and fall according to the distribution and abundance of prey (Harper 2004 and references). As the numbers of prey fluctuate, a population of predators responds numerically by changes in its demographic patterns or functionally through changes in behavior. The *numerical response* describes how the predator *population* responds to changes in prey density (e.g. Pech et al. 1992), manifested by changes in a predator's patterns of survival, reproduction, immigration, and emigration (Harper 2004 and references). Feral cats, for example, often disperse during a decline in density of their principal prey (Norbury et al. 1998a). A functional response might involve switching to alternative prey under such conditions (Harper 2004 and references), ordinarily when populations of the favored prey plummet. The *functional response* describes the effect of *individual* predators from the standpoint of prey eaten per predator at differing prey densities (e.g. Pech et al. 1992). The *total response* is the product of the numerical and functional responses.

Liberg (1984b) reported a functional response by free-ranging house cats during winter in southern Sweden when they switched from the favored prey (rabbits) to short-tailed voles and water voles (*Arvicola amphibius*). These cats were also opportunistic in 1977 when the rabbit population exploded. Rabbits then composed 93% of all prey by mass, and daily winter consumption of rabbit meat was 162 g/cat/d, 6 times more than in the previous 2 seasons. After 1977 the rabbits declined, average annual intake of all prey computed daily dropped from ~294 to 170 g/cat/d, and the cats lost mean body mass ($\bar{x} = 4.5$ kg to $\bar{x} = 3.8$ kg). Although feral animals represented 15–20% of the local cat population, their combined take of prey about equaled that of all the house cats combined, which over the seasons averaged 66 g/d, probably because most were fed regularly and had no need to hunt. Cats in semiarid eastern Australia showed a numerical response to rabbits (Pech et al. 1992). Fitzgerald and Karl (1979) apparently saw a functional response in the forested Orongorongo Valley, New Zealand, where rabbits contributed up to 40% by mass to the diet of feral cats in some seasons despite low numbers. When rabbits declined at Macquarie Island after release of the myxoma virus, cats, which had previously responded numerically to the rising rabbit population, increased their predation on seabirds (Copson and Whinam 2001).

Feral cats at Stewart Island/Rakiura, New Zealand, preyed mainly on rats (Harper 2004). Rat abundance varied seasonally, being lowest from early summer to early autumn. However, the appearance of rat parts in cat feces showed little change in seasonal frequency and no significant difference by mass (81.2% of prey taken overall). The cats continued to hunt and consume rats even when their abundance declined, indicating no prey-switching and thus no functional response until the rat population fell below some critical level. The relative abundance of rats and both frequency of bird parts and biomass of birds in the diet showed a strong negative correlation, indicating that consumption of birds rose as rats became less abundant, having finally reached that critical low. Bird parts occurred in 26.9% of scats but added only 13.2% to the diet by mass. Cats also emigrated when rats diminished in number (numerical response). Some that stayed behind starved to death, one specimen having lost 40% of its original body mass. Harper (2004: 25) wrote: "If it is assumed that the relative abundance of rats relates directly to density, and prey mass is linked to the number of individuals caught by cats, then there is little evidence that cats on Stewart Island/Rakiura 'prey switch'. ... "

Not assessed in these reports is the effect of population size on the growth and size distribution of the rats themselves. Rats in urban populations grow faster as their numbers rise (Davis 1951). During times of low reproduction and a decreasing population (e.g. winter and summer), male rats are heavier but females are similar all year, suggesting that with diminishing numbers the largest males survive. Any cats preying on them would then be forced to switch to other prey or become less size-selective in the rats they kill.

Konecny (1987b: 29), like others before him (e.g. Jones and Coman 1981, Marshall 1961, McMurry and Sperry 1941), reasoned that "Without the ability to emigrate during resource shortages, cats should be opportunistic and feed on whatever foods are most available. ... " His observations of feral cats on the Galápagos Islands showed this. However, they also ate birds, lizards, and grasshoppers in relatively uniform proportions despite variations in the seasonal abundance of each group, and Konecny noted, "Clearly, cats are not tracking the apparent populations in their consumption of prey items." Feral cats in central New South Wales showed little evidence of a functional response, continuing to focus on rabbits even after their numbers had been decimated by rabbit calicivirus disease and drought and the population had fallen 90%

(Molsher 2001, Molsher *et al.* 1999). Predation on rabbits actually rose, although house mice also increased in importance.

Prey-switching is often mandatory, especially on oceanic islands where seabirds are migratory. At Little Barrier Island, New Zealand, feral cats fed mainly on specimens of Cook's petrel (*Pterodroma cookii*) during their courting, nesting, and rearing phase (September–March) and switched to Polynesian rats, or kiores (*Rattus exulans*), from April–August (Imber 1975). The cats at Juan de Nova Island, Mozambique Channel, switched from insects and rats to sooty terns as soon as terns began arriving to nest and breed (Peck *et al.* 2008).

In arid South Australia, rabbits were the most important prey when their counts exceeded $10/km^2$, but at lower densities other vertebrates (e.g. rodents, birds, and reptiles) increased in the stomach contents from 1 to 2.5/cat (Read and Bowen 2001), indicating prey-switching. A functional response to rabbits was evident at Yathong Nature Reserve in semi-arid western New South Wales (Catling 1988). Feral cats switched to invertebrates, carrion, birds, and reptiles in times of drought when rabbits became scarce.

In contrast, Church *et al.* (1994: 748) reported laboratory cats demonstrating "anti-apostatic" food selection (i.e. increased preferences for rare foods) when presented with foods of mixed types differing in sensory properties. They attributed this to "learned and/or innate preferences for rare food items, based upon the nutritional benefits when a mixed diet may offer." These authors claimed (p. 749) that "anti-apostatic selection makes sense as a foraging strategy when foragers are required to select their diet from a nutritionally diverse array of food types, as urban feral cats are." I doubt this logic. The laboratory is not a suitable place for testing real-world foraging hypotheses. Opportunism is difficult to assess in confined spaces, numerical and functional responses impossible. Novel foods can be selected preferentially for several reasons, novelty not excluded (Mugford and Thorne 1980). Evidence is unequivocal that the eating habits of cats depend heavily on prior predatory experience and habituation to specific foods (Kane 1989; also see Chapter 6). Novelty appears to be less important to free-ranging cats than to cats kept confined.

The composition of prey organisms also seems of minor importance. Cats are generally insensitive about whether one food might be healthier than another (Kane 1989), basing their selections in the laboratory on odor and taste (i.e. flavor), and texture (Chapter 7). Although fats add flavor and cats show preferences for certain ones, I found no evidence of their choices being driven by dietary considerations. As Kane (1989: 151) wrote, "fat content per se, independent of flavor components and its effect on consistency, has little effect on dietary preference of the cat." Laboratory cats revealed no significant bias when choosing between diets containing 15% and 45% fat (Kane *et al.* 1987).

As mentioned, cats can kill prey near their own body size (e. g. adult European rabbits), but usually not without a struggle (Jones 1989). Large prey are sometimes too intimidating to kittens (Kuo 1930) and even adult cats (Adamec *et al.* 1980a). Small toys elicit more object play than big ones (Hall and Bradshaw 1998). Big toys were avoided if the cats had recently eaten, but elicited limited interest and contact in cats starved 16 h.

Just about any animal ranging from 5 to 1000 g makes suitable prey for a cat on the prowl (Apps 1986b). Brown pelican (*Pelecanus occidentalis*) chicks, which often attain 3.5–4.0 kg, are among the largest prey of free-ranging cats yet recorded (Anderson *et al.*

1989). More commonly, prey animals are ~1% of a cat's own body size, unusually small for a carnivoran (Pearre and Maass 1998). One investigator (Apps 1986b: 120) summarized the situation by stating that where a range of potential prey organisms is available, "Approximately 80–90 per cent of [the] diet is made up of items smaller than 100 g and 70–80 per cent of it is mammals." In an extensive analysis of the modern decline of Western Australia's vertebrate fauna, Burbidge and McKenzie (1989: 143) reported that "Extinctions and declines are virtually confined to non-flying mammals with mean adult body weights between 35 g and 4200 g." And of Australia generally, consumption of native prey by free-ranging cats comprises, at minimum, 48 species of mammals, 177 of birds, 46 of reptiles, 5 of amphibians, and many invertebrates, most <100 g (Dickman 1996a). Peacock *et al.* (2011: 826), citing Burbidge and McKenzie (1989) and other sources, noted that Australian mammals within the "critical weight range" of 35–5500 g have experienced the highest extinction rates. Feral cats were partly to blame.

At Dassen Island, birds outnumbered rabbits by a factor of 140, yet they composed just 20% of the diet, in part because the available species were large and aggressive, and cats were reluctant to attack them (Apps 1986b). Size is obviously a factor in prey selection. When the cat eradication program on Marion Island was underway and the population began to decline, predation on birds dropped too, and the favored prey became small mammals (Bloomer and Bester 1990). This suggests that prey size had been above optimal (Pearre and Maas 1998).

9.10 The motivation to hunt

How much of a cat's motivation to hunt is instinctive and how much is learned by watching the mother? Kuo's (1930) extensive laboratory experiments offer intriguing insight. Kuo started with 59 newborn kittens and divided them into groups of 30 and 29. He reared one group on a mostly carnivorous diet (beef, pork, milk, fish, and cooked rice), the other on a mainly vegetarian diet (milk, vegetables, beans, and cooked rice). Some cats were not fed 12 h before an experiment, the others given food and tested immediately after eating. Some kittens were raised in complete isolation almost from birth, others kept with a rat-killing mother, and still others placed alone shortly after birth, most of the time in a cage with a rodent, either an albino laboratory rat, a wild rat, or a mouse. The effects of these conditions were then tested. Kittens raised with rodent-killing mothers could see their mothers killing one of the three types of rodent (albino rat, wild rat, or mouse) but not eat it, kill and eat it, or share the kill with a kitten; other kittens were shielded from seeing, hearing, or smelling evidence of predation.

Of kittens raised in isolation ($n = 20$), 45% later killed one or more kinds of rodent before age 4 months compared with 85% raised in the "rat-killing environment." In other words, nearly half killed without having witnessed a kill previously. Although some kittens of the group able to watch their mothers killed more than one kind of rodent, all of them killed the type they had observed their mother kill, and the first predatory act came, on average, 11 d earlier in life than for kittens of the isolated group. Kittens raised only with a rodent for company ($n = 18$) displayed interesting behavior too. As time passed the familiar rodent was removed and replaced sequentially with one of the other kinds. Three of the kittens killed one of these rodents but never of the kind that had been its original cage-mate.

Kittens up to age 4 months that failed to kill a rodent (11 kittens raised in isolation plus 15 raised with rodents; $n = 26$) watched adult cats kill rodents and immediately afterward were placed with live rodents themselves. Nine from the isolation group became rodent-killers, apparently through emulative learning (Chapter 6) and subsequently reinforced by the killing act. Of the kittens raised with rodents, only one from the latter group of 15 became a killer. Thus familiarity and habituation to traditional prey starting at an early age is sufficient to inhibit later predation. However, three individuals from the isolation group had killed and eaten rodents without having observed another cat doing either, and several kittens reared in the rodent-killing environment did too, also without having seen their mothers in the act. Thus rodent-killing behavior can develop with or without observation and reinforcement, although stimuli associated with predation appear to be strong incentives.

Of the 59 kittens, 40 were rodent-killers, but only 17 ate their kills. Of the 40 that killed, half were reared on the carnivorous diet and half on the vegetarian diet. However, of the 20 vegetarian rodent-killers only 4 (20%) were rodent-*eaters*; 13 individuals (65%) killed rats but refused to eat them. A vegetarian diet discourages kittens from eating animals, but not from killing them. Moreover, most vegetarian kittens refused to eat meat when offered it at age 3–4 months.

As mentioned earlier, some kittens were starved prior to these experiments, others fed to satiation. Neither condition affected either rodent-killing or rodent-eating behavior, indicating that both behaviors (or their absence of expression) were not driven by state of hunger.

Prey size affected predation. The albino rats were much larger than the wild rats, which were much larger than the mice. No kitten younger than 3 months took on an albino rat, but mice and wild rats were killed by kittens as young as 40 d. In every instance a kitten that killed an albino rat also killed the smaller rodents. In evaluating his results, Kuo (1930: 19) concluded that kittens: (1) lack an "innate preference" for killing certain kinds of rodents, (2) preferentially kill the same prey as their mothers, (3) tend not to kill the species or type of rodent to which they have become habituated from early age, (4) avoid attacking large prey until becoming larger (or older) themselves, and (5) consistently kill prey smaller than the largest prey they become capable of killing. Thus "(1), (2), and (3) all point to the importance of environmental influence, while (4) and (5) indicate the importance of the age and size of the cat as well as the size of the rat as determining factor."

In his later work Kuo (1938) separated kittens from their mothers shortly after birth and reared them together in groups of 4 or 5, each group in the company of a pair of adult white rats. The kittens allowed the rats to eat with them from the same bowl, climb over them, and snatch food from their mouths. Seven of the 17 kittens tried to initiate play with rats. When the kittens reached 9 months they and their rats were separated then tested after 1 month apart for their reactions to an adult rat. The tests were repeated over 4 months, and the rats used were the original cage mates.

In all instances the cats were indifferent and did not show evidence of a killing response. However, no cat formed an attachment to a rat, as had been the case with isolated kittens (Kuo 1930), and separation did not result in behavior indicating anxiety. Nor did kittens attempt to "defend" their rats, in contrast to single kittens reared with rats. Instead, there was clear evidence of attachment to littermates, demonstrated by interaction (playing, eating, and sleeping together), and a kitten separated from the other kittens became anxious, the presence of rats providing no consolation.

The rat pairs ignored the kittens. They played together, mated and reproduced, and females became aggressive if a kitten ventured too near the nest. At such times the kittens were intimidated. Without any previous predatory experience, and at ages 2–4 months, 12 of the 17 kittens killed and ate newborn rats when the mother was away from her nest. Ordinarily the babies were removed from the nest and eaten in another part of the cage, typical predatory behavior in cats (Section 9.7). Predation on newborns eventually became a pattern, one group of kittens consuming 5 rat litters during their stay of 9 months with adult rats. Mother rats never interfered if witnessing these acts, but returned any uneaten carcasses to the nest. Some rats joined kittens at the feast, helping consume their own young. At no time did the cats' behavior toward the adult rats change.

Because the newborn rats were hairless, Kuo (1930) tested whether the absence of hair was a predatory stimulus. Test rats were placed in two groups, one with normal hair, the other shaved. Their ages were the same: 1 d, 7 d, 14 d, 30 d, 60 d, and 90 d. The kittens were ~5 months old and tested separately. Two rats, one with hair and the other hairless, were presented. Eleven of the 12 kittens that had killed newborn rats killed and ate the shaved rats regardless of their ages, but they reacted to unshaven rats the same as they did to their adult cage mates. Only one killed unshaven rats.

During tests of separation, just one cat killed its adult rat cage mate when reunited after the 4-month separation; the other 16 remained indifferent. In another series of experiments these cats were permitted to watch a known rat-killing cat perform the act of predation. Over time, 6 of the 16 killed rats themselves; others tried to attack rats but retreated after being bitten. These cats would carry dead rats killed by other cats, growl, and play with the carcasses, but not eat them.

If the results of Kuo's two experiments (Kuo 1930, 1938) are considered together the outline of a pattern comes into focus. Single kittens raised in a "rat-killing" environment are likely to become rat-killers themselves; for those reared in isolation the odds are ~0.5. However, single kittens raised with a rat form bonds with their cage mates, displaying anxiety when separated and moving to "defend" their rats if perceiving danger. These kittens never attacked their rats even after having witnessed predation on rats by other cats. Kittens reared in groups with pairs of rats formed attachments to littermates but not to the rats. These cats were more likely to kill rats spontaneously or after observing an experienced rat-killer perform predatory acts.

Results reinforce the idea that predation is a combination of instinctual and learned behavior. The refusal of adult cats reared alone with rats to kill their cage mates is difficult to interpret. Such a situation is purely artificial, designed to test the limits of interspecies interaction, not its normal course of development. The results doubtfully apply in the lives of free-ranging cats but demonstrate that social behavior is an important factor at a certain stage of development, the effects persisting through the age when social behavior is subsumed sequentially in object play and eventual separation from littermates (Chapter 6).

Others have carried out similar experiments. Rogers (1932) also showed the importance of emulative learning in predation by cats. Some kittens were left with their mothers when not being tested until day 42, others until day 76. Hungry kittens were observed individually with a white rat starting at day 1. In the mother's absence a rat seemed to comfort them. Very young kittens put their heads over the backs of the rats, became less restless, and vocalized less. Orientation toward the rats during week 1 had been in response to thermal and haptic stimulation, but it became visual when the kittens could open their eyes at ~day 8, and this was obvious by day 13. On day

10 kittens tapped the rats when the rats nibbled on them. Mutual stimulation eventually became aggressive by both species, especially when the rats were large, but the behavior still seemed harmless (kitten and rat would lie down together afterward). At the end of 8 weeks the kittens had not shown predatory behavior toward rats either large or small. They were then starved for 48–77 h but still showed no inclination to kill companion rats. At age 68 d a kitten caught a mouse placed in its cage. It growled, killed the mouse, and ate it, then lay down with a rat put in its cage. Other kittens oriented toward mice without killing them. They played with dead mice without eating them. Moreover, forced 100-h fasts did not induce predation on either mice or rats, even live mice with blood and beef juice squirted onto their ears and bodies.

Mugford (1977: 34) mentioned having conducted feeding trials similar to Kuo's except that fixed diets were started at weaning instead of at birth. Data were not provided. Results were different in that no cat formed a consistent preference for its rearing diet and as an adult chose instead "a novel alternative." In these cats, which were offered commercial foods, "massed feeding of a single nutritionally balanced diet … induces a transient depression of its relative palatability." The new food is preferred for a time, then interest in it wanes and the original food, if actually more palatable, becomes preferred once again.

Adamec (1980a) conducted experiments based on Kuo's general design but added competition from littermates as a variable. Their sample size was much smaller than Kuo's and the results less reliable. Kittens were reared under conditions that provided variable early experiences. By age 10 months, exposure to enforced hunger and dead prey produced the most aggressive adults, measured as willingness to attack rats. The next most effective combination of experiences inducing rat-killing behavior were watching an adult cat kill a rat (observation) in combination with hunger, exposure to dead prey, and solitary early exposure to live prey. Ranking third was the "normal" composite feline experience of observation, hunger, exposure to dead prey, and group competition. This last factor, in combination with exposure to dead prey but without observation, was no more effective than lack of any of these experiences. The effect of competition varied by individual, inhibiting the predatory behavior of some cats while facilitating it in others.

10 Management

10.1 Introduction

The previous nine chapters provide sufficient background on the behavior, biology, and ecology of free-ranging domestic cats to deepen the effectiveness of control programs. The more wildlife biologists know about the nutrition, prey choice, habitat preferences, sensory biology, space use, and reproduction of cats the more likely management programs can be tailored to specific locations and achieve the desired objectives.

Free-ranging cats affect wildlife directly through predation and indirectly by intimidation and competition with native predators for the same resources, and by serving as reservoirs of parasites and infectious diseases that are potentially transmissible to native species and humans (Castillo and Clarke 2003, Dickman 1996b, Duffy and Capece 2012, Jones 1989, Preisser *et al.* 2005). This last factor is not discussed. Free-ranging cats exert other effects that are less well known. Their consumption of herbivorous prey can even disrupt normal seed dispersal: cats in the Canary Islands eat frugivorous lizards, and seeds retained by the lizards later pass unchanged through the cats' digestive systems (Nogales *et al.* 1996).

Predation is perhaps the most visible impact of free-ranging cats, although intimidation might be more important. At some locations, cats and other exotic predators exert little apparent impact on the native fauna or flora (Amarasekare 1994). As discussed (Section 10.2), their presence nonetheless induces subtle effects like raising the alertness of small animals and altering their behavior (e.g. Stone *et al.* 1994). At worst they cause dangerous population declines in targeted prey species (Ebenhard 1988), even pushing them over extinction's precipice (Blackburn *et al.* 2004, Fitzgerald and Veitch 1985). I treat the terms *exotic, alien,* and *invasive species* as synonyms. A slightly modified definition devised for plants applies as well to animals: nonindigenous forms that spread without human assistance in natural and semi-natural habitats inducing substantial changes in composition, structure, or ecosystem processes (Safford and Jones 1998).

Sometimes exotic predators are considered aids in maintaining ecological stasis, as when insular cats prey on rats that in turn eat the eggs and hatchlings of birds (e.g. Ebenhard 1988, Fitzgerald *et al.* 1991, Tidemann *et al.* 1994), but this putative benefit masks a darker, less optimistic side. Where cats are sympatric with breeding seabirds they often prey on the adults, which has a far more devastating effect on populations than the killing of eggs and chicks (Sections 10.6 and 10.8). In an ideal world these places would be free of both rats and cats, the birds left to their own regulatory devices.

Cats have recently been eliminated from many oceanic and continental islands where they had become apex predators, some of the bigger locations being Ascension Island

---

Free-ranging Cats: Behavior, Ecology, Management, First Edition. Stephen Spotte.
© 2014 John Wiley & Sons, Ltd. Published 2014 by John Wiley & Sons, Ltd.
Companion Website: www.wiley.com/go/spotte/cats

(Ratcliffe *et al.* 2009), Marion Island (Bester *et al.* 2002), and Macquarie Island (Dowding *et al.* 2009). In previous chapters I discussed aspects of the biology and ecology of the cats that had once lived at these locations as if time had stopped and the eradication programs never commenced. Why? Because information in the older literature contributes a valuable historical record of what happens when alien predators and prey are released into otherwise undisturbed habitats without thought of the consequences. In most instances the devastation can never be repaired completely. These earlier investigations have heuristic value.

Free-ranging cats offer challenges to those who attempt to control them. Having spread around the world, they are alien everywhere. Although domesticated, they survive capably when turned outdoors. Once on its own such a cat, although free, remains imprisoned inside its evolutionary heritage. Obligate carnivory and high nitrogen turnover force it to kill relentlessly. In offsetting these liabilities are the cat's protean adaptability, polyestrous reproductive cycle, production of multiple offspring, promiscuous mating system, solitary life-style, elusiveness and mobility, and refined sensory modalities adapted to hunting. The ordinary cat is the hypostasis of predation. Here I review present control methods and offer one I believe is unique. I explain why feral and stray cats should be eradicated, not simply neutered and released, by describing their effects on prey populations. It was stated earlier (Chapter 9) that cats favor rodents, but this hardly makes their predation on the birds inhabiting big islands and continents trivial. As birds generally decline worldwide, cats contribute to the carnage (Beckerman *et al.* 2007, Woods *et al.* 2003). For summaries of methods and relative success in eradicating cats from islands, see Burbidge and Morris (2002), Nogales *et al.* (2004), Parr *et al.* (2000), Wood *et al.* (2002), and especially Campbell *et al.* (2011).

The two obvious choices for control of free-ranging cats are to interrupt their reproductive cycles or reduce the number of individuals. The first indirectly accomplishes the second but requires a generation to show results. Advocates of trap–neuter–release, or TNR (Section 10.3) concede this. Levy and Crawford (2004: 1358) wrote that "the extended survival of feral cats following sterilization indicates that natural attrition would result in a slow rate of population decline." Consequently, TNR efforts are unlikely to reduce predation on wildlife in the short term until these cats have died. Chemical repression of reproduction (Section 10.4) would be only partly effective as a population control measure for the same reason even if a permanent method could be devised. Killing unowned free-ranging cats is the most direct, rapid, and effective way of regulating their numbers, although the process can be expensive, labor-intensive, involves extensive monitoring, and is only completely effective where later recruitment is unlikely (e.g. oceanic islands) or practical to control.

10.2 Effect of free-ranging cats on wildlife

Feral and stray cats are well-known killers of wildlife. In contrast, predation by a well-fed house cat is often subtle and largely unnoticed unless its owner pays attention. Over 18 months a free-ranging pet cat at Rose Lake Wildlife Experiment Station, Michigan, United States, killed 1688 prey animals (Bradt 1949). At another Michigan location free-ranging house cats killed 0.7–1.4 birds/week of >23 species representing 12.5% of species breeding locally (Lepczyk *et al.* 2003). During a British survey conducted over 5 months (April through August 1997), Woods *et al.* (2003) logged 14 370 prey items brought home by 986 cats belonging to 618 households.

The kills comprised mammals (69%), birds (24%), amphibians (4%), reptiles (1%), fishes (<1%), invertebrates (1%), and unidentified items (1%). A subsample of 696 of these cats showed that 634 returned home with at least one prey item, and the mean number per cat was 11.3. Free-ranging house cats in southern Sweden obtained 15–90% of their food from natural prey (Liberg 1984b). Depredation is magnified in urban areas where wildlife is already diminished by habitat destruction, often leaving cat and human populations closer reflections of each other than either is to the density of available prey (Sims *et al.* 2008).

Because hunger and the motivation to hunt are loosely linked (Chapter 7), stomach and scat analyses might not yield accurate data on the species of prey killed or their numbers (Warner 1985). Counting prey animals left at the doorstep is always an underestimate of the actual number killed or mortally wounded because not every captured animal is taken home (Churcher and Lawton 1987, Woods *et al.* 2003). Some are eaten at the kill site, abandoned after being killed, or escape.

The release of cats onto islands, usually in attempts to control rodents, has produced devastating consequences worldwide. In a meta-analysis, Medina *et al.* (2011) concluded that cats are responsible for at least 14% of the global extinctions of insular mammal, bird, and reptile species and the principal threat to nearly 8% of those now critically endangered. Worldwide, cats have been involved directly in the extinction of at least 33 species of birds alone (Dauphiné and Cooper 2009, Nogales *et al.* 2004). The cat's efficiency as a hunter is illustrated by an event in the 1990s when a single cat killed every deer mouse (*Peromyscus guardia*) in a unique subpopulation at Estanque Island, México, causing its extinction (Vázquez-Domínguez *et al.* 2004).

Klopfer (1973: 4) wrote, "The race between the adaptations of the predator for capturing the prey and those of the prey for escaping a predator may be viewed as a race whose finish line is constantly moved ahead of the contestants." This is true only if evolution is allowed to proceed uninterrupted, but human meddling usually rigs the race to favor one or the other. Releasing exotic animals onto remote islands loosens the constraints of predation, parasitism and infectious diseases, and competition (Royle 2004), shifting evolution in other directions, not uncommonly at the expense of the indigenous fauna. Birds that breed on oceanic islands often lack the life-history, behavioral, or morphological traits to evade or defend themselves and their young against alien terrestrial predators (Atkinson 2001, Cooper *et al.* 1985, Fritts and Rodda 1998, Moors and Atkinson 1984, Peck *et al.* 2008). Insular seabirds tend to be numerous, clumped, noisy, helpless and clumsy on land or even flightless, easily detected, and nest in burrows or on open ground (Faulquier *et al.* 2009), all attributes heightening their vulnerability to introduced cats, rats, and mustelids. With reference to cats Moors and Atkinson (1984: 671) wrote: "Probably no other alien predator has had such a universally damaging effect on seabirds."

At mainland regions rodents typically make up the bulk of animals killed by feral cats, but on islands where seabirds gather they often become the predominate prey (van Aarde 1980). Islands are especially vulnerable to loss of diversity caused by invasion of alien species. Of birds that have become extinct since the start of the seventeenth century, 176 species and subspecies (93%) inhabited oceanic islands (reference in Say *et al.* 2002a). Most extinctions of insular animal species have been caused by introduced predators, 26% a result of cat predation (King 1985). Cats contributed substantially to the disappearance of 45 species of birds in Hawai'i and continue to exert a major impact on those remaining (e.g. Duffy and Capece 2012, Hu *et al.* 2001, Kowalsky *et al.* 2002, Snetsinger *et al.* 1994; also Table II of the online Appendix). Feral cats at Stewart Island/Rakiura were responsible for near extinction of the kakapo (*Strigops habroptilus*) prior to removal of the remaining 61 birds to cat-free islands, and the weka (*Gallirallus australiis scotti*), and cats there extirpated the brown teal (*Anas auklandica chlorotis*), an endemic duck (Harper 2004 and references).

Despite their grim climates, subantarctic islands have proved particularly vulnerable to cats. Marion Island, located at 46°54′S, 37°45′E in the southern Indian Ocean, is typical. Its interior is buried permanently under fields of snow and ice. Annual mean temperature is 5°C, and strong westerly winds are common. Gales occur on 10 d/y, gusting to 160 km/h. Humidity is high, and mean annual precipitation as rain, sleet, and snow is 2576 mm (Bester *et al.* 2002).

In 1949 staff members at the South African Meteorological Station imported 5 intact house cats, presumably to control mice, and allowed them to range freely and reproduce (van Aarde 1979). Prey was abundant during most of the year, and despite harsh conditions and the often high incidence of early postnatal and juvenile mortality the population rose 23.3% annually when averaged over the ensuing 26 y (van Aarde 1978). A feral cat was reported 14 km from the station in 1951, and by 1974 progeny of the initial founders occupied the entire periphery of the 290-km^2 island. According to van Aarde (1980) and Watkins and Cooper (1986), by 1965 the cats had exterminated the common diving petrel (*Pelecanoides urinatrix*). With adequate food and shelter and no predators the cat's capacity to pullulate is nearly boundless. Ten years later the population reached ~2139, and was increasing at 26% annually (van Aarde 1979), each cat now responsible for ~213 burrowing petrel deaths every year (van Aarde 1980). In 1979 the ~3405 cats were killing ~450 000 burrowing petrels and threatening their survival (Bester *et al.* 2000; van Aarde 1978, 1979). The estimated annual carnage was 455 119 seabirds comprising 7 species.

Macquarie Island was discovered by sealers in 1810 and within 10 y feral cats had become established (Brothers *et al.* 1985). Their presence was largely responsible for extinction of the endemic parakeet (*Cyanoramphus novaezelandiae erythrotis*) and banded rail (*Rallus phillippenis*) prior to 1900 (Taylor 1979, 1985) and more recently the drastic decline of burrowing petrels (Brothers 1984). Based on crude assessments of energy needs the 375 or so cats thought to occupy Macquarie Island at one time annually killed ~46 000 Antarctic petrels (*Pachyptila desolata*), 11 000 white-headed petrels (*Pterodroma lessonii*), 57 000 rabbits (van Aarde 1980), and probably exterminated the local gray petrel (*Procellaria cinerea*) population (Jones 1977). Cats at both Macquarie and Marion islands have since been eradicated.

The daily energy requirement of a nonpregnant, free-ranging adult cat living in a cool climate can be ~2131 kJ/d (Matias and Catry 2008). For a gestating female this rises to 2508 kJ/d, and for a lactating female to 6271 kJ/d (Matias and Catry 2008). New Island in the Falkland Islands lies at 51°43′18″S, 61°18′01″W. Average monthly temperature in the Falklands ranges from 2°C in winter to 9°C in summer. The feral cats prey on several species of mammals and birds. Petrels are present only during 5 months when they nest, but composed ~20% of the diet during austral summers of 2004/2005 and 2005/2006. Although introduced mammals (e.g. rabbits) were the principal prey, consider the basis for the mortality inflicted by cats on indigenous thin-billed petrels (*Pachyptila belcheri*). Assume one bird provides ~916 kJ. Assume also a cat population of 30–160 individuals, one-third of them pregnant and then lactating during 2 of the 5 months petrels are in residence. By simple extrapolation at least

4753 and perhaps 25 332 petrels were killed in 2 summers monitored by Matias and Catry (2008). The Kerguelens offer still another example. The feral cat population on 6600-km² Grand Terre, the largest island of the group, exploded after the 1956 release of just 2 cats at Port-aux-Français (references in Pontier et al. 2005). Feral cats occupied 20% of Grand Terre by 1974, their estimated number of 2500 up from 1750 just 3 y earlier, a mean annual growth rate of 33%. According to Pascal (1980), by 1977 the approximately 3500 feral cats occupying the islands were killing ~1.2 million seabirds annually.

Continental and oceanic islands of the lower latitudes have not been spared either. At Isla Isabela, México, feral cats killed 23–33% of nesting sooty terns, causing the population to drop precipitously from 150 000 to ~1000 over 13 y (Rodríguez et al. 2006). In 1995 a single cat killed 1-3 chicks per nest in 9 of 61 monitored nests of Heermann's gulls (*Larus heermanni*).

Using Nagy's equation (Nagy 1987), Keitt et al. (2002) estimated that a 2.7-kg feral cat on Natividad Island, México, needed to eat 328 g/d (range 190–452 g/d). If 90% of a cat's intake was black-vented shearwaters (*Puffinus opisthomelas*) of which half of each bird was consumed (~200 g, excluding feathers and large bones), the mortality would be 1.5 birds/d, or 45/month. The number of cats preying on the shearwater colony has an obvious effect on its population growth (Fig. 10.1). As the figure illustrates, population growth ($\lambda = 1.006$ in the absence of cats) declines with increasing predation. Assuming a shearwater population of 150 000, just 20 cats cause λ to fall ~5% in the first year, which is below the simulated rate of recruitment needed to maintain stability. Thereafter, if predation stays constant and assuming no outside recruitment of birds, the cats are consuming an ever larger proportion of a diminishing number. The calculated persistence (i.e. survival time) is 21 y before local extinction, an estimate that changes little even if immature birds are killed at twice the rate of adults.

According to Faulquier et al. (2009), a mere 10 feral cats could kill 900 Barau's petrels annually at Réunion Island in the southern Indian Ocean. They reasoned that during a breeding season lasting 210 d the mean number of birds killed by just one cat would average ~22 fledglings and 70 adults. By 1974 the cats at Little Barrier Island, New Zealand, were exerting a heavy toll on Cook's petrels, killing an estimated 7000–10 000 annually during the nesting season, and in one area of the island predation on

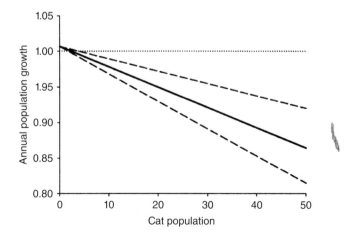

Fig. 10.1 Annual population growth, λ (±95% confidence intervals) for a population of 150 000 black-vented shearwaters (*Puffinus opisthomelas*) depending on the number of cats preying on the colony. Horizontal dotted line represents a stable population. Source: Keitt et al. 2002, figure 1, p. 220. Reproduced with permission of John Wiley & Sons.

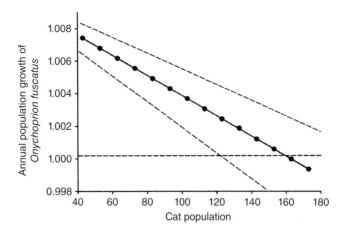

Fig. 10.2 Matrix population model showing population growth (±95% confidence interval) for a population of 2 million female sooty terns (*Onychoprion fuscatus*) plotted against the number of feral cats at Juan de Nova Island, Mozambique Channel. Horizontal dashed line represents a stable population of terns (annual population growth of 1.01 in the absence of cat predation). Source: Peck et al. 2008, figure 5, p. 70. Reproduced with permission of John Wiley & Sons.

adult and juvenile black petrels (*Procellaria parkinsoni*) was so fierce that not a single chick fledged over austral summer 1974–1975 (Girardet et al. 2001).

The estimated population of 10 feral cats at Juan de Nova Island in the Mozambique Channel killed a projected 416 sooty terns weekly, or 6 birds/d/cat during the nesting season, 22% of which were "surplus kills" and not eaten (Peck et al. 2008). At its previous maximum number (53 cats), 2205 terns were being killed weekly, equal to 0.1% of the breeding population in a season. The number of cats had a detectable influence on the modeled population growth (Fig. 10.2). Removing a single cat from each nesting colony lowered the short-term predation 10-fold. Even without cat predation the projected annual population growth ($\lambda = 1.01$) was nearly flat, barely keeping ahead of the annual mortality, but a situation typical of seabirds in natural conditions (Keitt et al. 2002, Russell 1999). A sensitivity analysis of the deterministic model illustrated the effect of changing reproductive success and adult survival on λ. Adult survival affected λ more than reproductive success (Fig. 10.3), as confirmed by elasticity values (adult survival = 0.53, reproductive success = 0.06).

Apps (1981, cited in Cooper et al. 1985) considered that in 1979 and 1980 the diet of each cat at Dassen Island, South Africa, comprised 133 rabbits, 36 jackass penguins, 57 birds of other species, and 24 house mice. Of birds, 33 penguins and 44 others were scavenged carcasses. By March 1982 there were >120 cats killing ~2000 birds annually. One particular individual was believed to have killed 51 terns in 3 months during 1971 and 1972 (Cooper 1977). The main predators of ground-nesting birds at the Upper Waitaki Basin, South Island, New Zealand, are feral cats, hedgehogs (*Eriaceus europaeus*), and ferrets (*Mustela furo*), all exotic, but only cats preyed indiscriminately on adults, chicks, and eggs (Sanders and Maloney 2002).

Predation exerts a direct impact on prey populations, but indirect effects also accrue, and in some respects are even more detrimental. Feeding stations set up for stray cats by animal welfare advocates apparently alter the distributions of birds and small mammals in the surrounding area. Such a station established east of San Francisco, California, reduced the numbers of western harvest mice (*Reithrodontomys megalotis*) and deer mice (*Peromyscus* sp.), as shown by Hawkins et al. (1999). Birds present during the breeding season, like California quails (*Callipepla californicus*) and California thrashers (*Toxostoma redivivum*), became less common, and prevalence of the house mouse rose. Cats attracted to the food thus changed the natural species compositions of resident

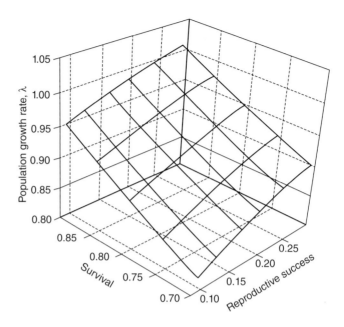

Fig. 10.3 Sensitivity analysis of the deterministic matrix model for sooty terns (*Onychoprion fuscatus*) illustrating the effect of changing reproductive success and adult survival on growth rate of the population at Juan de Nova Island, Mozambique Channel. Note the comparatively greater effect of adult survival. Source: Peck *et al*. 2008, figure 6, p. 71. Reproduced with permission of John Wiley & Sons.

rodents and birds in the vicinity and encouraged house mice to expand their ranges. In another study, Sanders and Maloney (2002) demonstrated that just the occasional *presence* of cats causes birds to abandon their nests.

Fear of cats increases wariness in Galápagos Islands lizards (*Tropidurus* spp.), as shown by Stone *et al*. (1994), and depresses prey populations of other animals by altering their behavior, including foraging patterns and use of space (Beckerman *et al*. 2007, Lima 1998). The effects can exceed those of actual predation (Preisser *et al*. 2005). The mere increased risk of predation can reduce clutch size in birds (Lima 1987). Because reproduction always includes a cost of heightened mortality, risk of predation then limits future reproductive success in regional populations. These predator-induced costs have been labeled *trait-mediated interactions* (*TMIs*) to distinguish them from *density-mediated interactions* (*DMIs*) in which predators lower prey population densities through direct consumption (Preisser *et al*. 2005). The costs of TMIs are exacted in many ways, such as reduced energy intake, excessive energy invested in defensive structures and modified "strategies," compromised reproductive success, greater vulnerability to other predators, or emigration. In a meta-analysis comparing values of DMIs and TMIs from the literature, Preisser *et al*. (2005: 501) found that, "On average, the impact of intimidation on prey demographics was at least as strong as direct consumption (63% and 51% the size of the total predator effect, respectively)." The consequences? "Predators can thus strongly influence resource density even if they consume few prey items." This indicates that "the costs of intimidation, traditionally ignored in predator–prey ecology, may actually be the dominant facet of trophic interactions." These are astonishing conclusions: *fear of predation is potentially more harmful at the population level than competition and perhaps predation itself.*

Even if direct mortality caused by cats were low (<1% of a prey's local population), a tiny reduction in fecundity induced by their presence (e.g. <1 prey offspring/y/cat) can reduce the abundance of a sensitive species of songbird as much as 95% (Beckerman *et al*. 2007). This occurs when cat densities are extraordinarily high. Cats throughout the United Kingdom, at an estimated 500/km^2 (Table 2.1), outnumber many songbirds on the basis of landscape area, sometimes by ratios of 35 or more (Beckerman et al. 2007). However, a sublethal effect can exist when only a few cats are around. At Baker Island, Phoenix Islands, central Pacific Ocean, the presence of just 2 feral cats forced most seabirds to abandon their historical nesting sites and emigrate to Howland Island (King 1973).

The cost of emigration, especially in *k*-selected species like seabirds, is often high. Besides removing individuals from the breeding population, it subjects those displaced to added stresses of adapting to a different habitat, maybe in a location where most space has already been allocated to residents and their immediate descendants. The effect becomes magnified as the trophic chain lengthens. In the meta-analysis of Preisser and colleagues (2005), TMI accounted for 58% of predator effects on the prey's resources in two-level food chains, but 85% in chains of three levels, at which point the trophic cascade turns into an avalanche. Therefore, TMIs dominate trophic cascades.

As discussed, intimidation affects prey populations by reducing either their densities (e.g. through emigration) or activities. Restricting foraging space to safe locations is an example of the latter (Spotte 2007: 75–85). Each year free-ranging cats kill birds and small mammals by the hundreds of millions. Summing their carcasses is one way of assessing the carnage, but the consequences of intimidation, although seldom apparent outwardly, would seem just as damaging.

The apparently fewer extinctions on large islands might be illusory. Extensive land masses can support larger populations of native species, making them less prone to extinction, and they offer more places of refuge from extinction-promoting forces (Blackburn et al. 2004). Furthermore, the "filter effect" allows the more resistant forms to survive, giving a false picture of fewer losses overall. Humans introduced exotic mammals onto the world's islands, either purposely or inadvertently. Blackburn et al. (2004) confirmed the probability that a species of bird extirpated from each of 220 oceanic islands correlates directly with the number of exotic predatory species of mammals established after European colonization. However, the proportions of species threatened today are independent of the numbers of predatory species introduced, indicating that the most susceptible birds have already vanished.

Stomach content analyses of 316 feral and stray cats in arid South Australia revealed identifiable remnants of 62 vertebrate species (8 mammals, 16 birds, 37 reptiles, and 1 amphibian) plus 11 invertebrate taxa comprising insects, spiders, centipedes, and scorpions (Read and Bowen 2001). Among the reptiles consumed were three species of venomous snakes (*Simoselaps bertholdi*, *S. fasciolatus*, and *Suta suta*).

Such predation often goes unnoticed at continental locations, partly because the prey species are thought to be vermin or possess little commercial value. At the 45 km^2 Revinge area of southern Sweden, Liberg (1984b) estimated that feral and house cats combined fluctuated from 84 to 121 over 2 y in the mid-1970s (all seasons combined). Within this time the prey killed comprised approximately 171 000 short-tailed voles (*Microtus agrestis*), or 18% of the total number, 20 000 wood mice (*Apodemus sylvaticus*), or 24.4% of the annual production, and 90 000 European rabbits representing 3.6% of the population.

Feral cats are believed to be responsible for the extinction of many of Australia's smaller mammals (<200 g) through direct predation (Dickman 1996a), especially those occupying open habitats (Dickman 1996b). Predation intensifies in arid regions during droughts when prey becomes concentrated near sources of water. Habitat fragmentation from clearing land is another contributing factor because it restricts prey animals to remnants of undisturbed landscapes, concentrating their numbers and increasing vulnerability. As discussed elsewhere (Sections 10.3 and 10.6), the availability of exotic alternative prey (e.g. rabbits) actually exerts a negative effect on native species by supporting larger populations of predators (Dickman 1996b, Didham et al. 2007, Smucker et al. 2000). The same factors modulating predatory opportunities for cats apply to red foxes and dingoes (Canis lupus), two other exotic carnivorans widespread in Australia, but the native animals they kill are generally >200 g (Dickman 1996b).

Eradication of cats from oceanic islands can have a demonstrably positive effect on seabird populations (e.g. Cooper et al. 1995, Keitt and Tershy 2003, Rodríguez et al. 2006), but this is not always the case with other animals, nor is it universally true of seabirds or terrestrial birds (Girardet et al. 2001). Results are usually clearer in simplified ecosystems. At Natividad Island, México, the mortality of black-vented shearwaters declined >90% after eradication of the 25 feral cats, each of which had been killing 36.7 birds/month (Keitt and Tershy 2003). In the first few years after eradication of cats at Marion Island the populations of great-winged petrels (*Pterodroma macroptera*) and blue petrels (*Halobaena caerulea*) rebounded quickly, but the growth of white-chinned petrel colonies stayed flat (Cooper et al. 1995). Results at other places have been equivocal too with positive differences seen in the numbers of some previous prey species but not others (Girardet et al. 2001). Seabirds, however, present a special management case as extreme k-selected species (Bonnaud et al. 2010, Russell 1999). Their long generation times, delayed maturity (e.g. Croxall and Rothery 1991: 273, Table 13.1), and low rates of recruitment mean that rapid recovery is not always possible, some populations taking years to show positive results (Cooper et al. 1995, Jones 2002). Some species do not even breed annually. The gray-headed albatross (*Diomedea chrysostoma*) at Bird Island, South Georgia, is mostly – but not consistently – a biennial breeder, <1% reproducing 2 y in succession (Prince et al. 1994). Most adults (68%) return to breed 2 y later, 11% the third year, and 5% not until the fourth year. Moreover, >50% of adults that failed to rear a chick reproduced the following year, but 23% postponed reproduction still another year.

Sooty shearwaters (*Puffinus griseus*) do not reappear at breeding colonies until 3–5 y after fledging, reproducing for the first time at age 5–7 y (Richdale 1963). Any advantage of delayed reproduction must be weighed against cohort survivorship over time, and this is difficult to measure. To monitor adult survival requires large samples before significant change can be detected. As pointed out by Croxall and Rothery (1991), survival from year to year of 1000 banded wandering albatrosses (*Diomedea exulans*) retains a standard error of 0.8% assuming 100% recapture. However, the birds are so long-lived that a 1–2% annual mortality can be statistically significant. Nonetheless, the results of cat eradication are immediate using other measures. At Isla Isabela, México, where 23–33% of the sooty tern population was typically found dead at the end of the nesting season, this fell to 5% and eventually to <2%, and the proportion of reproductive individuals dropped significantly (Rodríguez et al. 2006). In all, tern deaths attributable to cats declined permanently by >80% despite an apparent increase in the rat population.

In addition, accurate assessments of survival for one year must wait another 2–3 y in species that breed irregularly. Mature sooty shearwaters might not even breed every

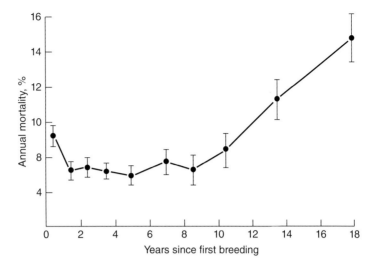

Fig. 10.4 Mean (±SEM) percentage annual mortality of breeding short-tailed shearwaters (*Puffinus tenuirostris*) banded at Fisher Island, Australia, in relation to the time since they first reproduced. Source: Bradley *et al.* 1991, figure 4, p. 57. Reproduced with permission of John Wiley & Sons.

year, sometimes returning to the colony or sometimes not during the hiatus. Birds that have lost mates do not reproduce until finding replacements, which can extend into the next season (Croxall and Rothery 1991, Oro *et al.* 2004). The more successful breeders of some seabirds (e.g. short-tailed shearwaters, *Puffinus tenuirostris*) appear to live longer, and continued breeding with the same mate substantially improves reproductive success (Wooller *et al.* 1992). Add to these factors the negative effects of an annual mortality that increases relative to the time of first breeding (Fig. 10.4) and the necessity of protecting the adults in breeding populations becomes obvious.

Still other factors work against seabird populations. In sooty shearwaters courtship and preparation of a burrow follow arrival at the colony. The single egg laid each season is incubated ~53 d by both parents. The chick, which is also fed by both parents, does not fledge until 3 months after hatching, a prolonged period of vulnerability in colonies stalked by predators. Not surprisingly, studies of seabirds continue for decades (Bradley *et al.* 1991) as older investigators retire and transfer their duties to younger colleagues. This has the added advantage of placing rare events and effects of short-term environmental perturbations in perspective, but also leads to unusual distractions like monitoring the rate at which stainless steel leg bands weather to project whether they can outlast the birds wearing them (Bradley *et al.* 1991). The difficulty is knowing whether a population's slow recovery originates from the expected life-history factors or if the causes are confounded by overfishing (e.g. Croxall and Rothery 1991: 281, Fig. 13.3, Wooller *et al.* 1992), failure of fishing gear to exclude seabirds from the incidental catch and climate change (Wooller *et al.* 1992), epizootics, and local weather events.

Uncertain results in the population growth of secondary avian prey have not been limited to seabirds. Similar results were seen in passerine populations over 15 y at Little Barrier Island spanning 1976–1980 when cats were being eradicated (Girardet *et al.* 2001). Transect data showed that the numbers of some species rose, others stayed flat, and still others actually decreased. As noted by Girardet *et al.* (2001: 25), "Counts give only an index of abundance (reliant on conspicuousness), not a census." As an example they noted that "bellbirds may call more often than stitchbirds, and this will make them more conspicuous even if they are not more abundant." Extrapolation of census data can only produce estimates that hint at cause. Girardet *et al.* (2001: 26) believed that eliminating cats at Little Barrier Island would assist populations of ground-nesting seabirds. However, failure to detect differences in the numbers of forest birds indicated either that cats had exerted little effect on them or that rats were responsible for depressing populations after the cats were gone. Clearly, when results are equivocal any thoughts of cause are reduced to guessing, especially in multi-predator systems.

10.3 Trap–neuter–release (TNR)

The welfare of feral and stray cats is incompatible with wildlife conservation for the obvious reason that preserving and extending the lives of cats sustains their presence, offering more time to kill wildlife. *Trap–neuter–release* (*TNR*) describes programs in which animal welfare proponents and veterinarians cooperate to capture and neuter stray cats and then release them, typically at locations near the capture sites. In addition to being neutered, cats in TNR programs are usually fed and watered regularly by volunteer caretakers, vaccinated, treated for parasites and infectious diseases, and given veterinary care if they later become sick or injured (e.g. Andersen *et al.* 2004, Hughes and Slater 2002, Hughes *et al.* 2002, Longcore *et al.* 2009, Nutter *et al.* 2004, Zaunbrecher and Smith 1993). Some TNR advocates have even built "houses" in which aggregations of strays can shelter (Hammond 1981), and soliciting donations to buy cat food is common practice (Fig. 10.5). Programs of TNR followed by subsequent "managed care" (e.g. Slater 2001) are usually successful if success is defined as extension of health and longevity (Hughes *et al.* 2002, Levy and Crawford 2004), and here a problem arises. Habitat protection alone is often inadequate to assist wildlife in the presence of exotic predators, at which point intervention and management must be

Fig. 10.5 Sign posted in an alley soliciting donations to buy food for stray cats in Venice, Italy. Source: Stephen Spotte.

considered (Safford and Jones 1998). As I show here, TNR and wildlife conservation are motivated by conflicting goals.

The stated objective of TNR (Levy et al. 2003: 42) is not necessarily to reduce the number of free-ranging cats, as sometimes claimed; rather, it is "to halt reproduction without causing harm to the cats." Limiting predation on wildlife is not a consideration because reducing the breeding potential of an aggregation of cats does not affect the incidence of predation by those still alive. Predation over time is actually extended by lengthening average life-spans of the cats through provisioning and veterinary care. Assuming the stated objective is achieved, benefits to wildlife could not accrue until the last neutered cat has died and predation then ends, a process that can take years. In one aggregation of English strays, 70% were still alive after 5 y. In a monitored TNR group in Florida, 83% of the cats had been in residence >6 y, about as long as pet cats survive (Levy et al. 2003). Most had been neutered as adults and were of unknown age. All individuals of an aggregation neutered in the late 1970s at Fitzroy Square, Regents Park, London, had died by 1990 (Remfry 1996), but total mortality like this is rare. Aggregations usually inflate instead of shrinking, ordinarily from immigration of both neutered and intact cats. Availability of food – the factor making certain locations in nature attractive to feral cats – has the same effect on strays, and vacancies at feeding sites are seldom left empty for long (Remfry 1996). In addition, neutering appears to increase the stability of cat aggregations (Rees 1981), which means less chance of a population reduction at TNR sites through emigration.

Advocates of TNR mislead by presenting their programs as eradication's suitable alternative, which is true only when the purpose is prolonging the lives of stray cats without regard to how wildlife is affected. Advocates also present TNR as if most citizens approve when the truth is that approval is given or withheld depending on the audience. Wallace and Levy (2006: 279), for example, described TNR as "an increasingly popular alternative to mass euthanasia." Their statement embodies two logical fallacies. The first is appeal to the popular even though popularity in this case refers to those who are already believers. The second begs the question while incorporating circular reasoning, its sometime corollary; that is, our statement is true because we say so.

Advocates who claim to make their case based on experimental findings fare little better. Robertson (2008: 366) wrote, "Data support the success of TNR in reducing cat populations. ... " Most studies have shown merely stasis or actual increases in the number of strays. Because few of these efforts meet the necessary zero-tolerance requirements on immigration and reproduction, blanket statements like Robertson's are false.

Proponents of TNR claim that free-ranging cats (1) harm wildlife only on islands, not continents; (2) fill a natural "niche," making them wildlife too and therefore deserving of protection, not persecution; (3) are not contributors to the decline of native animals; (4) are minor vectors or reservoirs of parasites and infectious diseases; and (5) when managed properly will eventually die without reproducing, their aggregations failing to grow because the resident cats are "territorial" and repel invasion by outsiders. All are false, easily and quickly refuted by the scientific literature, and impossible for any informed person to take seriously (Longcore et al. 2009). They demonstrate what happens when narrow concerns about animal welfare – a single exotic species, in this case – divagate from the much broader and serious environmental issues. The first claim, for example, ignores habitat fragmentation during urban and suburban

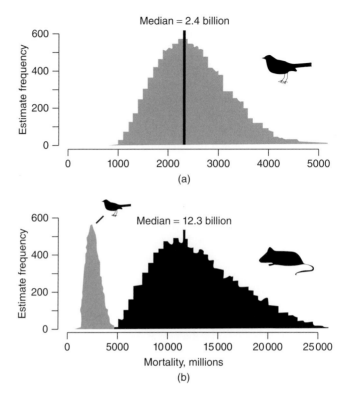

Fig. 10.6 Probability distributions of free-ranging cat predation throughout mainland areas of the contiguous United States: (a) bird mortality; (b) mammal mortality. Source: Loss *et al.* 2013. Reproduced with permission of Nature Publishing Group.

development, clearing of land for farming and ranching, and expansion of industries, all of which isolate remaining natural areas into "islands," which has the effect of increasing free-ranging cat populations while simultaneously diminishing spaces for wild animals, making them more vulnerable to predation.

Maintaining aggregations of stray cats – neutering, feeding, and treating their maladies – offers not a single redeeming feature to wildlife conservation, but inflicts undeniable harm on the environment. Outdoor cats kill indigenous animals. Nature receives no benefit from their presence, but suffers greatly for it. In the United States alone, free-ranging cats kill billions of animals, many of which might otherwise survive and reproduce. Such numbers are staggering (Fig. 10.6). Loss et al. (2013) estimated that cats in the contiguous United States annually destroy 1.4–3.7 billion birds, 69% of the mortality attributable to cats without owners. Of the 6.7–20.7 billion mammals killed by free-ranging cats, 89% succumb to feral and stray cats.

Improbable though it seems, TNR advocates believe that the impact of free-ranging cats remains to be assessed. Levy and Crawford (2004: 1354) stated: "Debate about the true impact of free-roaming cats on the environment, on feline welfare, and as a reservoir of feline and zoonotic diseases is ongoing, often emotional, and fueled largely by a lack of sound scientific data on which to base credible conclusions." This can only be read as a wistful entelechy twisted back on itself. Any hint that objectivity rises above all else, in which instance TNR advocates might gladly become abjurers, seems unlikely. At what point are the data declared "sound," and who decides? Ignoring the possibility that their own assessments might be incorrect, these authors (p. 1357) sought to establish even more TNR programs and described one "in which outside agencies were solicited to accept cats for release in appropriate environments in exchange for stipends of up to $25,000 for 50 cats." A similar disconnect with environmental reality has led TNR's proponents to deny that eradication efforts can actually control numbers of free-ranging cats. Slater (2004: 1351) wrote: "To the best of my knowledge, no location has ever achieved long-term control of free-roaming cats by use of this method." The facts tell a different story: 87 campaigns of eradication starting in the 1970s have successfully exterminated free-ranging cats at 83 locations (Campbell *et al.* 2011), and the positive results have ramped up efforts on a worldwide scale.

Neither provisioning nor neutering lowers the scavenging and predatory tendencies of free-ranging cats and might produce the opposite effect. Veterinary care is an additional factor helping to sustain unowned cats at artificially high numbers and insulating them from the normal numerical or functional responses to fluctuating prey densities (Chapter 9) endured by native predators (Coleman and Temple 1993). Ample evidence exists that predation and satiation in the cat are decoupled and function more or less independently, and even well-fed house cats continue to scavenge and hunt (e.g. Brickner-Braun *et al.* 2007, Warner 1985). House cats in Israel, although provisioned by their owners, still foraged on garbage and killed wildlife (Brickner-Braun *et al.* 2007). Prey included 12 species of mammals, 26 of birds, 18 kinds of reptiles, and 1 amphibian. In terms of stomach volume, mammals dominated (75%), followed by amphibians (10%), birds (9%), and reptiles (6%).

Well-fed strays do not relinquish hunting either (Davis 1957, Hutchings 2003, Natoli *et al.* 2001). Because satiated cats in TNR programs have no incentive to relocate, expand their home ranges, or prey-switch they continue hunting a favored prey organism even as its numbers diminish locally. This is similar to hyperpredation (see later) and confirms that any human activity or human-induced situation encouraging stray cats to maintain large populations actually increases their predation on native wildlife. According to Meckstroth *et al.* (2007: 2391), "Supplemental foods help support higher feral cat densities, do not replace wild prey, and are counter-productive to managing wildlife." Their investigation showed how free-ranging cats in the vicinity of South San Francisco Bay, California, obtained most of their nutrition from commercial cat food set out regularly by citizen volunteers, although small rodents nonetheless supplied 28% of the daily energy requirements. The recommendation was that feeding strays be prohibited everywhere.

Neutering alone does not quell the urge to hunt (e.g. Churcher and Lawton 1987, Flux 2007) and in some cases might intensify it. Robertson (1998) reported that neutered house cats were more likely than intact cats to kill wildlife. A castrated male house cat at Metula, northern Israel, brought home 48 items of prey over 12 months that included 4 species of mammals, 5 of birds, 8 of reptiles, and various insects (Bricker-Braun *et al.* 2007). Neutering sometimes limits a stray cat's movements (Rees 1981), but not always. It had no influence over movement patterns, home-range size, or extent of overlap in home ranges of cats on Santa Catalina Island, California, during their intermittent shifts between the stray and feral states (Guttilla and Stapp 2010). That every cat returned to its original capture site stays there is unlikely (Guttila and Stapp 2010). For those that do stay the hunting impact of the aggregation becomes concentrated (Schmidt *et al.* 2007), probably a moot point in the end: a cat that emigrates merely shifts its predatory activities to another location.

The probability is high that intact cats attracted by food will immigrate into TNR-managed aggregations and reproduce. This has two effects on the aggregation:

(1) increase its number, and (2) sustain it through another generation. Although the cat is a carnivoran it has one thing in common with rabbits and rats, two other prolific breeders that cause management problems: its pseudopregnant luteal phase (Chapter 4) is only about half as long as the duration of gestation. This offers an advantage over dogs and other carnivoran competitors because the briefer diestrus allows quick recycling back into proestrus and estrus (Paape *et al.* 1975). In addition, folliculogenesis can begin 7–10 d after pseudopregnancy. Any potential diminution of reproductive capacity in a population of free-ranging cats caused by teratospermia might be compensated by the females, for which 4–5 seasonal pseudopregnancies are possible (Paape *et al.* 1975). In contrast, a feral or stray dog that misses producing a litter during the breeding season waits through ~5 months of anestrus after corpora luteal regression before proestrus commences once again (Spotte 2012: 150–173). Seldom mentioned is that providing food and veterinary care to unsterilized cats helps prime them to reproduce all year instead of seasonally (Chapter 4). The result is to make intact cats in "managed care" aggregations even more prolific.

Wildlife biologists and TNR advocates obviously view the control of strays from opposite poles. Leaving healthy feral or stray cats in the habitat while simultaneously trying to mitigate the damage they cause is unlikely to achieve both objectives. I scoured the literature seeking at least one synthesis, a single, carefully documented anchoring point from which I might pivot in a circle and see the logic of both points of view. What I found instead was a one-way transfer of red herrings across the divide as TNR advocates purposely muddled the issues with unsupported claims intended to distract a largely neutral public. For their part, biologists have continued documenting the carnage caused by free-ranging cats, and most seem to favor eradication over TNR. Like pressure waves from a falling tree, their data propagate outward in steady, predictable pulses, while those in denial inhabit a world of self-inflicted deafness.

Arguing that stray cats are, or are not, the moral equal of the animals they kill, if not a red herring is at least one tinted a deep pink. The common rebuttal that cats, although exotic, occupy a defined ecological niche is another (e.g. Tantillo 2006). In the parts of Eurasia and Africa where free-ranging cats and wildcats are sympatric they compete for food and hybridize (Beaumont *et al.* 2001, Biró *et al.* 2005, Driscoll *et al.* 2011, Germain *et al.* 2008, Hubbard *et al.* 1992, Pierpaoli *et al.* 2003, Randi *et al.* 2001). Neither is necessarily evidence of either niche-sharing or niche overlap.

Hughes *et al.* (2002: 286) confused the meaning of ecological niche with territoriality, writing that TNR is "a more effective means than euthanasia of controlling feral cat populations because cats continue to occupy the environmental niche [*sic*], thus making it less likely that new cats will immigrate to the colony [*sic*]. ... " The domestic cat is considered a domestic species, *Felis catus,* distinct from its wild ancestors and not a subspecies of European or African wildcats, making it alien wherever it occurs (Dauphiné and Cooper 2009). This being the case, feral and stray domestic cats on large land masses wedge themselves into ecosystems, dislodging indigenous predators of similar size. Meanwhile, freeloading at locations made attractive and convenient by humans insulates them against disease, predation, and the stress of surviving as actual wild creatures.

Advocates of TNR would have us believe cats are a legitimate form of wildlife and therefore deserving of similar protection (references in Longcore *et al.* 2009). Gorman and Levy (2004) contended that the long-term presence of free-ranging cats lessens their ecological impact. This is certainly false considering they can outnumber native predators of their own size, often by orders of magnitude (Coleman and Temple 1996, Dauphiné and Cooper 2009, Woods et al. 2003). For example, the cat is the dominant predator in the field–forest mosaics of east-central Poland (Goszczyński et al. 2009), and in North America the absence of coyotes (e.g. Quinn 1997) or similar predators higher in the trophic chain results in predation by cats and cat-sized predators becoming largely unidirectional. Crooks and Soulé (1999), in examining a partly developed landscape in coastal southern California, found that a typical 20-ha fragment of forest held 35 free-ranging house cats compared with 1–2 pairs of native predators of similar size. Collectively, they killed, at minimum, 525 birds annually, some species of which did not exceed 10 individuals inside that limited space.

Debates about moral equivalency and niche occupancy distract from the real issue, which is conservation of biodiversity. Its importance for the survival of all species, including us, is beyond question (Rockström et al. 2009). Arguing how habitat destruction, global warming, and ocean acidification are more dangerous to the environment than a few cats is still another red herring intended as a distraction. Conservation's objective is to mitigate *all* factors contributing to loss of biodiversity, not only those that seem largest, timely, or popular. Animal welfare advocates want to save cats while extending their individual longevities; opponents propose saving wildlife by eradicating cats, not lengthening their survival so they can kill still more wildlife. Cat advocates see this viewpoint as insensate. Cats indeed suffer at the hands of biologists, but no less than birds and squirrels suffer at the paws of unrestrained cats.

Other issues aside, TNR does not live up to its claim of consistently eliminating aggregations of free-ranging cats (Castillo and Clarke 2003, Jessup 2004, Stoskopf and Nutter 2004, Winter 2004, Zaunbrecher and Smith 1993). The complete removal of unowned cats, even by attrition, was never the purpose of TNR anyway. Its true objective has always been to keep the cats alive as long as possible through provisioning, veterinary care, and favorable legislation. As Zaunbrecher and Smith (1993: 451) wrote, "The effectiveness of the [TNR] program was demonstrated by the low turnover and improved health of the colony [sic] over the 3-y period." Statements like this, coupled with thorough critiques of TNR (e.g. Dauphiné and Cooper 2009, Duffy and Capece 2012, Jessup 2004, Longcore et al. 2009, Winter 2004), provide ample verification.

Trap–neuter–release, by design, operates at cross purposes with wildlife conservation by not actively reducing the number of cats at aggregation sites, as does any auxiliary intervention that extends longevity and encourages philopatry. Immigration of both neutered and intact individuals is usually uncontrolled. New cats arrive by various means. Some might be born at feeding sites and stay, others simply appear, and still others are dumped there by owners who evidently feel less guilty about abandoning their pets knowing they will be fed. The result is a group that grows instead of shrinks (Levy and Crawford 2004, Levy et al. 2003, Natoli et al. 2006).

When new arrivals outnumber those dying natural deaths and emigrating, the only relief comes from regular adoption (Levy and Crawford 2004, Levy et al. 2003), which is tangential to TNR efforts, not a result of them. Considering adoptions the equivalent of natural mortality for accounting purposes is misleading. Andersen et al. (2004: 1875) wrote, "Adoption programs are similar in effect to euthanasia because these cats are permanently removed from the free-roaming population." This assumption is logical for purposes of data analysis, although we have no evidence of its actual truth. Adoptions subtract cats from specific TNR programs, although whether they disappear permanently in free-ranging form is problematic. Cats accustomed to living outdoors often adapt poorly to confinement, even those "socialized" to humans. Most likely, adopted free-ranging strays become free-ranging pets, in which case their impact on wildlife remains unchanged. Others abandon their new homes (Remfry 1981: 77, Table 2). Adoption is a shell game in either case, dispersing the aggregation to out-of-sight locations devoid of any guarantee that damage to the environment has been ameliorated. Gorman and Levy (2004: 174) claimed TNR to result "in a long-term reduction of the feral cat population. ... " Levy and Crawford (2004) said much the same thing. Whether adoption actually reduced the number of free-ranging cats was left unaddressed.

During monitoring of strays at two south Florida parks, total numbers fell 20% in the first year from the initial 81, but by the end of the study had risen to 88, refuting the claim (e.g. Mahlow and Slater 1996), as Levy et al. (2003: 45) stated, "that an established colony [sic] of cats will defend its territory and prevent the immigration of new arrivals." Cats are not territorial, and an aggregation of strays would fail to defend the space it occupies under any circumstances (Chapter 2). In Italy, where the killing of free-ranging cats is illegal and strays and their caretakers are registered, the number of aggregations just in Rome increased from 76 in 1991 to 965 in 2000 despite a TNR program in place (Natoli et al. 2006). A survey of 103 of these aggregations showed a decrease in 55, stability of numbers in 20, and an increase in 28. However, the cats were counted, not followed, and whether they died or emigrated was unknown.

Success defined in terms of TNR has meant that enhanced longevity continues to expose wildlife to longer periods of predation, usually at increasing predator-to-prey ratios. Mendes-de-Almeida et al. (2011), for example, initiated a program in 2004 of sterilizing females >6 months. The starting aggregation was estimated at ~40 strays (both sexes combined). By 2006 that number had been reduced to 26 and to 17 by 2008, admirable because the numbers fell, but the effect of the slow decline on wildlife was undoubtedly greater than if the cats had simply been killed.

Aggregations of provisioned strays can be expected to grow unless the sterilization program is vigorous and consistently implemented (Tennent and Downs 2008). Longevity of both individuals and the aggregation will likely be extended (Guttilla and Stapp 2010) if only healthy cats are neutered and the sick ones killed (Hughes and Slater 2002, Zaunbrecher and Smith 1993).

The TNR concept has been subjected to modeling, which has given inconsistent results because different questions were asked. Budke and Slater (2009) modeled groups of strays for estimating the minimum number requiring sterilization annually to maintain stable numbers when the useful objective should have been to estimate the rate of decline after the entire group has been sterilized. In another attempt, matrix modeling predicted that 71–94% of an aggregation must be neutered before its numbers decline (Andersen et al. 2004, Foley et al. 2005), assuming all are released healthy and no further immigration occurs. As the model of Andersen et al. (2004: 1871) demonstrated, "All possible combinations of survival and fecundity values of free-roaming cats led to predictions of rapid, exponential population growth." Growth could be controlled only by killing ≥50% of all cats each year or neutering >75% of the intact proportion.

Elasticity analyses proved the first option superior (i.e. population growth, λ, was more sensitive to changes in survival than to fecundity). A 25% decrease in fecundity brought λ to 1.59; a reduction in survival of 25% reduced it to 1.36. However, both values of λ are not simply positive but indicate substantial overall growth; that is, growth well above the equilibrium level of none at all, or 1.00. A 50% decrease in annual survival led to a predicted population decline of 10%/y. In contrast, a 75% reduction in annual fecundity at both juvenile and adult stages triggered an increase in population growth. Highest elasticity values were for juvenile survival followed by adult survival, illustrating that λ is most sensitive to the survival of young cats and not adults. This is explained by certain life-history traits of cats reviewed in previous chapters. For example, they have relatively short lives, breed at an early age, can reproduce 2–3 times annually, are polyestrous and polytocous, and experience high postnatal and juvenile mortality unless attended by humans. In other words, cats differ from many seabirds for which λ shows high values for adults but not juveniles (Sections 10.2 and 10.6).

Jones and Downs (2011) reported similar findings after modeling urban strays on five campuses of the University of KwaZulu-Natal, South Africa. Modeling showed the population to drop by 10 cats with each 16% increase in the number sterilized, yet the overall density would still rise by 1 cat/ha. At zero sterilization the population was anticipated to double every 5 y; at 100% it would fall by half over the same time period. A no-growth level ($\lambda = 1.03$, total $n = 186$ with an increase of 6 cats in 5 y) required a projected rate of continuous sterilization of >90% annually. If this rate of neutering fell to 55% the population would increase to 1217 in 5 y.

Every TNR program steps onto a treadmill of ever increasing speed, the specific velocity modulated by kitten survival, adult longevity, rates of emigration and immigration, births, deaths, adoptions, and sterilization. Changes in any of these factors can affect whether a given group of strays increases or decreases. "Managing" aggregations encourages immigration, discourages emigration, and usually causes the number of births to rise. Provisioning and veterinary care increase kitten survival and extend longevity of adults, both of which mitigate natural mortality, and the combination of adoption and sterilization has seldom proven adequate to hold back the rising tide.

Contrary to the opinions of some (e.g. Gorman and Levy 2004) the extended presence of free-ranging cats does not cushion their harmful impact on wildlife and allow them to fade harmlessly into the ecological background. Quite the opposite. Providing food and veterinary care lengthens individual survival, increases cohort survivorship, and culminates in competitive advantages over indigenous predators of similar size. Freed from natural controls on morbidity and mortality, and having slipped the shackles of functional and numerical responses, they can hunt prey of ever increasing scarcity without suffering adverse consequences. To Longcore *et al.* (2009: 889), "This is a form of hyperpredation, similar to what occurs on oceanic islands where an exotic prey species (e.g., rats) supports an exotic predator (e.g., cats) that then devastates native prey. ... " The *hyperpredation effect*, as defined by Quillfeldt *et al.* (2008), occurs "when one or several prey species ... introduced into an environment in which a predator has also been introduced ... sustain high predator numbers, such that local prey, less adapted to high levels of predation, could suffer a population decline and possibly even extinction."

The results are most pronounced at locations with abbreviated food chains, such as islands and fragmented continental habitats where space, habitat variation, and species diversity are limited. Examples abound, making shambles of the notion that abundant introduced prey reduces predation on indigenous species. The presence of rabbits did little to lower predation by feral cats on seabirds at Macquarie Island (e.g. Brothers and Bone 2008). Similarly, TNR coupled with regular feeding is unlikely to diminish predation on wildlife by stray cats while introducing the potential to sustain or increase it.

In its classic form, hyperpredation can be said to occur when a fecund exotic prey species is introduced into a habitat containing a fecund predator. With food abundant the predator multiplies and inflicts damage on essentially defenseless indigenous prey species (e.g. Brothers 1984; Courchamp *et al.* 1999a, 1999b; Fritts and Rodda 1998; Jones 1977; Pech *et al.* 1995; Smith and Quin 1996; Taylor 1979). By feeding stray cats, humans substitute food for exotic prey into this sequence, providing the nutrition for cats to multiply and subsequently prey even more heavily on native animals. The cat food in this example is what Davis (1957: 466) called a "buffer," writing that "if a buffer is present the predators can turn to it when the prey become scarce and thus the predators remain and are ready when the prey begins to increase." He continued, "The buffer thus may keep the predators on hand to check the increase of prey and thereby the predators may hold down the prey population."

Davis tested his idea using free-ranging strays on a farm near Baltimore. He released even more cats into the population, maintained a continuing census of brown rats, and alternately fed the cats or withheld their food. He continued the experiment over several months, examining cat scats for evidence of cat food and prey parts. Supplementary feeding had no effect on predation. The function of the buffer, which can be considered secondary prey, is to sustain predators during times when the primary prey is scarce. In these experiments the objective was to see whether the cats continued to hunt even when alternative food was available. They did. Pigeons and mice were alternative buffers, and their consumption indicated a partial functional response. However, when rats, the primary prey, began to reproduce in spring, predation on them intensified, and their numbers did not increase.

Davis surmised that in natural systems an appropriate buffer has three characteristics: (1) it should be available when populations of the primary prey decline, (2) it must be palatable, and (3) its reproductive rate must be adequate. In this instance the principal buffer was palatable cat food, and the frequency of replenishing the food bowls the equivalent of its reproductive rate. For "control" to occur (i.e. for cats to reduce the number of rats), the predator population must remain high even when individuals of the primary prey are scarce. At the population level these cats responded numerically to supplementary feeding. In other words, the actual response was artificial (new cats were introduced to increase their numbers), but supplementary food kept them from either starving or emigrating. From a practical standpoint the results indicate that for rat control using cats, keep lots of cats around, not just a few, and feed them. The belief that a hungry cat makes a superior ratter is false.

In the absence of territorial behavior by a resident aggregation of strays, other cats come and go freely, making a sham of the notion that existing aggregations repel invaders at feeding sites, thus keeping the original number from increasing. Finally, eradication is unquestionably the most effective method of population control, but it directly opposes TNR's objective, which is to save free-ranging cats and extend their lives by keeping them healthy even at the risk of abandoning logic. Consider a statement by Hughes *et al.* (2002): "If eradication is successful and food and shelter are not eliminated, new cats from surrounding areas will move in to fill the vacant niche [*sic*]. … " This reprises an earlier statement by Hammond (1981: 89): "If cats are removed it is likely that more will move in to take advantage of the food and shelter that attracted the original cats." What these authors actually meant is that if eradication is successful, new cats from surrounding areas will move in *because* food and shelter have not been eliminated. Any useful program to eradicate an aggregation of cats would simultaneously remedy the cause of the aggregation and prevent its recurrence.

Predictably, advocates of TNR (e.g. Centonze and Levy 2002, Gorman and Levy 2004, Hughes *et al.* 2002, Levy *et al.* 2003, Robertson 2008, Scott *et al.* 2002) generally favor protecting free-ranging cats without concern for the ecological damage they cause. Scott *et al.* (2002: 205), for example, wrote: "It [TNR] is controversial because reports vary as to its effectiveness for controlling cat populations and whether the welfare of feral [*sic*] cats is served by neutering followed by return to the environment instead of euthanasia." This is indeed a narrow view of nature by limiting the concept of "welfare" to cats while excluding any thought of their potential prey. Then any TNR program undeniably promotes *cat* welfare, but that more wild creatures are saved if ownerless cats are removed instead of released ought to be equally clear.

Advocates of TNR have taken the peculiarly bigoted stance that control measures other than TNR are unsophisticated and somehow inferior. Levy *et al.* (2011: 1517), for example, wrote: "In parts of the world where progressive animal control resources are not available, measures such as poisoning and shooting are still used to control cats. … " Such statements and the attitudes behind them show a startling lack of compassion for animals other than pets, reducing their lives as individuals – and even their survival as species – to irrelevancy. Duffy and Capece (2012: 174) observed that "there is a profound asymmetry between the cat and endangered endemic species: the death of a cat is the loss of an individual of a planet-wide species, but in an endangered species an individual that dies may be irreplaceable."

This is indeed an odd juxtaposition considering people feed stray cats largely to enjoy bonding with them (Centonze and Levy 2002, Levy and Crawford 2004). The cats, in other words, cease to be generic and are transformed into individuals. But why should the life of a gray squirrel someone recognizes and names, perhaps feeds, and tames to the point it eats from his hand, be worth less than the life of the feral or stray cat that kills it? And ultimately, why should another squirrel's life be valued less just because it remains wild and anonymous in the woods? Jessup (2004: 1378), in a strong refutation of TNR (which he termed "trap, neuter, and reabandon") pointed out that "Wild animals are not only killed by cats but are also maimed, mauled, dismembered, ripped apart, and gutted while still alive, and if they survive the encounter they often die of sepsis because of the virulent nature of the oral flora of cats."

Saving feral and stray cats conveniently ignores the conservation issue, which is superbly documented and irrefutable: free-ranging cats inflict relentless, widespread harm on nature. Were they without blame, control programs would be superfluous. Doubters have only to consult the online Appendix for partial lists of vertebrate species known to have been killed or scavenged by free-ranging cats around the world. The information places in stark perspective the opposing side of the TNR question, which asks if *wildlife* might be better served were feral and stray cats killed instead of captured, neutered, and released. The answer is an unqualified yes.

10.4 Biological control

Howell (1984: 111) defined *biological control* as "the influence of either parasites, predators or pathogenic microorganisms on the population density of another organism, which as a result, is maintained at a lower level than would have occurred in their absence. … " The release of myxoma virus to control wild rabbits in Australia, New Zealand, and Europe is a well-known example; release of feline panleucopenia (FPL) virus to control feral cats is another. If the control organism is an infectious agent it must have several properties to be efficacious: (1) high host susceptibility, (2) high host specificity, (3) high virulence, (4) the capacity to spread rapidly, and (5) resistance to environmental inactivation during periods outside the host (Howell 1984). In susceptible cats, FPL is usually fatal in <175 h, and kittens are especially vulnerable (Howell 1984). Transfer is through direct contact but also contact with contaminated objects or bodily excretions (e.g. feces, urine, vomit, saliva). As an effective control agent, FPL meets all requirements. Morbidity can approach 100%, and the virus resists many chemical agents and survives over a wide temperature range, making it effective nearly everywhere (van Rensburg et al. 1987). Modeling predicts such biological agents to be useful tools for cat eradication when applied in combination with other methods (Oliveira and Hilker 2010).

The initial rapid drop in population leads to reduced encounters of the survivors, which naturally slows further spread of the virus. This is a limiting factor especially in extensive landscapes. More important, surviving females can pass colostral immunity to their kittens, raising the possibility of life-long immunity in these individuals and eventual stability within the population (Howell 1984). Monitoring during the feral cat eradication program at Marion Island revealed a reduction in antibody titers of FPL by 1982, or 5 y after its initial release in 1977. The feral cat population had grown exponentially to >3400 by 1977 (van Aarde 1978). Release of the virus produced an average drop in population of 29% annually during 1977–1981, which had declined to 8%/y by 1982 when the total population was then ~615, and to ~531 in 1983, indicating stabilization of negative growth (van Rensburg et al. 1987).

Comparison with 1975 census date prior to release of the FPL virus revealed a shift in age structure, subadult cats decreasing and adults increasing. Age distributions of the 1975 and 1982 populations were statistically similar, although the ratio of subadults (age-class 1) to adults (age-classes 2–9) were statistically different (Fig. 10.7). Prenatal mean litter sizes were similar in 1975 and 1982 at ~4.5, and surviving females in both years produced about 2 litters annually. However, litter sizes at weaning decreased significantly ($\bar{x} = 2.66$ in 1975 to $\bar{x} = 1.65$ in 1982). Thus 93.6% of all females collected

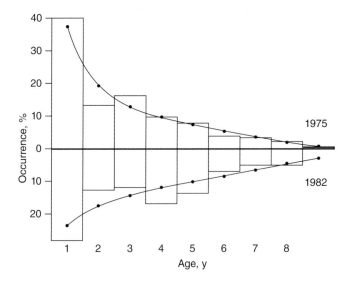

Fig. 10.7 Age structure of the 1975 and 1982 feral cat populations at Marion Island before and after release of the feline panleucopenia (FPL) virus in 1977. Source: van Rensburg et al. 1987, figure 2, p. 67. Reproduced with permission of John Wiley & Sons.

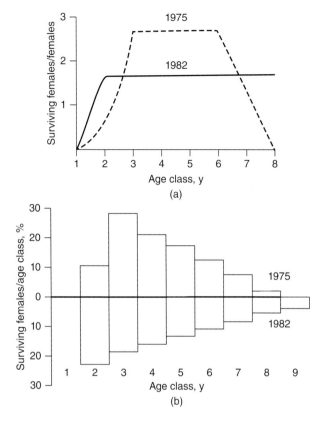

Fig. 10.8 Age-specific fecundity of the 1975 and 1982 feral cat populations at Marion Island before and after release of the feline panleucopenia (FPL) virus in 1977. Source: van Rensburg et al. 1987, figure 3, p. 68. Reproduced with permission of John Wiley & Sons.

during the breeding season (mid-September to mid-March, mean date of birth = 30 December) were reproducing, and mean age-specific fecundity was 1.54 surviving female births per female per season for all reproductive age classes. Age-specific fecundity was lower during 1975 for age classes 2, 8, and 9, and higher for 3–7 than in 1982 (Fig. 10.8a), and age-class 2 had the highest reproductive output of all classes in 1982; in 1975, age-class 3 was highest (Fig. 10.8b).

The age-related shifts at Marion Island and low survival of kittens and subadults aged 4–12 weeks indicated loss of maternally conferred resistance, probably from the mothers' low titers (van Rensburg et al. 1987). The result would be seasonal spikes in mortality resulting from the seasonal breeding pattern. Titers will eventually attenuate once the FPL virus has been released into a population of free-ranging cats. However, seasonal epizootics are still likely when the kittens become susceptible at 2–3 months.

Viruses are obviously useful for controlling unwanted animals (Tyndale-Biscoe 1993). Depending on the objective, species-specific viruses can be loosed into unwanted populations to cause intentional mortality; alternatively, recombinant viruses can be prepared that induce sterilization (i.e. *immunosterilization*). Induction of mass mortality is the favored method when the immediate purpose is to reduce population size, induced sterilization when the immediate goal is reduced fecundity with a lower population being the ultimate objective. The first procedure triggers an epizootic, usually by releasing laboratory-infected conspecifics into a susceptible population. In the second procedure, reproductive proteins that bring egg and sperm together at fertilization are isolated, the genes encoding them sequenced, and the antigens delivered to target species through baits or other means. Cats are fecund (Chapter 4), and when eradication is impractical or impossible the logical alternative is to lower reproductive potential. Immunosterilization offers the advantages of being humane and environmentally harmless, and if products contained in baits that could be broadcast were developed they would be less expensive than surgically neutering cats one at a time (Courchamp and Cornell 2000).

Immunosterilization might be rendered problematic when the target species – foxes or rabbits, for example – are both fecund and social. Tyndale-Biscoe (1993) argued that this situation requires a decision on whether to render the population simultaneously sterile and impotent or just sterile. The second choice does nothing to upset the social order. Because the reproductive success of a social species is often modulated by social rank, which is in turn driven by comparative concentrations of reproductive hormones, lower-ranking animals are often prevented from breeding, the equivalent of social castration. Chemically castrating individuals without regard to their social status can then result in high-ranking animals being displaced, nullifying any influence over those previously below them and allowing them to breed.

Whether such reasoning actually makes a difference in practice has yet to be tested in any social species, but even if it does its application to free-ranging cats, which are asocial, would seem irrelevant. The current approach of surgical neutering without regard to rank in a putative hierarchy (Chapter 1) is undoubtedly more effective than attempting to identify any so-called "dominant" individuals in an aggregation of strays and sterilizing only them. That existing TNR programs are nonspecific can be interpreted as indirect acknowledgement that dominance hierarchies are either nonexistent in free-ranging cats or exert no substantial influence over their reproductive potential.

Efforts to chemically control the reproductive potential of cats have mostly relied on manipulating steroids that influence hormonal functioning of the gonads, but these might prove less effective than non-hormonal approaches like manipulating genes expressed in the testes (Kean 2012). For example, *JQ1* inhibits division of cancerous cells by interfering with bromodomain proteins, including a testes-specific form. Knocking out this substance in mice stops spermatogenesis. The compound H2-gamendazole shows promise for human male sterilization (Kean 2012).

The ideal immunosterilizing agent would be species-specific, rendering recipients permanently sterile after just one or very few applications. When delivery involves a biological agent, control is relinquished, the process is likely irreversible, and extensive pre-release testing must assure no collateral damage to the environment (Courchamp and Cornell 2000). A product that meets these requirements has yet to be developed. Creagh (1992) explained the principles and use of immunocontraception in wildlife management, Munson (2006) reviewed contraception in cats, and Levy et al. (2004: 1128) noted that "an ideal cat contraceptive would provide permanent sterilization." These last authors then added a qualifier that widespread use of immunocontraceptive vaccines for free-ranging cats "will require the development of products that provide long-term immunity with a single treatment in a high proportion of animals." Such products would still be substandard by not sterilizing *all* treated animals permanently. Where wildlife is endangered by cats, a single pair left unsterilized is too many. Three of 9 individuals in the experiments of Levy and colleagues designed to test GnRH immunocontraception of males did not respond to treatment, demonstrating sperm counts similar to sham-treated controls. Of GnRH-treated females, Levy et al. (2011: 1518) stated: "Treatment must be efficacious in a high proportion of the population, but it is not essential that every treated animal be rendered infertile." As emphasized above, simply sterilizing the "majority" in an aggregation of cats is an empty exercise, failing in the end to stop population growth or protect wildlife. Nothing less than total, permanent sterilization is acceptable and then only if eradication is not possible. When true regulation of free-ranging cats is the objective, there can be no such entity as a "partly sterilized population."

10.5 Poisoning and other eradication methods

The impact cats exert on nature is undeniably negative (see online Appendix), and urgent action must sometimes take priority over detailed projections of the outcome if nothing is done. Cats can wipe out entire local populations and even species, and each loss leaves the ecosystem into which they intrude less diverse and more vulnerable to disturbances. As Rockström *et al.* (2009: 474) wrote, "Today, the rate of extinction of species is estimated to be 100 to 1,000 times more than what could be considered natural."

Scientists have advised that prior to implementing a program to eradicate a predator a thorough assessment of its diet should be undertaken (Meckstroth *et al.* 2007) and its place in local food webs evaluated (Matias and Catry 2008, Zavaleta *et al.* 2001). However, detailed studies of ecosystem effects take time, both to conduct research and to process the results. As Simberloff (2003: 83) emphasized, research in population biology is not always necessary or even useful to control an alien species, and eradication using "brute-force chemical and mechanical techniques" is often just as effective.

Poisoning is the most effective method of eradicating cats, especially when combined with trapping, shooting, and monitoring. Poisoned baits are generally distributed, the active ingredient historically inserted in sooty tern flesh (Rauzon 1985), dried meat (Risbey *et al.* 1997), polymerized fishmeal (Eason and Frampton 1991), raw fish (Domm and Messersmith 1990, Ratcliffe *et al.* 2009), nonpolymerized fishmeal (Risbey *et al.* 1997), dead day-old chicks (Bester *et al.* 2002, Brothers 1982), and kangaroo-meat sausage (Algar *et al.* 2002, Risbey *et al.* 1997). Any palatable flesh-based substance suffices.

Baiting and bait-induced trapping are more effective if initiated at times when prey density is low (Algar *et al.* 2002, Short *et al.* 1997, Twyford *et al.* 2000, Veitch 1985) and cats are hungrier and less discriminating (Short *et al.* 1997). Acceptance of poisoned baits demonstrated strong inverse correlation with the abundance of rabbits. Cats took >60% of baits they approached when the rabbit index was <1.0/km, but this declined to 34% at 3.0/km. Therefore, monitoring changes in the abundance of principal prey species can be important in eradication programs (Short *et al.* 1997).

Knowledge of the prey's natural history is also useful. In arid Australia the ideal time to set out poisoned baits for feral cats is a month after the first substantial rain of the year, or before young rabbits become independent and available for predation (Short *et al.* 1997). Rabbits mate within 1–2 weeks of rain (Poole 1960), have a 30-d gestation, and the neonates require 21 d from birth until emergence from their burrows. Poole (1960: 33) wrote, "Rains falling from mid October to the third week in November [austral spring] further increased pasture growth, and by November 15, 84 per cent. of the fertile females were pregnant." In contrast, Mykytowycz (1960) reported that up to 81% of litters might be lost by resorption or abortion during austral mid-winter (July). Thus the timing of mating, and even the subsequent viability of the embryos, depends ultimately on rainfall and improvement in forage for development of newly weaned offspring (Poole 1960).

Poisoned baits are often distributed as the first stage in a program of eradication to achieve a quick "knock-down" of the population (e.g. Algar *et al.* 2002). This is especially effective over large land masses. On small islands poisoning after trapping and shooting has also proved effective (Domm and Messersmith 1990).

A poisoning campaign can be more effective if preceded by acclimation; that is, if baits without poison are set out first and replenished several times before start of the program (Phillips *et al.* 2005, Twyford *et al.* 2000). Consumption of poisoned baits is likely to be highest during the first application and then decline as the initial kill takes effect and survivors develop aversions (Rodríguez *et al.* 2006). Changing baits to introduce novelty is sometimes effective (Rodríguez *et al.* 2006); in fact, novelty is not usually a deterrent to acceptance, but exceptions have been reported (Molsher 2001). Other methods (e.g. trapping, shooting) should be implemented quickly after the initial knock-down (Veitch 1985).

Most poisons tried to date are nonselective and cause collateral kills of nontargeted wildlife (e.g. Broome 2009, Gooneratne *et al.* 1995, McIlroy and Gifford 1992). This could potentially be used to advantage, and secondary poisoning of cats can sometimes be achieved by first targeting other alien species like rodents or rabbits (Alterio 1996, Broome 2009, Risbey *et al.* 1997, Rodríguez *et al.* 2006). Effects, however, have been inconsistent (Heyward and Norbury 1998). In addition, releasing former prey of cats (e.g. rats) from predation can have an adverse effect on native plants and animals (Ratcliffe *et al.* 2009).

Incidences of natural secondary poisoning have been reported. Peacock *et al.* (2011) found historical evidence of feral cats in Australia dying after consuming bronzewing pigeons (*Phaps chalcoptera* and *P. elegans*) and marsupials that had eaten parts of "desert poison bushes" (*Gastrolobium* spp.), the leaves, flowers, cotyledons, and seeds of which contain the toxin monofluoroacetate, the same basic compound used to formulate the toxicant 1080 (see later). Plants of the genus *Gastrolobium* contain 1.6–107 mg/kg "total fluorine" depending on species, and on the plant part and its age (Peacock *et al.* 2007). Native fauna that feed on these plants regularly are immune to the effects but retain the toxicant in their tissues.

The most popular poison used to kill cats has been sodium monofluoroacetate, commonly called 1080, a fine white powder soluble in water. The substance is nearly odorless and tasteless and less likely to be rejected than other poisons. It degrades harmlessly in the environment, but while still active is nonselective, killing nontargeted birds and mammals that eat it (Bester *et al.* 2002). Ingestion of 1080 by cats induces vomiting followed by death ~20 h later (McIlroy 1981). A dosage of 1080 administered at 0.6 mg/g bait (equivalent to 1.3 mg/kg M), causes death within 24 h, and 2 mg/g bait was recommended as being both humane and lethal (Eason and Frampton 1991). Tests by McIlroy (1981) showed the LD_{50} for cats to be 0.4 mg/kg M (± 0.31–0.52 95%CI, $n = 20$). Vomiting the poison before it takes effect can allow cats to survive. Those that learn from the experience later avoid baits (Ratcliffe *et al.* 2009). For this reason the anti-emetic Pileran (metoclopramide) is sometimes added to the poisoned mix at 5 mg/mL of total solution (Phillips *et al.* 2005).

Other toxicants used less commonly are brodifacoum and PAPP. Brodifacoum is a 4-hydroxycoumarin vitamin K antagonist and, like 1080, an anticoagulant. Its LD_{50} for rodents and cats is ~0.25 mg/kg M, and the toxic effect is cumulative. It is nontoxic to invertebrates and reptiles (Rodríguez *et al.* 2006). Cats are comparatively more susceptible than most mammals to PAPP – *p*-aminopropiophenone, or 1-(4-aminophenyl)-1-propanone,ethyl *p*-aminophenyl ketone. The lethal dosage is 20–34 mg/kg M, death occurring within 37–246 min (Eason *et al.* 2010). In contrast to 1080 and brodifacoum, PAPP interferes with oxygenation of the blood by converting hemoglobin to methemoglobin and inducing methemoglobinemia (Bright and Marrs 1983). Birds are less susceptible than carnivorans to PAPP's effects, giving it at least some degree of specificity (Eason *et al.* 2010).

Attempts to poison cats sometimes fail (e.g. Risbey *et al.* 1997). Causes have been insufficient exposure of baits to the cat population (Risbey *et al.* 1997) and their unnatural appearance (Risbey *et al.* 1997). The latter can sometimes be ameliorated by simple habituation; that is, exposing cats to the same baits without poison until resistance is overcome, as mentioned previously.

The success of a poisoning campaign is largely dependent on a bait's texture and palatability. Day-old chicks injected with 1080, for example, are quickly accepted (e.g. Bester *et al.* 2002). Bait degradation reduces palatability, and loss to nontargeted species can render poisoning campaigns less effective. Some compounds (e.g. 1080) are degraded by rain (McIlroy *et al.* 1988, Rodríguez *et al.* 2006), a problem in wet climates, and poisoned baits deployed at tropical and subtropical locations can be lost to land crabs. The species *Coenobita compressus* and *Johngarthia lagostoma* are well-known bait thieves, but immune to the effects of rodenticides and 1080 (Aguirre-Muñoz *et al.* 2008, Ratcliffe *et al.* 2009). Other losses of 1080 are attributable to consumption by fly larvae (maggots), defluorination by microorganisms, and leakage after injection of the toxic solution into the baits (McIlroy *et al.* 1988).

Lures have been used in efforts to enhance bait uptake and increase trapping efficiency. These can be olfactory, visual, or auditory. Their efficacy has been inconsistent and the evaluations incomplete. Olfactory lures, the most commonly tested, are based on food or "social" odors. Of the substances tried (e.g. cat urine, cat feces, cat anal-sac derivatives, commercial wild-animal lures, fish oils, catnip, matatabi), none has worked consistently (Algar *et al.* 2002, Clapperton *et al.* 1994, Edwards *et al.* 1997, Molsher 2001, Wood *et al.* 2002). Scents concocted from cats not in residence might be more attractive (Wood *et al.* 2002), but this hypothesis has not been tested. Of 13 scent-based lures tried by Edwards *et al.* (1997), only sun-dried prawn extract and the anal-sac preparations of male and female conspecifics attracted feral cats. From what we know about the cat's sensory systems (Section 9.6), visual lures are unlikely to attract interest unless they move, and their poor performance (Edwards *et al.* 1997) seems to bear this out. Even so, no experienced free-ranging cat would likely attack an artifactual object, such as a bait covered in bird feathers, much less ingest it.

Auditory lures, mostly recorded cat vocalizations, have yielded mixed results (Algar *et al.* 2002, Chawkins 2012). Broadcasting the ultrasounds of sympatric rodent prey might prove more effective, especially those of neonate or adolescents, but this approach has not been suggested previously and any success is therefore hypothetical. Rhodamine B (50 mg in meat baits) has been applied successfully as a systemic bait marker for free-ranging cats and can be used to evaluate acceptance (Fisher *et al.* 1999) during preliminary deployment of nonpoisonous baits.

Trapping often yields inconsistent results (e.g. Bester *et al.* 2002), although part of the failure can be blamed on low-quality baits. In addition, cats become trap-shy under persistent pressure (Algar *et al.* 2002, Twyford *et al.* 2000). This culminates in greater effort and diminishing success as a trapping campaign proceeds. In rare cases the opposite occurs, and a particular cat is caught repeatedly (Molsher 2001). In one instance just a single cat was captured over >450 trap-nights; trapping success elsewhere is usually <5% (e.g. Berruti 1986, Bloomer and Bester 1992, Jones 1977, Twyford et al. 2000). However, trapping accounted for 78% of the cats removed at Heirisson Prong, Shark Bay, Western Australia (Short et al. 1997). Trapping effectiveness obviously varies by location. When live-traps produce diminishing returns switching to gin, or leg-hold, models can be effective (Ratcliffe et al. 2009).

Dogs, either to hunt and kill cats directly, flush them out, or identify their locations (Bester et al. 2002, Wood et al. 2002) have not been consistently effective. Dogs trained to detect cats are sometimes helpful by identifying promising locations for trapping, setting baits, or providing "the impetus to continue the trapping effort when no other indications of remaining cats were evident" (Broome 2009: 18). Jack Russell terriers are among the most motivated breeds (Wood et al. 2002). Shooting is an effective control method only if the landscape is flat and the field of view unobstructed (Twyford et al. 2000), and human habitation is unaffected.

Ideally, a plan to eradicate cats should consider the negative ecological consequences from all known perspectives. These can be sizeable if other introduced pests are not targeted simultaneously, although cause–effect relationships that seem obvious are not necessarily true. It was the killing of all cats at Macquarie Island that some believe triggered an explosion of European rabbits, devastating the island's vegetation (Bergstrom et al. 2009a, 2009b). Dowding et al. (2009) were doubtful this was the cause, and Brothers and Bone (2008: 143) agreed: "Any perception that cat eradication was somehow a catalyst for the rabbit population increase that subsequently occurred is erroneous and convincingly disputed by the historical trends in rabbit abundance on the island irrespective of cat numbers through that time. ... " In their opinion, cat predation was never adequate to *regulate* the rabbit population, considering that no more than ~250 adult cats ever inhabited the island at a given time (Brothers et al. 1985). Cat removal on Raoul in the Kermadecs allowed the Pacific rats to increase which, according to Imber et al. (2003) and Rayner et al. (2007), adversely affected the native Cook's petrel.

Johnson et al. (2011: 182) wrote: "Existing tools cannot achieve reductions in cat populations over large areas without presenting a hazard to wildlife." The need to narrow the range of target specificity is indeed a major problem in managing free-ranging cats. I suggest trying a different approach. The ideal compound used in cat control would be a true "silver bullet" targeted specifically at cats and degrade harmlessly in the environment, but it need not be an actual compound when the absence of one might work as well. The domestic cat's Achilles' heel is its crucial need of dietary arginine. Cats die within a few hours if arginine is omitted from just one meal (Chapter 7). Because this requirement is unique to the cat, arginine-free baits could be devised and distributed without collateral kills of birds or mammals other than cats. If a cat ate an arginine-free bait and then consumed other prey the effect might be lessened.

Amino acid deficiencies are unusual in feral and stray cats so long as protein is available and its composition adequate to sustain gluconeogenesis (Chapter 8), but cats have a crucial weakness when dietary arginine is missing, and the result is striking (Chapter 7). *The complete absence of arginine from a single meal given to a hungry cat can induce hyperammonemia in <1 h and evidence of ammonia poisoning in 2–5 h. Death is inevitable*. Devising baits without arginine might work as well as setting out poisoned baits, and omitting arginine would not affect a bait's acceptance.

Like poisons, arginine-free baits should be most effective if the cat population is hungry because prey has become scarce, meaning that distribution might work better at certain times of low prey densities. Baits that include arginine should be deployed several days beforehand. This might condition cats to visit the stations, encourage feeding to satiation, and minimize chances that other prey would be ingested before arginine deficiency takes effect. Prey animals contain enough arginine to nullify effects of a deficiency if the amount of tissue ingested is large enough. As in poisoning campaigns, rate of encounter is critical, and the number of baits must be adequate to cover each cat's home range, encouraging uptake preferentially to the surrounding prey options. For example, at one site in the Galápagos Islands extra poisoned baits were set around a garbage dump where cats were known to aggregate.

The advantages of arginine as a replacement for poison are obvious. To repeat, arginine deprivation targets cats exclusively, meaning there would be no collateral kills of *any* nontargeted species, which is always a possibility with 1080 (McIlroy 1981, Murphy *et al.* 1999), brodifacoum (Dowding *et al.* 2009, Shah 2001), and other nonspecific toxicants. Arginine is an amino acid, and its degradation in the environment would be harmless.

Research into induction of teratospermia is another avenue of investigation that might produce in a useful control measure by compromising the reproductive potential of males. We know that pleiomorphic spermatozoa routinely fail to fertilize ova *in vitro* and assume their participation *in vivo* is nonexistent (Pukazhenthi *et al.* 2001). Teratospermic species like the cheetah, which produce spermatozoa that are >70% malformed still breed and father viable young. Pukazhenthi *et al.* (2001: 432) wrote: "This ambiguity is probably explained not on the basis of total malformed sperm, but rather on the proportion of the most serious defects." If true, then even insular populations of domestic cats that might test out teratospermic could continue to breed and flourish, so long as the defects were not too serious and enough males still produced normal sperm in adequate quantities during the breeding season (Chapter 7). Nonetheless, it seems worthwhile to initiate basic research into methods of inducing teratospermia in free-ranging cats for the purpose of limiting male fertility.

10.6 Integrated control

When another species of exotic prey is present and abundant it provisions the apex carnivore, allowing its population to expand and become even more harmful to scarce indigenous animals (Courchamp *et al.* 1999a, 1999b). The red-crowned parakeet (*Cyanoramphus novaezelandiae*) existed in uneasy equilibrium with Macquarie Island's cats until the release of European wild rabbits (Taylor 1979, 1985), after which the cats multiplied and the parakeets became locally extinct.

Because cats exert a beneficial effect on islands where mice, rats, and rabbits have also been introduced, removing them can allow their prey to proliferate, causing more damage (Bergstrom *et al.* 2009a, 2009b; Cook and Yalden 1980; Dowding *et al.* 2009; Glen and Dickman 2005; Matias and Catry 2008, Pontier *et al.* 2005). To avoid such "trophic cascade effects," many authorities have suggested targeting all invasive species simultaneously (e.g. Bonnaud *et al.* 2007, Creagh 1992). For example, it is not always obvious whether eradicating cats on oceanic islands might not stimulate increases in rat and rabbit populations (Apps 1983, Karl and Best 1982, Tidemann *et al.* 1994), which then intensifies predation on smaller vertebrates or bird eggs (Courchamp and Cornell 2000). The complex web of relationships among native and introduced animals on islands, coupled with the large land masses of some, makes identification of effective management strategies difficult (Pech *et al.* 1995).

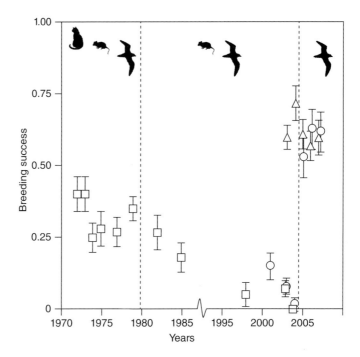

Fig. 10.9 Fraction of burrows of Cook's petrels (*Pterodroma cookii*) in which chicks were fledged during successive predator regimes at Little Barrier Island, New Zealand. Circles, squares and triangles represent different sites on the island ($\bar{x} \pm$ SEM). Far left column, before eradication of cats and Pacific rats; middle column, after eradication of cats; far right column, after eradication of rats. Source: Rayner *et al.* 2007. Reproduced with permission of the National Academy of Sciences.

Events at Macquarie Island exemplify the trophic cascade effects that occur by attempting to eliminate pest species one at a time. Release of the myxoma virus in the late 1970s knocked down the rabbit population, inducing the cats to increase predation on seabirds. Fewer rabbits resulted in local extinction of the weka (*Gallirallus australis*), another introduced predator. Vegetation rebounded, including tall tussock (*Poa foliosa*), which in turn allowed black rats to spread. A similar series of events has occurred elsewhere. Rayner *et al.* (2007) demonstrated that cat eradication at Little Barrier Island lowered the reproductive success of Cook's petrels (measured as the fraction of chicks that fledged in monitored burrows), which subsequently rose after later eradication of the Pacific rats (Fig. 10.9). Rats had risen to apex status after the cats disappeared, and reproductive success of the petrels continued to decline until the rats were eventually eliminated too.

Courchamp *et al.* (1999a, 1999b) suggested that subtracting an apex predator like the feral cat might raise the incidence of mortality by secondary predators (e.g. rats in this case), the so-called "mesopredator release effect" in which population growth of a smaller predator (the "mesopredator") has been etiolated by a larger apex predator. Once "released" from predation by a decline in numbers of the top predator the mesopredator population grows, exerting a sometimes damaging effect on prey numbers (e.g. Crooks and Soulé 1999, Ritchie and Johnson 2009). However, as pointed out by Peck *et al.* (2008), such predictions are founded on models that fail to account for age structure of the prey and for disparities in growth rate to changing survival or breeding success.

Mesopredator release can seem straightforward when applied to abbreviated terrestrial ecosystems such as those typifying islands or Australia where the dominant predators are exotic. For example, Lundie-Jenkins *et al.* (1993) saw an increase in the feral cat population in the Tanami Desert, Northern Territory, after a culling of dingoes, and in a controlled field experiment Moseby et al. (2012) reported dingoes killing both foxes and cats. Others (e.g. Smith and Quin 1996) have not found evidence of a negative correlation, and any trend might depend at least partly on the characteristics of the habitat (Glen and Dickman 2005). Designating which species constitute apex predators, mesopredators, or prey depends entirely on the local food web. Feral cats might be apex predators on islands where rats serve as mesopredators, although cats would rank lower in continental systems where coyotes replace them at the top (MacCracken 1982, Quinn 1997, Soulé et al. 1988). The designations shift again when the chain extends beyond three links including the prey. In such instances the mesopredator hypothesis in its original form no longer holds, and instead the apex predator releases the smallest predator in the chain (Levi and Wilmers 2012). For example, in a gray wolf → coyote → fox → ground squirrel chain, it is the fox that is spared predation.

Exploitation competition is sometimes difficult to separate from apparent release effects. Red foxes supposedly suppress feral cats in rural Australia, with dingoes suppressing both. A cat census after a poisoning program for foxes and dingoes resulted in more cats, but the difference could not be separated from effects of above-average rainfall and the subsequent increased prey abundance (references in Glen and Dickman 2005). Molsher et al. (1999) reported that when neither foxes nor feral cats were controlled their home ranges actually overlapped. Conilurine rodents have declined sharply in Australia as a result of predation by foxes and feral cats. Where dingoes occur rodent numbers are higher; where dingoes and foxes are absent, cats prey on them (Smith and Quin 1996) but the effects are difficult to separate from those of habitat modification.

The release of mid-sized predators like feral cats represents just one of many possible situations even in insular habitats (Roemer et al. 2002, Zavaleta et al. 2001). For example, elimination of a large herbivore can have adverse consequences on predators. A recent example is the Galápagos Islands where elimination of goats (*Capra hircus*) altered the vegetative landscape to such an extent that the population of indigenous hawks (*Buteo galapagoensis*) plummeted (Rivera-Parra et al. 2012). At Sarigan Island, Northern Mariana Islands, eradication of feral goats and pigs released an alien vine (*Operculina ventricosa*), which grew explosively (Zavaleta et al. 2001). Its presence had not even been detected until then, a striking example of "niche opportunity" when an invasive species remains at low densities until an upset in the ecosystem provides an opening to multiply (Allington et al. 2012, Shea and Chesson 2002). At California's Channel Islands, introduced pigs provided sufficient food (piglets) for golden eagles (*Aquila chrysaetos*) to colonize from the mainland (Roemer et al. 2002). Once there they preyed on the two indigenous carnivorans, island foxes (*Urocyon littoralis*) and island spotted skunks (*Spilogale gracilis*). Eagles supplanted foxes and skunks as apex predators, which then became the prey. The skunks, being nocturnal, survived better than the foxes, populations of which dipped toward local extinction. Models showed that in the absence of eagles, foxes were likely to drive the skunks toward local extinction. At Macquarie Island the presence of rabbits stimulated an increase in the population of great skuas (*Stercorarius antarctica*). As a result, skuas extended their range, which exposed more petrels to predation. After the myxoma virus was introduced and the rabbit population fell abruptly, skuas turned to preying and scavenging on penguins, and petrel predation declined. These observations led Brothers and Bone (2008: 135) to conclude, "As a result of pest management actions, there may be readjustments between indigenous species that do not necessarily result in greater abundance for each of those species."

When seabirds are the prey, small differences in annual survivorship strongly influence reproduction (Fig. 10.10). The predicted breeding life is 17 y for species with a constant annual adult survival of 96%, but half this long if survivorship drops to 92%. The lowest survivorship shown in the figure might represent kittiwakes (*Rissa* spp.) and their estimated 80% annual survival. Notice the abrupt decline in survival after breeding 12 y. The middle curve could represent the short-tailed shearwater. Its annual survival is 90% and falls sharply after ~20 y. The top curve is the wandering albatross, with its 94% annual survival of breeders. As of 1992 the first cohorts of wandering albatrosses had been breeding >20 y without signs of a downturn in survivorship. The data predict this species to breed 30 y before attaining a proportion of survivors (~10%) showing a substantial drop like the other two species.

The short-tailed shearwater deserves a few more words because it offers a good example of how life-history affects survivorship and ultimately adult mortality (Bradley *et al.* 1991). In this species the death rate increases with age and breeding experience (Fig. 10.10). Reproductive success is enhanced by a familiar partner and diminishes if that bird dies and another partner is acquired. Maturity is delayed, and short-tailed shearwaters first breed at 4–15 y ($\bar{x}$ = 7.2 y). Meanwhile, cohorts can be devastated during any given year by storms that flood burrows, drowning chicks or eggs. The shifting of ocean currents can temporarily make food to unavailable to foraging adults, which then abandon their young to starve. The species is transequatorial, and perhaps a half-million birds are killed each year by the salmon fishery as they migrate north through the Pacific basin during the nonbreeding season, a gauntlet affecting young and old alike. The short-tailed shearwater's strict philopatry makes it unlikely to breed elsewhere, even an adjacent island, rendering it especially vulnerable to disturbances by cats at natal sites.

Awkerman *et al.* (2006: 483) wrote, "Long-lived species with slow annual reproductive rates and low intrinsic mortality are especially vulnerable to increased mortality from extrinsic sources. … " Demographic studies and the use of periodic matrix models show how the rate of population growth for such animals is affected most by adult survival, with breeding success, or fraction of chicks fledged per egg laid (Jones 2002, Oro *et al.* 2004), and other parameters proving less sensitive indicators. (In at least one

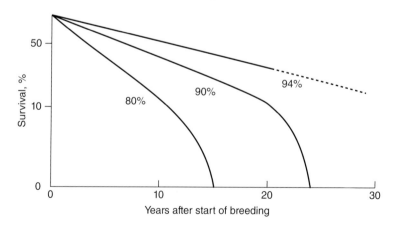

Fig. 10.10 Seabirds having a constant annual adult survival of 96% typically breed for 17 y, but a species in which survival is 92% has a breeding life-span of half this time. The consequences of small differences in survival are therefore large. Source: Wooller *et al.* 1992. Reproduced with permission of Elsevier.

instance breeding success has been defined as burrow occupancy; see Cooper *et al.* 1995, van Rensburg and Bester 1988.) Because most seabirds fit this pattern (Awkerman *et al.* 2006, Croxall and Rothery 1991, Peck *et al.* 2008) their populations are influenced more by altered adult survival than fluctuations in breeding success defined conventionally. Where cats and rats occur sympatrically the cats are more likely to kill adult birds. Rats prey on eggs and chicks instead, indicating that population growth is probably suppressed more by cats (Croxall and Rothery 1991). Newell's (Townsend's) shearwater (*Puffinus auricularis*) can live ~36 y (Ainley *et al.* 2001). The northern royal albatross (*Diomedea epomophora*) might live until 60 y (Robertson 1993, Tickell 2000: 37). According to Robertson (1993), a female banded at unknown age as a breeding adult in 1937 at Taiaroa Head, South Island, New Zealand, and nicknamed "Grandma" was still alive at the time of his publication. More recently, a 62-year-old Laysan albatross (*Phoebastria immutabilis*) nicknamed "Wisdom" hatched a viable chick at Midway Island in the central Pacific (Ho 2013).

Additional vulnerability is built into the life histories of the Procellariiformes (albatrosses, petrels, and shearwaters). For Galápagos waved albatrosses (*Phoebastria irrorata*), which can survive at least 38 y (Douglas and Fernández 1997), the care of both parents is necessary to rear the single offspring successfully: when one dies during a nesting season the egg or chick is doomed (Croxall and Rothery 1991), and this is true of albatrosses in general. If males and females have disparate rates of mortality then reproductive success is controlled ultimately by the surviving population of the limiting sex (Awkerman *et al.* 2006). Direct capture for human consumption and incidental mortality by commercial fishing also takes a toll of adults (Awkerman *et al.* 2006, Russell 1999), as is the case for albatrosses everywhere (Wooller *et al.* 1992).

Seabirds are especially vulnerable to extrinsic factors for several other reasons. The young develop slowly, prolonging their time of susceptibility to predation on land, and those that survive mature late, often not reproducing until >5 y (Tickell 2000: 337). The Procellariiformes typically breed once a year, some less often (e.g. biennially or even less frequently) because of the long chick-rearing period (Russell 1999). Ordinarily, the wandering albatross is a biennial breeder. However, some pairs might reproduce only every 3–4 y, and failure one year might lead them to defer another 2–3 y (Croxall and Rothery 1991). The Procellariiformes are permanently monogamous, lay a single egg that is not replaced if lost, and demonstrate strong philopatry (Rauzon *et al.* 2011), returning to the same nesting site and even the same burrow year after year.

10.7 Preparation for eradication programs

During integrated campaigns not just cats but other alien species (e.g. mice, rats, rabbits) are eliminated simultaneously to prevent trophic cascades that sometimes occur after removing only the apex predator (e.g. Courchamp *et al.* 1999a, 1999b; Medina and Nogales 2009; Rayner *et al.* 2007; Veitch 1985). These consequences include the upsetting of trophic food webs (Courchamp *et al.* 1999b, Roemer *et al.* 2002, Zavaleta *et al.* 2001), increases in numbers of rodents or lagomorphs (e.g. Rodríguez *et al.* 2006) and other unwanted species, and landscapes damaged by uncontrolled herbivory after the cats have been killed (Bergstrom *et al.* 2009a, 2009b). Targeting prey but not predator often induces an undesirable predatory effect. For example, a precipitous decline in rabbits after the introduction of rabbit calicivirus can have a harmful effect on the native fauna when feral cats address a shortage of their principal prey with a functional response. Such events can make the results of cat eradication programs equivocal (Dowding *et al.* 2009, Girardet *et al.* 2001). The purpose should be the restoration of ecosystems (Aguirre-Muñoz *et al.* 2011). With this in mind, few would disagree that efforts are best planned and implemented at the ecosystem level (Broome 2009, Courchamp *et al.* 2003).

The most important facet of a cat eradication program is the commitment of all involved, and loss of dedication or funding at any step along the way results in failure to achieve the goal. Culling cats, rabbits, and rodents intermittently and stopping before eradication is achieved can have undesirable consequences if the populations rebound rapidly afterward (Apps 1986b). Past eradication programs on large islands have taken years to complete. Cats at Marion Island, which is 290 km^2, were eradicated over 19 y. In this case the extended time can be blamed on several factors: the difficult landscape and climate, funding delays, bureaucratic entanglements, and the need to test different control methods.

Every eradication program must be modified to fit the circumstances. Preparation often includes both human and ecological elements if humans, in addition to their pets and livestock, could be affected. It is useful to know such factors as the identities of main prey species, where cats are concentrated and their times of highest activity, age structure of the population and its rate of growth, and the risk to non-targeted species. Identifying preferred habits, for example, is useful because ridding those areas of cats encourages repopulation from suboptimal locations, increasing the effectiveness of baiting, trapping, and shooting (Bester *et al.* 2000). The methods chosen depend on local ordinances; size of the area to be cleared of cats; nature of the terrain; local weather; proximity of human habitation; potential danger to humans, pets, livestock, and wildlife; and how thoroughly the ecology of the targeted cat population has been described (e.g. dispersion pattern, prey preferences, prey availability, diel activity patterns, reproductive biology, demography and age distribution, population numbers). The human aspect includes public relations and educational components and must be designed so residents can see the conservation value of eradicating alien species.

Ideally, eradication of unwanted flora and fauna is accompanied by full government backing, local conservation education, community involvement, ecosystem restoration, and research. Tershy *et al.* (2002) described the results of an ongoing bi-national plan for >250 islands off northwestern México. Ratcliffe *et al.* (2009) reported results from a more limited program that successfully eradicated free-ranging cats at Ascension Island where they had devastated seabird populations since 1815, extirpating 11 of 12 species from the main island of 97 km^2 and relegating them to nearby rock stacks and islets.

A public information campaign at Ascension began 6 months prior to the start of the campaign. The government and local animal welfare agency were informed of its details and benefits and gave approval, provided certain safety issues were met and steps would be taken to protect house cats. There were public meetings, articles in the newspaper, emails to employers, and casual conversations with residents to confirm that everyone was informed about how the program would be implemented and monitored. Local people were hired and trained to assist, providing an economic benefit and integrating the project into the community. A visitor center was established. It contained educational displays and distributed leaflets describing the adverse environmental impact of cats and a field guide to the local seabirds. Experts who supervised the eradication program were outsiders – not part of the community – who would leave at the project's completion.

All future imported house cats and those currently in residence were required to be sterilized and microchipped, the latter at no cost to residents. Reflective collars were issued so that pets could be distinguished visually from feral cats both day and night. Biological controls were not considered as a management measure because of their effect on these cats, but owners were warned that any free-ranging pets were nonetheless in danger from poisoned baits (1080 injected into chunks of raw fish) and shooting. The land crab *Johngarthia lagostoma*, the only indigenous scavenger, is unaffected by 1080.

After the eradication of unowned cats, steps should be taken to reduce predation by house cats. Existing cats must be sterilized and, in ideal circumstances, importation of new ones banned. The imposing of "curfews" on free-ranging house cats has been suggested (e.g. Barratt 1997b, Stracey 2011) as a means of lowering their destruction of wildlife, and confining cats indoors at night can reduce predation on amphibians and reptiles (Woods et al. 2003). Equipping pet cats with collars containing bells occasionally spares the birds and mammals (Gordon et al. 2010, Nelson et al. 2005, Ruxton et al. 2002, Woods et al. 2003). Nelson et al. (2005) found that cats wearing such collars brought home 34% fewer mammals and 41% fewer birds than those fitted with plain collars. Two bells were more effective than one. Cats equipped with sonic devices on their collars returned with 38% fewer mammals and 51% fewer birds. Quick-release collars are available for owners concerned that their cats might become entangled in underbrush.

The landscape itself is an important consideration in devising the program. Terrain differs by location, and in rugged, heavily vegetated areas shooting might be both dangerous and impractical. The use of shotguns and rifles is most effective where vistas are open and the sightlines in most directions unobstructed (Twyford et al. 2000). A combination of methods usually works best. In rural areas and uninhabited islands where free-ranging house cats are not present, releasing biological agents like the FPL virus is an effective opening gambit because it causes an immediate population crash, especially in cats aged <1 y (Bester et al. 2002). The program is initiated by trapping and inoculating cats and then releasing them as "carriers." Secondary and tertiary programs (e.g. trapping, shooting) need to be in place and commence immediately afterward before the survivors and their progeny develop resistance, which occurred within 5 y when FPL was released into feral cats at Macquarie Island in 1977 (Bester et al. 2002). In rugged country, aerial drops of poisoned baits can be effective, especially if followed by trapping and shooting, both labor-intensive and therefore expensive. Meanwhile, a trapping program must remain in place to continuously monitor blood titers for viral immunity.

10.8 "Secondary" prey management

Years of scat and stomach-content analysis in many parts of the world show clearly that a cat's first choice of prey is a small mammal, making birds "secondary" prey in most situations. Islands with nesting seabirds are exceptions, especially if bottom-up management to lower predator density has involved culling the principal mammalian prey (e.g. rabbits). Where free-ranging cats have become troublesome on large islands and continents, total eradication might not be possible at this time, leaving assertive management as the only option. The deterministic approach taken by Jones (2002) has much to offer. Jones modeled the conservation management of sooty shearwaters on mainland New Zealand in an effort to ascertain the population level at which a colony could sustain numerical stability as secondary prey under threat of predation. In New Zealand, as on other large land masses, cats prefer mice, rats, and rabbits, ordinarily expressing a functional response (Chapter 9) and shifting to birds when mammalian prey becomes scarce. If the secondary prey is a seabird the main objective should be protecting the adults even through temporary dips in reproductive success, however defined.

Predation pressure on secondary prey tends to be negatively density-dependent (Pech et al. 1995), resulting in possible extinction of small populations without fatally affecting the predator's population status. In contrast, large populations of secondary prey hover above a safety net manifested by their own abundance and the continued availability of primary prey, which limits mortality enough to allow population growth. When risk of predation is related inversely to population numbers, as it is in this example, at some theoretical point the secondary prey attains a degree of insulation permitting predator control management to slacken. In the words of Jones (2002: 2), "An 'escape threshold' prey density would therefore exist where unassisted population growth occurs as introduced predator impacts become relatively less significant."

Life-histories are dynamic, and Jones (2002) modeled stages of the sooty shearwater's in these equations:

$$n^e_{t+1} = n^A_t \times (1 - \text{ABS}) \times (1 - \text{SKIP}) \times 0.5 \tag{10.1}$$

$$n^5_{t+5} = n^e_t \times S^e_{(m)} \times S^{ch}_{(m)} \times (S^{juv})^{4.5} \times 1/R \tag{10.2}$$

$$n^6_{t+1} = [n^5_t \times (1 - \text{ABS}) \times S^5_{(m)}] + [n^5_t \times \text{ABS} \times S^5_{(max)}] \tag{10.3}$$

$$n^A_{t+1} = [n^6_t \times (1 - \text{ABS}) \times S^6_{(m)}] + [n^6_t \times \text{ABS}) \times S^6_{(max)}]$$
$$+ [n^A_t \times (1 - \text{ABS}) \times S^A_{(m)}] + [n^A_t \times \text{ABS} \times S^A_{(max)}] \tag{10.4}$$

where n^i_t = number of individuals of age class i at time t, e = eggs, ch = chicks, juv = juvenile birds of 6 months to 4 y, 5 and 6 = pre-breeders (age 5 to 6-y-old), A = adults aged 7 + y, $n^{total}_t = n^e_t = n^5_t + n^A_t$, $S^i_{(m)}$ = probability of survival of individuals of class i under management level m or experiencing maximum survival ($S^i_{(max)}$) while absent from the colony, ABS = the proportion of adults or pre-breeders absent in any year (assumed equal for both classes, SKIP = the proportion of adults present at the colony but skipping breeding in any given year, R = the proportion of birds aged 5 y comprising natal fledglings returning to the colony (i.e. immigration rate), and λ (population growth rate) = $n^{total}_t + 1/n^{total}_t$. The exponent 4.5 represents an estimated "reappearance" rate at 5 y and differs from survival (S) by including emigration; in other words, all birds that do not return eventually to breed are necessarily dead. Immigration data are lacking for sooty shearwaters. In colonies of a related species, the short-tailed shearwater, natal recruits compose 45% of all breeders (Bradley et al. 1991). Jones used this value for returning sooty shearwaters of age 5 y.

For assumptions and applications other than described here, see Jones (2002). Essentially, perturbations were induced to test their effects on λ. In this case the procedure involved elasticity analysis, comparing how proportional changes in parameters affected λ and assuming that the association between λ and the test parameter is also linear for large changes. In this case the mean value of each parameter was reduced 5% while keeping the rest fixed with the objective of isolating which factors exerted the most effect and could then be used to devise management tools. The elasticity of λ, e_p to a tested parameter, p, is

$$e_p = (\Delta\lambda/\lambda)/(\Delta p/p) \tag{10.5}$$

Two important assumptions are that (1) the daily individual risk of dying from predation (1 − daily survival) represents a "baseline" by assuming no predators and remains invariable regardless of population size, and (2) a daily risk of mortality resulting from predation that varies according to the number of birds present. From earlier work (Jones 2001, in Jones 2002) it was seen that risk of predation in a secondary prey population, n, varied inversely with population density such that e^β = relative risk resulting from a unit change in n.

When tested empirically, $\beta = -0.004$, or $e^\beta = 0.996$. Risk of mortality from predation at $n+1 = 0.996 \times$ mortality risk from predation at n. Jones used the predation risk estimation from the unprotected colony ($n = 30$) to start, subsequently estimating the daily risk for eggs in colonies of increasing numbers, assuming no management program. Another assumption was that β was the same for eggs and chicks, thus giving a similar distribution. The daily risk of death from predation could then be varied by colony size, which included the presence of adults and pre-breeders, but also factoring in the intensity of the management program in which predator control was relaxed or ramped up. This was accomplished by $1 - I$ where I = the proportional reduction in predation resulting from predator control. Reducing predation 80%, for example, gives daily risk estimates 0.2 times estimates without management intervention.

To obtain the daily mortality risk by colony size, Jones combined the estimate of daily risk with the baseline value resulting from natural mortality (i.e. the absence of predators). To obtain discrete survivorship by stage involved increasing daily survival probabilities to the pertinent power of the number of days in the stage examined (egg or chick). The model was tested for initial populations of 30–1000 and predator control intensities of 0.0–1.0. Survivorship by life-history stage (e.g. eggs, chicks, proportion of adults and pre-breeders present) varied with population size and management intensity. Estimates of λ were based on all individuals in the population, both present and absent.

Using mean values the model returned a predicted population growth rate of 1.044 (95% confidence intervals 0.967, 1.130). *Elasticity values confirm that adult survival in the colony ($S^A_{(m)}$) was clearly the most influential parameter.* Although survival of birds from fledging to return at age 5 y ranked second in importance, it was resistant to predator control, indicating that even intense efforts were unlikely to alter it much. The survival of pre-breeders within the colony was also important. As mentioned, population growth rate responded to both predator management intensity and colony size. Raising the level of predator removal enhanced the well-being of all colonies regardless of size.

Lower adult survival influences population growth adversely in both the present and future. In the long term it reduces lifetime reproductive output (Wooller *et al.* 1992); short-term effects reduce survival of eggs and chicks because bi-parental care is not provided. The survival of pre-breeders assures a reservoir of breeder-in-waiting, insulating the colony from excessive adult mortality caused by storms and other environmental catastrophes. Immigration is necessary too because like the other factors just discussed it provides additional elasticity to the population, making it more flexible to perturbations, including predation.

Models are only as accurate and precise as the assumptions used in their design and the mathematical constants and other data entered into the equations that define them. Models like this one can add necessary quantitative and predictive value to strategic decisions needed to implement effective predator control programs where secondary prey require protection. They show that even small populations can survive provided management efforts are intense and persistent. If primary prey and predator are both pests (e.g. rabbits and feral cats) an integrated program might be the best approach (Jones 2002). Simply removing the primary prey leaves the secondary prey more vulnerable to a functional response by the predator. As mentioned, thorough knowledge of the life-histories of all species involved – including seasonal prey densities and the causes of their fluctuations – is useful before initiation a management program.

References

Adamec, R. E. (1976). The interaction of hunger and preying in the domestic cat (*Felis catus*): an adaptive hierarchy? *Behavioral Biology* **18**: 263–272.

Adamec, R., C. Stark-Adamec, and K. E. Livingston. (1980a). The development of predatory aggression and defense in the domestic cat (*Felis catus*). I. Effects of early experience on adult patterns of aggression and defense. *Behavioral and Neural Biology* **30**: 389–409.

Adamec, R., C. Stark-Adamec, and K. E. Livingston. (1980b). The development of predatory aggression and defense in the domestic cat (*Felis catus*). II. Development of aggression and defense in the first 164 days of life. *Behavioral and Neural Biology* **30**: 410–434.

Adamec, R., C. Stark-Adamec, and K. E. Livingston. (1980c). The development of predatory aggression and defense in the domestic cat (*Felis catus*). III. Effects on development of hunger between 180 and 365 days of age. *Behavioral and Neural Biology* **30**: 435–447.

Adams, T. (1963). Body-temperature regulation in the normal and cold-acclimatized cat. *Journal of Applied Physiology* **18**: 772–777.

Ademolu, K. O., A. B. Idowu, and G. O. Olatunde. (2010). Nutritional value assessment of variegated grasshopper [*sic*], *Zonocerus variegatus* (L.) (Acridoidea: Pygomorphidae), during post embryonic development. *African Entomology* **18**: 360–364.

Adkins, Y., S. C. Zicker, A. Lepine, and B. Lönnerdal. (1997). Changes in nutrient and protein composition of cat milk during lactation. *American Journal of Veterinary Research* **589**: 370–375.

Adler, H. E. (1955). Some factors of observational learning in cats. *Journal of Genetic Psychology* **86**: 159–177.

Adolph, E. F. (1947). Tolerance to heat and dehydration in several species of mammals. *American Journal of Physiology* **151**: 564–575.

Aguirre-Muñoz, A., D. A. Croll, C. J. Donlan, R. W. Henry, M. A. Hermosillo, and G. R. Howald et al. (2008). High-impact conservation: invasive mammal eradications from the islands of western México. *Ambio* **27**: 101–107.

Aguirre-Muñoz, A., A. Samaniego-Herrera, L. Luna-Mendoza, A. Ortiz-Alcaraz, M. Rodríguez-Malagón, and F. Méndez-Sánchez et al. (2011). Island restoration in Mexico: ecological outcomes after systematic eradications of invasive mammals. In *Island Invasives: Eradication and Management*, C. R. Veitch, M. N. Clout, and D. R. Towns (eds.). Gland, Switzerland: International Union for the Conservation of Nature, pp. 250–258.

Ainley, D. A., R. Podolsky, L. Deforest, G. Spencer, and N. Nur. (2001). The status and population trends of the Newell's shearwater on Kaua'i: Insights from modeling. *Studies in Avian Biology* **22**: 108–123.

Akre, K. L., H. E. Farris, A. M. Lea, R. A. Page, and M. J. Ryan. (2011). Signal perception in frogs and bats and the evolution of mating signals. *Science* **333**: 751–752.

Alexander, L. G., C. Salt, G. Thomas, and R. Butterwick. (2011). Effects of neutering on food intake, body weight and body composition in growing female kittens. *British Journal of Nutrition* **106**: S19–S23.

Alexander, R. D. (1974). The evolution of social behavior. *Annual Review of Ecology and Systematics* **5**: 325–383.

Algar, D. A., A. A. Burbidge, and G. J. Angus. (2002). Cat eradication on Hermite Island, Montebello Islands, Western Australia. In *Turning the Tide: The Eradication of Invasive Species*. Occasional Paper of the IUCN, Species Survival Commission No. 27, C. R. Veitch and M. N. Clout (eds.). Gland, Switzerland: International Union for the Conservation of Nature, pp. 14–18.

Allin, J. T. and E. M. Banks. (1972). Functional aspects of ultrasound production by infant albino rats (*Rattus norvegicus*). *Animal Behaviour* **20**: 175–185.

Allington, G. R. H., D. N. Koons, S. K. M. Ernest, M. R. Schutzenhofer, and T. J. Valone. (2012). Niche opportunities and invasion dynamics in a desert annual community. *Ecology Letters* doi: 10.1111/3l3.12023, 9 pp.

Alterio, N. (1996). Secondary poisoning of stoats (*Mustela erminea*), feral ferrets (*Mustela furo*), and feral house cats (*Felis catus*) by the anticoagulant poison, brodifacoum. *New Zealand Journal of Zoology* **23**: 331–338.

Alterio, N. and H. Moller. (1997). Daily activity of stoats (*Mustela erminea*), feral ferrets (*Mustela furo*) and feral house cats (*Felis catus*) in coastal grassland, Otago Peninsula, New Zealand. *New Zealand Journal of Ecology* **21**: 89–95.

Alterio, N., H. Moller, and H. Ratz. (1998). Movements and habitat use of feral house cats *Felis catus*, stoats *Mustela erminea* and ferrets *Mustela furo*, in grassland surrounding yellow-eyed penguin *Megadyptes antipodes* breeding areas in spring. *Biological Conservation* **83**: 187–194.

Amarasekare, P. (1994). Ecology of introduced small mammals on western Mauna Kea, Hawaii. *Journal of Mammalogy* **75**: 24–38.

Andersen, M. C., B. J. Martin, and G. W. Roemer. (2004). Use of matrix population models to estimate the efficacy of euthanasia versus trap-neuter-return for management of free-roaming cats. *Journal of the American Veterinary Medical Association* **225**: 1871–1876.

Anderson, D. W., J. O. Keith, G. R. Trapp, F. Gress, and L. A. Moreno. (1989). Introduced small ground predators in California brown pelican colonies. *Colonial Waterbirds* **12**: 98–103.

Anderson, G. D. and P. R. Condy. (1974). A note on the feral house cat and house mouse on Marion Island. *South African Journal of Antarctic Research* **4**: 58–61.

Anderson, R. S. (1982). Water balance in the dog and cat. *Journal of Small Animal Practice* **23**: 588–598.

Anthony, J. A., D. D. Roby, and K. R. Turco. (2000). Lipid content and energy density of forage fishes from the northern Gulf of Alaska. *Journal of Experimental Marine Biology and Ecology* **248**: 53–78.

Appleby, M. C. (1983). The probability of linearity in hierarchies. *Animal Behaviour* **31**: 600–608.

Appleton, D. J., J. S. Rand, and G. D. Sunvold. (2001). Insulin sensitivity decreases with obesity, and lean cats with low insulin sensitivity are at greatest risk of glucose intolerance with weight gain. *Journal of Feline Medicine and Surgery* **3**: 211–228.

Appleton, D. J., J. S. Rand, J. Priest, G. D. Sunvold, and J. R. Vickers. (2004). Dietary carbohydrate source affects glucose concentrations, insulin secretion, and food intake in overweight cats. *Nutrition Research* **24**: 447–467.

Apps, P. J. (1983). Aspects of the ecology of feral cats on Dassen Island, South Africa. *South African Journal of Zoology* **18**: 393–399.

Apps, P. J. (1986a). Home ranges of feral cats on Dassen Island. *Journal of Mammalogy* **67**: 199–200.

Apps, P. J. (1986b). A case study of an alien predator (*Felis catus*) introduced on Dassen Island: selective advantages. *South African Journal of Antarctic Research* **16**: 118–122.

Aronson, L. R. and M. L. Cooper. (1966). Seasonal variation in mating behavior in cats after desensitization of glans penis. *Science* **152**: 226–230.

Ashmole, N. P. (1963). The biology of the wideawake or sooty tern *Sterna fuscata* on Ascension Island. *Ibis* **103b**: 297–364.

Atkinson, I. A. E. (2001). Introduced mammals and models for restoration. *Biological Conservation* **99**: 81–96.

Atwood, T. C. (2006). Behavioral interactions between coyotes, *Canis latrans*, and wolves, *Canis lupus*, at ungulate carcasses in southwestern Montana. *Western North American Naturalist* **66**: 390–394.

Avilés, L. (2002). Solving the freeloaders paradox: genetic associations and frequency-dependent selection in the evolution of cooperation among nonrelatives. *Proceedings of the National Academy of Sciences* **99**: 14268–14273.

Awkerman, J. A., K. P. Huvaert, J. Mangel, J. A. Shigueto, and D. J. Anderson. (2006). Incidental and intentional catch threatens Galápagos waved albatross. *Biological Conservation* **133**: 483–489.

Backus, R. C., N. J. Cave, and D. H. Keisler. (2007). Gonadectomy and high dietary fat but not high dietary carbohydrate induce gains in body weight and fat of domestic cats. *British Journal of Nutrition* **98**: 641–650.

Baerends-van Roon, J. M. and G. P. Baerends. (1979). *The Morphogenesis of the Behaviour of the Domestic Cat, with Special Emphasis on the Development of Prey-catching*. Amsterdam: North Holland, 116 pp.

Baker, M. A. and P. A. Doris. (1982). Control of evaporative heat loss during changes in plasma osmolality in the cat. *Journal of Physiology* **328**: 535–545.

Baker, P. J., A. J. Bentley, R. J. Ansell, and S. Harris. (2005). Impact of predation by domestic cats *Felis catus* in an urban area. *Mammal Review* **35**: 302–312.

Ballard, F. J. (1965). Glucose utilization in the mammalian liver. *Comparative Biochemistry and Physiology* **14**: 437–443.

Ballard, W. B., L. A. Ayers, P. R. Krausman, D. J. Reed, and S. G. Fancy. (1997). Ecology of wolves in relation to a migratory caribou herd in northwest Alaska. *Wildlife Monographs* (135): 47 pp.

Banks, D. H. and G. Stabenfeldt. (1982). Luteinizing hormone release in the cat in response to coitus on consecutive days of estrus. *Biology of Reproduction* **26**: 603–611.

Banks, D. R., S. R. Paape, and G. H. Stabenfeldt. (1983). Prolactin in the cat: I. Pseudopregnancy, pregnancy and lactation. *Biology of Reproduction* **28**: 923–932.

Barker, J. and R. C. Povey. (1973). The feline urolithiasis syndrome: a review and an inquiry into the alleged role of dry cat foods in its aetiology. *Journal of Small Animal Practice* **14**: 445–447.

Baron, A., C. N. Stewart, and J. M. Warren. (1957). Patterns of social interaction in cats (*Felis domestica*). *Behaviour* **11**: 56–66.

Barratt, D. G. (1997a). Home range size, habitat utilisation and movement patterns of suburban and farm cats *Felis catus*. *Ecography* **20**: 271–280.

Barratt, D. G. (1997b). Predation by house cats, *Felis catus* (L.), in Canberra, Australia. I. Prey composition and preference. *Wildlife Research* **24**: 263–277.

Barrett, P. and P. Bateson. (1978). The development of play in cats. *Behaviour* **66**: 106–120.

Bateson, P. (1981). Discontinuities in development and changes in the organization of play in cats. In *Behavioral Development. The Bielefeld Interdisciplinary Project*, K. Immelmann, G. W. Barlow, L. Petrinovich, and M. Main (eds.). Cambridge: Cambridge University Press, pp. 281–295.

Bateson, P., M. Mendl, and J. Feaver. (1990). Play in the domestic cat is enhanced by rationing of the mother during lactation. *Animal Behaviour* **40**: 514–525.

Bateson, P. and M. Young. (1979). The influence of male kittens on the object play of their female siblings. *Behavioral and Neural Biology* **27**: 374–378.

Bateson, P. and M. Young. (1981). Separation from the mother and the development of play in cats. *Animal Behaviour* **29**: 173–180.

Bateson, P., P. Martin, and M. Young. (1981). Effects of interrupting cat mothers' lactation with bromocriptine on the subsequent play of their kittens. *Physiology and Behavior* 27: 841–845.

Bateson, P., M. Mendl, and J. Feaver. (1990). Play in the domestic cat is enhanced by rationing of the mother during lactation. *Animal Behaviour* 40: 514–525.

Bauer, J. E. (1997). Fatty acid metabolism in domestic cats (*Felis catus*) and cheetahs (*Acinonyx jubatus*). *Proceedings of the Nutrition Society* 56: 1013–1024.

Beauchamp, G. K., O. Maller, and J. G. Rogers Jr., (1977). Flavor preferences in cats (*Felis catus* and *Panthera* sp.). *Journal of Comparative and Physiological Psychology* 91: 1118–1127.

Beaumont, M., E. M. Barratt, D. Gottelli, A. C. Kitchener, M. J. Daniels, and J. K. Pritchards et al. (2001). Genetic diversity and introgression in the Scottish wildcat. *Molecular Ecology* 10: 319–336.

Beaver, B. V. (1977). Mating behavior in the cat. *Veterinary Clinics of North America* 7: 729–733.

Beck, A. M. (1971). The life and times of Shag, a feral dog in Baltimore. *Natural History Magazine*, 80: 58–65.

Beck, A. M. (1973). *The Ecology of Stray Dogs: A study of Free-ranging Urban Animals*. Baltimore, MD: York Press, xiv + 98 pp.

Beckerman, A. P., M. Boots, and K. J. Gaston. (2007). Urban bird declines and the fear of cats. *Animal Conservation* 10: 320–325.

Bednekoff, P. A. (1997). Mutualism among safe, selfish sentinels: a dynamic game. *American Naturalist* 150: 373–392.

Bekoff, M. (1976). Animal play: problems and perspectives. In *Perspectives in Ethology, Vol. 2*, P. P. G. Bateson and P. H. Klopfer (eds.). New York: Plenum Press, pp. 165–188.

Bekoff, M. (1978). Social play: structure, function, and the evolution of co-operative social behavior. In *The Development of Behavior*, G. Burghart and M. Bekoff (eds.). New York: Garland Press, pp. 367–383.

Bekoff, M. and J. A. Byers. (1981). A critical reanalysis of the ontogeny and phylogeny of mammalian social and locomotor play: an ethological hornet's nest. In *Behavioral Development. The Bielefeld Interdisciplinary Project*, K. Immelmann, G. W. Barlow, L. Petrinovich, and M. Main (eds.). Cambridge: Cambridge University Press, pp. 296–337.

Beliveau, G. P. and R. A. Freedland. (1982). Metabolism of serine, glycine and threonine in isolated cat hepatocytes *Felis domestica*. *Comparative Biochemistry and Physiology* 71B: 13–18.

Bergstrom, D. M., A. Lucieer, K. Kiefer, J. Wasley, L. Belbin, and T. K. Pedersen et al. (2009a). Indirect effects of invasive species removal devastate World Heritage island. *Journal of Applied Ecology* 46: 73–81.

Bergstrom, D. M., A. Lucieer, K. Kiefer, J. Wasley, L. Belbin, and T. K. Pedersen, et al. (2009b). Management implications of the Macquarie Island trophic cascade revisited: a reply to Dowding et al. (2009). *Journal of Applied Ecology* 46: 1133–1136.

Berkeley, E. P. (1982). *Maverick Cats: Encounters with Feral Cats*. New York: Walker, 142 pp.

Bermingham, E. N., D. G. Thomas, P. J. Morris, and A. J. Hawthorne. (2010). Energy requirements of adult cats. *British Journal of Nutrition* 103: 1083–1093.

Bermingham, E. N., S. Kittelmann, G. Henderson, W. Young, N. C. Roy, and D. G. Thomas. (2011). Five-week dietary exposure to dry diets alters the faecal bacterial populations in the domestic cat (*Felis catus*). *British Journal of Nutrition* 106: S49–S52.

Bernstein, I. S. (1981). Dominance: the baby and the bathwater. *Behavioral and Brain Sciences* 4: 419–429.

Berruti, A. (1986). The predatory impact of feral cats *Felis catus* and their control on Dassen Island. *South African Journal of Antarctic Research* 16: 123–127.

Berry, C. S. (1908). An experimental study of imitation in cats. *Journal of Comparative Neurology* 18: 1–25.

Bester, M. N., J. P. Bloomer, P. A. Bartlett, D. D. Muller, M. van Rooyen, and M. Büchner. (2000). Final eradication of feral cats from sub-Antarctic Marion Island, southern Indian Ocean. *South African Journal of Wildlife Research* 30: 53–57.

Bester, M. N., J. P. Bloomer, R. J. van Aarde, B. H. Erasmus, P. J. J. van Rensburg, and J. D. Skinner et al. (2002). A review of the successful eradication of feral cats from sub-Antarctic Marion Island, Southern Indian Ocean. *South African Journal of Wildlife Research* **32**: 65–73.

Beynon, R. J. and J. L. Hurst. (2003). Multiple roles of major urinary proteins in the house mouse, *Mus domesticus*. *Biochemical Society Transactions* **31**: 142–146.

Biben, M. (1979). Predation and predatory play behaviour of domestic cats. *Animal Behaviour* **27**: 81–94.

Biourge, V., J. M. Groff, C. Fisher, D. Bee, J. G. Morris, and Q. R. Rogers. (1994). Nitrogen balance, plasma free amino acid concentrations and urinary orotic acid excretion during long-term fasting in cats. *Journal of Nutrition* **124**: 1094–1103.

Biró, Z., L. Szemethy, and M. Heltai. (2004). Home range sizes of wildcats (*Felis silvestris*) and feral domestic cats (*Felis silvestris* f. catus [sic]) in a hilly region of Hungary. *Mammalian Biology* **69**: 302–310.

Biró, Z., J. Lanszki, L. Szemethy, M. Heltai, and E. Randi. (2005). Feeding habits of feral domestic cats (*Felis catus*), wild cats (*Felis silvestris*) and their hybrids: trophic niche overlap among cat groups in Hungary. *Journal of Zoology (London)* **266**: 187–196.

Blackburn, T. M., P. Cassey, R. P. Duncan, K. L. Evans, and K. J. Gaston. (2004). Avian extinction and mammalian introductions on oceanic islands. *Science* **305**: 1955–1958.

Bloomer, J. P. and M. N. Bester. (1990). Diet of a declining feral cat *Felis catus* population on Marion Island. *South African Journal of Wildlife Research* **20**: 1–4.

Bloomer, J. P. and M. N. Bester. (1992). Control of feral cats on sub-Antarctic Marion Island, Indian Ocean. *Biological Conservation* **60**: 211–219.

Blottner, S. and K. Jewgenow. (2007). Moderate seasonality in testis function of domestic cat [sic]. *Reproduction in Domestic Animals* **42**: 536–540.

Bonanni, R., S. Cafazzo, C. Fantini, D. Pontier, and E. Natoli. (2007). Feeding-order in an urban feral domestic cat colony: relationship to dominance rank, sex and age. *Animal Behaviour* **74**: 1369–1379.

Bonds, A. B. (1974). Optical quality of the living cat eye. *Journal of Physiology* **243**: 777–795.

Bonnaud, E., K. Bourgeois, E. Vidal, Y. Kayser, Y. Tranchant, and J. Legrand. (2007). Feeding ecology of a feral cat population on a small Mediterranean island. *Journal of Mammalogy* **88**: 1074–1081.

Bonnaud, E., D. Zarzoso-Lacoste, K. Bourgeois, L. Ruffino, J. Legrand, and E. Vidal. (2010). Top-predator control on islands boosts endemic prey but not mesopredator. *Animal Conservation* **13**: 556–567.

Boudreau, J. C. (1974). Neural encoding in cat geniculate ganglion tongue units. *Chemical Senses and Flavor* **1**: 41–51.

Bourre, J-M., O. Dumont, M. Piciotti, M. Clément, J. Chaudière, and M. Bonneil et al. (1991). Essentiality of $\omega 3$ fatty acids for brain structure and function. *World Research in Nutrition and Dietetics* **66**: 103–117.

Boyd, J. S. (1971). The radiographic identification of the various stages of pregnancy in the domestic cat. *Journal of Small Animal Practice* **12**: 501–506.

Boyd, R. and J. B. Silk. (1983). A method for assigning cardinal dominance ranks. *Animal Behaviour* **31**: 45–58.

Bradford, K., R. W. Henry, A. Aguirre Muñoz, C. García, L. L. Mendoza, and M. Á. Hermosillo et al. (2006). El impacto de los gatos introducidos (*Felis catus*) en el ecosistema de Isla Guadalupe. http://www2.ine.gob.mx/publicaciones/libros/477/cap14.html (accessed 25 February 2014).

Bradley, J. S., I. J. Skira, and R. D. Wooller. (1991). A long-term study of short-tailed shearwaters *Puffinus tenuirostris* on Fisher Island, Australia. *Ibis* **133** (Suppl. 1): 55–61.

Bradshaw, J. and C. Cameron-Beaumont. (2000). The signalling repertoire of the domestic cat and its undomesticated relatives. In *The Domestic Cat: The Biology of Its Behaviour*, 2nd edn., D. C. Turner and P. Bateson (eds.). Cambridge: Cambridge University Press, pp. 67–93.

Bradshaw, J. W. S. (2006). The evolutionary basis for the feeding behavior of domestic dogs (*Canis familiaris*) and cats (*Felis catus*). *Journal of Nutrition Supplement* **136**: 1927S–1931S.

Bradshaw, J. W. S. and S. L. Brown. (1992). Social behaviour of cats. *Tijdschrift voor Diergeneeskunde* **117**: 54S–56S.

Bradshaw, J. W. S. and C. Thorne. (1992). Feeding behaviour. In *The Waltham Book of Dog and Cat Behaviour*, C. Thorn (ed.). Oxford: Pergamon Press, pp. 118–129.

Bradshaw, J. W. S., D. Goodwin, V. Legrand-Defrétin, and H. M. R. Nott. (1996). Food selection by the domestic cat, an obligate carnivore. *Comparative Biochemistry and Physiology* **114A**: 205–209.

Bradt, G. W. (1949). Farm cat as predator. *Michigan Conservation* **18**: 23–25.

Brickner-Braun, I., E. Geffen, and Y. Yom-Tov. (2007). The domestic cat as a predator of Israeli wildlife. *Israeli Journal of Ecology and Evolution* **53**: 129–142.

Bright, J. E. and T. C. Marrs. (1983). The induction of methaemoglobin by *p*-aminophenones. *Toxicology Letters* **18**: 157–161.

Bristol-Gould, S. and T. K. Woodruff. (2006). Folliculogenesis in the domestic cat (*Felis catus*). *Theriogenology* **66**: 5–13.

Bristol, S. K. and T. K. Woodruff. (2004). Follicle-restricted compartmentalization of transforming growth factor β superfamily ligands in the feline ovary. *Biology of Reproduction* **70**: 846–859.

Broome, K. (2009). Beyond Kapiti – a decade of invasive rodent eradications from New Zealand islands. *Biodiversity* **10**: 14–24.

Brothers, N. and C. Bone. (2008). The response of burrow-nesting petrels and other vulnerable bird species to vertebrate pest management and climate change on sub-Antarctic Macquarie Island. *Papers and Proceedings of the Royal Society of Tasmania* **142**: 123–148.

Brothers, N. P. (1982). Feral cat control on Tasman Island. *Australian Ranger Bulletin* **2**(2): 9.

Brothers, N. P. (1984). Breeding, distribution and status of burrow-nesting petrels at Macquarie Island. *Australian Wildlife Research* **11**: 113–131.

Brothers, N. P., I. J. Skira, and G. R. Copson. (1985). Biology of the feral cat, *Felis catus* (L.), on Macquarie Island. *Australian Wildlife Research* **12**: 425–436.

Brown, J. L. and G. H. Orians. (1970). Spacing patterns in mobile animals. *Annual Review of Ecology and Systematics* **1**: 239–262.

Buddington, R. K., J. W. Chen, and J. M. Diamond. (1991). Dietary regulation of intestinal brush-border sugar and amino acid transport in carnivores. *American Journal of Physiology* **261**: R793-R801.

Budke, C. M. and M. R. Slater. (2009). Utilization of matrix population models to assess a 3-year single treatment nonsurgical contraception program versus surgical sterilization in feral cat populations. *Journal of Applied Animal Welfare Science* **12**: 277–292.

Burbidge, A. A. and N. L. McKenzie. (1989). Patterns in the modern decline of Western Australia's vertebrate fauna: causes and conservation implications. *Biological Conservation* **50**: 143–198.

Burbidge, A. A. and K. D. Morris. (2002). Introduced mammal eradications for nature conservation on Western Australian islands: a review. In *Turning the Tide: The Eradication of Invasive Species*. Occasional Paper of the IUCN, Species Survival Commission No. 27, C. R. Veitch and M. N. Clout (eds.). Gland, Switzerland: International Union for the Conservation of Nature, pp. 64–70.

Burger, I. H., R. S. Anderson, and P. W. Holme. (1980). Nutritional factors affecting water balance in the dog and cat. In *Nutrition of the Dog and Cat*, R. S. Anderson (ed.). Oxford: Pergamon Press, pp. 145–158.

Burger, I. H., S. E. Blaza, P. T. Kendall, and P. M. Smith. (1984). The protein requirement of adult cats for maintenance. *Feline Practice* **14**(2): 8–14.

Burness, G. P. (2010). Elephants, mice, and red herrings. *Science* **296**: 1245–1247.

Burt, W. H. (1943). Territoriality and home range concepts as applied to mammals. *Journal of Mammalogy* **24**: 346–352.

Cafazzo, S. and E. Natoli. (2009). The social function of tail up in the domestic cat. *Behavioural Processes* **80**: 60–65.

Calhoon, R. E. and C. Haspel. (1989). Urban cat populations compared by season, subhabitat and supplemental feeding. *Journal of Animal Ecology* **58**: 321–328.

Call, J., M. Carpenter, and M. Tomasello. (2005). Copying results and copying actions in the process of social learning: chimpanzees (*Pan troglodytes*) and human children (*Homo sapiens*). *Animal Cognition* **8**: 151–163.

Campbell, K. J., G. Harper, D. Algar, C. C. Hanson, B. S. Keitt, and S. Robinson. (2011). Review of feral cat eradications on islands. In *Island Invasives. Eradication and Management*, C. R. Veitch, M. N. Clout, and D. R. Towns (eds.). Gland, Switzerland: International Union for the Conservation of Nature, pp. 37–46.

Campos, C. B., C. F. Esteves, K. M. P. M. B. Ferraz, P. G. Crawshaw Jr., and L. M. Verdade. (2007). Diet of free-ranging cats and dogs in a suburban and rural environment, south-eastern Brazil. *Journal of Zoology (London)* **273**: 14–20.

Carey, G. P., Z. Kime, Q. R. Rogers, J. G. Morris, D. Hargrove, and C. A. Buffington et al. (1987). An arginine-deficient diet in humans does not evoke hyperammonemia or orotic aciduria. *Journal of Nutrition* **117**: 1734–1739.

Caro, T. M. (1979). Relations between kitten behaviour and adult predation. *Zeitschrift für Tierpsychologie* **51**: 158–168.

Caro, T. M. (1980a). Predatory behaviour in domestic cat mothers. *Behaviour* **74**: 128–148.

Caro, T. M. (1980b). The effects of experience on the predatory patterns of cats. *Behavioral and Neural Biology* **29**: 1–28.

Caro, T. M. (1980c). Effects of the mother, object play and adult experience on predation in cats. *Behavioral and Neural Biology* **29**: 29–51.

Caro, T. M. (1981a). Sex differences in the termination of social play in cats. *Animal Behaviour* **29**: 271–279.

Caro, T. M. (1981b). Predatory behaviour and social play in kittens. *Behaviour* **76**: 1–24.

Caro, T. M. and D. A. Collins. (1987). Male cheetah social organization and territoriality. *Ethology* **74**: 52–64.

Caro, T. M. and M. D. Hauser. (1992). Is there teaching in nonhuman animals? *Quarterly Review of Biology* **67**: 151–174.

Carr, G. M. and D. W. Macdonald. (1986). The sociality of solitary foragers: a model based on resource dispersion. *Animal Behaviour* **34**: 1540–1549.

Carss, D. N. (1995). Prey brought home by two domestic cats (*Felis catus*) in northern Scotland. *Journal of Zoology (London)* **237**: 678–686.

Castillo, D. and A. L. Clarke. (2003). Trap/neuter/release methods ineffective in controlling domestic cat "colonies" on public lands. *Natural Areas Journal* **23**: 247–253

Castonguay, T. W. (1981). Dietary dilution and intake in the cat. *Physiology and Behavior* **27**: 547–549.

Catling, P. C. (1988). Similarities and contrasts in the diets of foxes, *Vulpes vulpes*, and cats, *Felis catus*, relative to fluctuating prey populations and drought. *Australian Wildlife Research* **15**: 307–317.

Center, S. A., K. L. Warner, J. F. Randolph, J. J. Wakshlag, and G. D. Sunvold. (2011). Resting energy expenditure per lean body mass determined by indirect calorimetry and bioelectrical impedance analysis in cats. *Journal of Veterinary Internal Medicine* **25**: 1341–1350.

Centonze, L. A. and J. K. Levy. (2002). Characteristics of free-roaming cats and their caretakers. *Journal of the American Veterinary Medical Association* **220**: 1627–1633.

Chamberlin, A., Y. Mitsuhashi, K. Bigley, and J. E. Bauer. (2011). Unexpected depletion of plasma arachidonate and total protein in cats fed a low arachiodonic acid diet due to peroxidation. *British Journal of Nutrition* **106**: S131–S134.

Chamero, P., T. F. Marton, D. W. Logan, K. Flanagan, J. R. Cruz, and A. Saghatelian et al. (2007). Identification of protein pheromones that promote aggressive behaviour. *Nature* **450**: 899–903.

Chase, I. D., C. Bartolomeo, and L. A. Dugatkins. (1994). Aggressive interactions and inter-contest interval: how long do winners keep winning. *Animal Behaviour* **48**: 393–400.

Chase, I. D., C. Tovey, D. Spangler-Martin, and M. Manfredonia. (2002). Individual differences versus social dynamics in the formation of animal dominance hierarchies. *Proceedings of the National Academy of Sciences* **99**: 5744–5749.

Chawkins, S. (2012). Complex effort to rid San Nicholas Island of cats declared a success. http://articles.latimes.com/2012/feb/26/local/la-me-adv-san-nicolas-20120227 (accessed 25 February 2014).

Chesler, P. (1969). Maternal influence in learning by observation in kittens. *Science* **166**: 901–903.

Childs, J. E. (1986). Size-dependent predation on rats (*Rattus norvegicus*) by house cats (*Felis catus*) in an urban setting. *Journal of Mammalogy* **67**: 196–199.

Christian, D. P. (1975). Vulnerability of meadow voles, *Microtus pennsylvanicus*, to predation by domestic cats. *American Midland Naturalist* **93**: 498–502.

Christiansen, I. J. (1984). *Reproduction in the Dog and Cat*. London: Baillière Tindall, x + 309 pp.

Church, S. C., J. A. Allen, and J. W. S. Bradshaw. (1994). Anti-apostatic food selection by the domestic cat. *Animal Behaviour* **48**: 747–749.

Church, S. C., J. A. Allen, and J. W. S. Bradshaw. (1996). Frequency-dependent food selection by domestic cats: a comparative study. *Ethology* **102**: 495–509.

Churcher, P. B. and J. H. Lawton. (1987). Predation by domestic cats in an English village. *Journal of Zoology (London)* **212**: 439–455.

Clapperton, B. K., R. J. Pierce, and C. T. Eason. (1992). *Experimental eradication of feral cats* (Felis catus) *from Matakohe (Limestone) Island, Whangarei Harbour*. Science and Research Series No. 54. Wellington, NZ: Department of Conservation, 11 pp.

Clevenger, A. P. (1995). Seasonality and relationships of food resource use of *Martes martes, Genetta genetta* and *Felis catus* in the Balearic Islands. *Revue Écologique (Terre et Vie)* **50**: 109–131.

Clum, N. J., M. P. Fitzpatrick, and E. S. Dierenfeld. (1996). Effects of diet on nutritional content of whole vertebrate prey. *Zoo Biology* **15**: 525–537.

Clutton-Brock, T. (2009). Structure and function in mammalian societies. *Philosophical Transactions of the Royal Society* **364B**: 3229–3242.

Clutton-Brock, T. and K. McAuliffe. (2009). Female mate choice in mammals. *Quarterly Review of Biology* **84**: 3–27.

Clutton-Brock, T. H. (1989). Mammalian mating systems. *Proceedings of the Royal Society of London* **236B**: 339–372.

Clutton-Brock, T. H. (2006). Cooperative breeding in mammals. In *Cooperation in Primates and Humans: Mechanisms and Evolution*, P. M. Kappeler and C. P. van Schaik (eds.). Berlin: Springer-Verlag, pp. 173–190.

Clutton-Brock, T. H. and D. Lukas. (2012). The evolution of social philopatry and dispersal in female mammals. *Molecular Ecology* **21**: 472–492.

Clutton-Brock, T. H. and G. A. Parker. (1995). Sexual coercion in animal societies. *Animal Behaviour* **49**: 1345–1365.

Cole, D. D. and J. N. Shafer. (1966). A study of social dominance in cats. *Behaviour* **27**: 39–53.

Coleman, J. S. and S. A. Temple. (1993). Rural residents' free-ranging domestic cats: a survey. *Wildlife Society Bulletin* **21**: 381–390.

Coleman, J. S. and S. A. Temple. (1996). On the prowl. *Wisconsin Natural Resources Magazine*, December issue. http://dnr.wi.gov/wnrmag/html/stories/1996/dec96/cats.htm (accessed 25 February 2014).

Coman, B. J. and H. Brunner. (1972). Food habits of the feral house cat in Victoria. *Journal of Wildlife Management* **36**: 848–853.

Concannon, P. W. (1991). Reproduction in the dog and cat. In *Reproduction in Domestic Animals*, 4th edition, P. T. Cupps (ed.). San Diego, CA: Academic Press, pp. 517–554.

Concannon, P. W. and D. H. Lein. (1983). Feline reproduction. In *Current Veterinary Therapy: Small Animal Practice*, Vol. 8, R. W. Kirk (ed.). Philadelphia, PA: W. B. Saunders, pp. 932–936.

Concannon, P., B. Hodgson, and D. Lein. (1980). Reflex LH release in estrous cats following single and multiple copulations. *Biology of Reproduction* 23: 111–117.

Concannon, P. W., D. H. Lein, and B. G. Hodgson. (1989). Self-limiting reflex luteinizing hormone release and sexual behavior during extended periods of unrestricted copulatory activity in estrous domestic cats. *Biology of Reproduction* 40: 1179–1187.

Connelly, M. E. and N. B. Todd. (1972). Age at first parity, litter size and survival in cats. *Carnivore Genetics Newsletter* 2: 50–52.

Cook, N. E., E. Kane, Q. R. Rogers, and J. G. Morris. (1985). Self-selection of dietary casein and soy-protein by the cat. *Physiology and Behavior* 34: 583–594.

Cook, L. M. and D. W. Yalden. (1980). A note on the diet of feral cats on Deserta Grande. *Bocagiana* 26: 1–4.

Cooper, J. (1977). Food, breeding and coat colour of feral cats on Dassen Island. *Zoologica Africana* 12: 250–252.

Cooper, J., P. A. R. Hockey, and R. K. Brooke. (1985). Introduced mammals on South and South West African islands: history, effects on birds and control. In *Proceedings of the Symposium on Birds and Man*, L. J. Bunning (ed.). Johannesburg, South Africa: Witwatersrand Bird Club, pp. 179–203.

Cooper, J., A. v. N. Marais, J. P. Bloomer, and M. N. Bester. (1995). A success story: breeding of burrowing petrels (Procellaridae) before and after the eradication of feral cats *Felis catus* at subantarctic Marion Island. *Marine Ornithology* 23: 33–37.

Cooper, J. B. (1944). A description of parturition in the domestic cat. *Journal of Comparative Psychology* 37: 71–79.

Copson, G. and J. Whinam. (2001). Review of ecological restoration programme on subantarctic Macquarie Island: pest management progress and future directions. *Ecological Management and Restoration* 2: 129–138.

Corbett, L. K. (1978). A comparison of the social organization and feeding ecology of domestic cats (*Felis catus*) in two contrasting environments in Scotland. *Carnivore Genetics Newsletter* 3: 269. [Abstract]

Cornwallis, C. K., S. A. West, K. E. Davis, and A. S. Griffin. (2010). Promiscuity and the evolutionary transition to complex societies. *Nature* 466: 969–972.

Cosgrove, J. J., D. H. Beermann, W. A. House, B. D. Toddes, and E. S. Dierenfeld. (2002). Whole-body nutrient composition of various ages of captive-bred bearded dragons (*Pogona vitteceps*) and adult wild anoles (*Anolis carolinensis*). *Zoo Biology* 21: 489–497.

Cottam, Y. H., P. Caley, S. Wamberg, and W. H. Hendriks. (2002). Feline reference values for urine composition. *Journal of Nutrition* 132: 1754S-1756S.

Courchamp, F. and S. J. Cornell. (2000). Virus-vectored immunocontraception to control feral cats on islands: a mathematical model. *Journal of Applied Ecology* 37: 903–913.

Courchamp, F., M. Langlais, and G. Sugihara. (1999a). Cats protecting birds: modelling the mesopredator release effect. *Journal of Animal Ecology* 68: 282–292.

Courchamp, F., M. Langlais, and G. Sugihara. (1999b). Control of rabbits to protect island birds from cat predation. *Biological Conservation* 89: 219–225.

Courchamp, F., J-L. Chapuis, and M. Pascal. (2003). Mammal invaders on islands: impact, control and control impact. *Biological Reviews* 78: 347–383.

Creagh, C. (1992). New approaches to rabbit and fox control. *Ecos* 71: 18–24.

Crémieux, J., C. Veraart, and M. C. Wanet-Defalque. (1986). Effect of deprivation of vision and vibrissae on goal directed locomotion in cats. *Experimental Brain Research* 65: 229–234.

Crespi, B. J. and D. Yanega. (1995). The definition of eusociality. *Behavioral Ecology* 6: 109–115.

Crissey, S. D., J. A. Swanson, B. A. Lintzenich, B. A. Brewer, and K. A. Slifka. (1997). Use of a raw meat-based diet or a dry kibble diet for sand cats (*Felis margarita*). *Journal of Animal Science* 75: 2154–2160.

Crooks, K. R. and M. E. Soulé. (1999). Mesopredator release and avifaunal extinctions in a fragmented system. *Nature* **400**: 563–566.

Croxall, J. P. and P. Rothery. (1991). Population regulation of seabirds: implications of their demography for conservation. In *Bird Population Studies: Relevance to Conservation and Management*, C. M. Perrins, J-D. LeBreton, and G. N. Hirons (eds.). Oxford: Oxford University Press, pp. 272–296.

Cuthbert, R. (2003). Sign [sic] left by introduced and native predators feeding on Hutton's shearwaters *Puffinus huttoni*. *New Zealand Journal of Zoology* **30**: 163–170.

Dards, J. (1981). Habitat utilisation by feral cats in Portsmouth Dockyard. In *The Ecology and Control of Feral Cats*, O. Jackson (chmn.). South Mimms, Potters Bar, Hertfordshire (UK): Universities Federation for Animal Welfare, pp. 30–49.

Dards, J. L. (1978). Home ranges of feral cats in Portsmouth dockyard. *Carnivore Genetics Newsletter* **3**: 242–255.

Dards, J. L. (1983). The behaviour of dockyard cats: interactions of adult males. *Applied Animal Ethology* **10**: 133–153.

Dauphiné, N. and R. J. Cooper. (2009). Impacts of free-ranging domestic cats (*Felis catus*) on birds in the United States: a review of recent research with conservation and management recommendations. In *Proceedings of the Fourth International Partners in Flight Conference: Tundra to Tropics: Connecting Birds, Habitats and People*, T. D. Rich, C. Arizmendi, D. W. Demarist, and C. Thompson (eds.). McAllen, TX: Partners in Flight, pp. 205–219.

Davies, N. B. (1991). Mating systems. In *Behavioural Ecology: An Evolutionary Approach*, 3rd edn., J. R. Krebs and N. B. Davies (eds.). London: Blackwell Scientific, pp. 263–294.

Davis, D. E. (1951). The relation between level of population and size and sex of Norway rats. *Ecology* **32**: 462–464.

Davis, D. E. (1957). The use of food as a buffer in a predator-prey system. *Journal of Mammalogy* **38**: 466–472.

Dawood, A. A. and M. A. Alkanhal. (1995). Nutrient composition of Najdi-Camel meat. *Meat Science* **39**: 71–78.

Dawson, A. B. (1946). Postpartum history of the corpus luteum of the cat. *Anatomical Record* **95**: 29–51.

Dawson, A. B. (1952). The domestic cat: *Felis catus* Linnaeus, 1758; *Felis domestica* Gemlin, 1788. In *The Care and Breeding of Laboratory Animals*, E. J. Farris (ed.). New York: John Wiley & Sons, Inc., pp. 202–233.

Dawson, A. B. and H. B. Friedgood. (1940). The time and sequence of preovulatory changes in the cat ovary after mating or mechanical stimulation of the cervix uteri. *Anatomical Record* **76**: 411–429.

Deag, J. M., C. E. Lawrence, and A. Manning. (1987). The consequences of differences in litter size for the nursing cat and her kittens. *Journal of Zoology (London)* **213B**: 153–179.

Deag, J. M., A. Manning, and C. E. Lawrence. (1988). Factors influencing the mother-kitten relationship. In *The Domestic Cat: The Biology of Its Behaviour*, D. C. Turner and P. Bateson (eds.). Cambridge: Cambridge University Press, pp. 23–39.

Dean, L. G., R. L. Kendal, S. J. Shapiro, B. Thierry, and K. N. Laland. (2012). Identification of the social and cognitive processes underlying human cumulative culture. *Science* **335**: 114–118.

de Boer, J. N. (1977a). The age of olfactory cues functioning in chemocommunication among male domestic cats. *Behavioural Processes* **2**: 209–225.

de Boer, J. N. (1977b). Dominance relations in pairs of domestic cats. *Behavioural Processes* **2**: 227–242.

DeLaurier, A., A. Boyde, M. A. Horton, and J. S. Price. (2006). Analysis of the surface characteristics and mineralization status of feline teeth using scanning electron microscopy. *Journal of Anatomy* **209**: 655–669.

Denny, E., P. Yakovlevich, M. D. B. Eldridge, and C. Dickman. (2002). Social and genetic analysis of a population of free-living cats (*Felis catus* L.) exploiting a resource-rich habitat. *Wildlife Research* **29**: 405–413.

de-Oliveira, L. D., A. C. Carciofi, M. C. C. Oliveira, R. S. Vasconcellos, R. S. Bazolli, and G. T. Pereira *et al.* (2008). Effects of six carbohydrate sources on diet digestibility and postprandial glucose and insulin responses in cats. *Journal of Animal Science* **86**: 2237–2246.

Derenne, P. (1976). Notes sur la biologie du chat haret de Kerguelen. *Mammalia* **40**: 532–595.

Derenne, P. and J. L. Mougin. (1976). Données écologiques sur les mammifères introduits de l'Ile aux Cochons, Archipel Crozet (46°06'S, 50°14'E). *Mammalia* **40**: 22–53.

Devillard, S., L. Say, and D. Pontier. (2003). Dispersal pattern of domestic cats (*Felis catus*) in a promiscuous urban population: do females disperse or die? *Journal of Animal Ecology* **72**: 203–211.

Devillard, S., L. Say, and D. Pontier. (2004). Molecular and behavioural analyses reveal male-biased dispersal between social groups of domestic cats. *Ecoscience* **11**: 175–180.

Devillard, S., H. Santin-Janin, L. Say, and D. Pontier. (2011). Linking genetic diversity and temporal fluctuations in population abundance of the introduced feral cat (*Felis silvestris catus*) on the Kerguelen archipelago. *Molecular Ecology* **20**: 1541–5153.

De Wilde, R. O. and T. Jansen. (1989). The use of different sources of raw and heated starch in the ration of weaned kittens. In *Nutrition of the Dog and Cat: Waltham Symposium 7*, I. H. Burger and J. P. W. Rivers (eds.). New York: Cambridge University Press, pp. 259–266.

Dickman, C. R. (1996a). *Overview of the Impacts of Feral Cats on Australian Native Fauna*. Canberra: Australian Nature Conservation Agency, 93 pp. (unpaginated).

Dickman, C. R. (1996b). Impact of exotic generalist predators on the native fauna of Australia. *Wildlife Biology* **2**: 185–195.

Didham, R. K., J. M. Tylianakis, N. J. Gemmell, T. A. Rand, and R. M. Ewers. (2007). Interactive effects of habitat modification and species invasion on native species decline. *Trends in Ecology and Evolution* **22**: 489–496.

Dilks, P. J. (1979). Observations on the food of feral cats on Campbell Island. *New Zealand Journal of Ecology* **2**: 64–66.

Domm, S. and J. Messersmith. (1990). Feral cat eradication on a barrier reef island, Australia. *Atoll Research Bulletin* (**338**): 1–4.

Donoghue, A. M., L. A. Johnston, K. L. Goodrowe, S. J. O'Brien, and D. E. Wildt. (1993). Influence of day of oestrus on egg viability and comparative efficiency of in vitro fertilization in domestic cats in natural or gonadotrophin-induced oestrus. *Journal of Reproduction and Fertility* **98**: 85–90.

Douglas, H. D. and P. Fernández. (1997). A longevity record for the waved albatross. *Journal of Field Ornithology* **68**: 224–227.

Douglas, T. C., M. Pinnino, and E. S. Dierenfeld. (1994). Vitamins E and A, and proximate composition of whole mice and rats used as feed. *Comparative Biochemistry and Physiology* **107A**: 419–424.

Dowding, J. E., E. C. Murphy, K. Springer, A. J. Peacock, and C. J. Krebs. (2009). Cats, rabbits, *Myxoma* virus, and vegetation on Macquarie Island: a comment on Bergstrom et al. (2009). *Journal of Applied Ecology* **46**: 1129–1132.

Dreux, P. (1970). La population des Chats de la Péninsule Courbet (Ile de Kerguelen). Un exemple de l'effet du foundateur. *Terres Australes et Antarctiques Françaises* (52–53): 45–46.

Drews, C. (1993). The concept and definition of dominance in animal behaviour. *Behaviour* **125**: 283–313.

Driscoll, C., N. Yamaguchi, S. J. O'Brien, and D. W. Macdonald. (2011). A suite of genetic markers useful in assessing wildcat (*Felis silvestris* ssp.) – domestic cat (*Felis silvestris catus*) admixture. *Journal of Heredity* **102** (Suppl. 1): S87–S90.

Driscoll, W. W. and J. W. Pepper. (2010). Theory for the evolution of diffusible external goods. *Evolution* **64**: 2682–2687.

Duffy, D. C. and P. Capece. (2012). Biology and impacts of Pacific island invasive species. 7. The domestic cat (*Felis catus*). *Pacific Science* **66**: 173–212.

Dugatkin, L. A., M. Perlin, and R. Atlas. (2003). The evolution of group-beneficial traits in the absence of between-group selection. *Journal of Theoretical Biology* **220**: 67–74.

Duvaux-Ponter, C., K. Rigalma, S. Roussel-Huchette, Y. Schawlb, and A. A. Ponter. (2008). Effect of a supplement rich in linolenic acid, added to the diet of gestating and lactating goats, on the sensitivity to stress and learning ability of their offspring. *Applied Animal Behaviour Science* **114**: 373–394.

Eason, C. T. and M. C. Frampton. (1991). Acute toxicity of sodium monofluoroacetate (1080) baits to feral cats. *Wildlife Research* **18**: 445–449.

Eason, C. T., E. C. Murphy, S. Hix, and D. B. MacMorran. (2010). Development of a new humane toxin for predator control in New Zealand. *Integrative Zoology* **5**: 31–36.

Ebenhard, T. (1988). Introduced birds and mammals and their ecological effects. *Swedish Wildlife Research (Viltrevy)* **13**(4): 1–78.

Eberhard, T. (1954). Food habits of Pennsylvania house cats. *Journal of Wildlife Management* **18**: 284–286.

Edwards, G. P., K. C. Piddington, and R. M. Paltridge. (1997). Field evaluation of olfactory lures for feral cats (*Felis catus* L.) in central Australia. *Wildlife Research* **24**: 173–183.

Edwards, G. P., N. de Preu, B. J. Shakeshaft, I. V. Crealy, and R. M. Paltridge. (2001). Home range and movements of male feral cats (*Felis catus*) in a semiarid woodland environment in central Australia. *Australian Ecology* **26**: 93–101.

Egan, J. (1976). Object-play in cats. In *Play – Its Role in Development and Evolution*, J. Bruner, A. Jolly, and K. Sylva (eds.). New York: Basic Books, pp. 161–165.

Eisert, R. (2011). Hypercarnivory and the brain: protein requirements of cats reconsidered. *Journal of Comparative Physiology* **181B**: 1–17.

Elgar, M. A. and P. H. Harvey. (1987). The lions share. *Trends in Ecology and Evolution* **2**: 57–58.

Erofeeva, M. N. M., and S. V. S. Naĭdenko. [Spatial organization of felid populations and some traits of their reproductive strategies.]. (2011) *Zhurnal Obshceĭ Biologii* **72**: 284–297. [Russian with English abstract]

Errington, P. L. (1936). Notes on food habits of southern Wisconsin house cats. *Journal of Mammalogy* **17**: 64–65.

Espinosa-Gayosso, C. V. and S. T. Álvarez-Castañeda. (2006). Status of *Dipodomys insularis*, an endemic species of San José Island, Gulf of California, Mexico. *Journal of Mammalogy* **87**: 677–682.

Evans, C. A., W. E. Carlson, and R. G. Green. (1942). The pathology of Chastek paralysis in foxes: a counterpart of Wernicke's hemorrhagic polioencephalitis. *American Journal of Pathology* **18**: 79–91.

Everett, G. M. (1944). Observations on the behavior and neurophysiology of acute thiamin deficient cats. *American Journal of Physiology* **141**: 439–448.

Ewer, R. F. (1959). Suckling behaviour in kittens. *Behaviour* **15**: 146–162.

Ewer, R. F. (1961). Further observations on suckling behaviour in kittens, together with some general consideration of the interrelations of innate and acquired responses. *Behaviour* **17**: 247–260.

Ewer, R. F. (1969). The "instinct to teach." *Nature* **222**: 698.

Fagen, R. (1974). Selective and evolutionary aspects of animal play. *American Naturalist* **108**: 850–858.

Fagen, R. M. (1976). Exercise, play, and physical training in animals. In *Perspectives in Ethology, Vol. 2*, P. P. G. Bateson and P. H. Klopfer (eds.). New York: Plenum Press, pp. 189–219.

Fagen, R. M. (1978). Population structure and social behavior in the domestic cat (*Felis catus*). *Carnivore Genetics Newsletter* **3**: 276–281.

Fanatico, A. C., L. C. Cavitt, P. B. Pillai, J. L. Emmert, and C. M. Owens. (2005). Evaluation of slower-growing broiler genotypes grown with and without outdoor access: meat quality. *Poultry Science* **84**: 1785–1790.

Faulquier, L., R. Fontaine, E. Vidal, M. Salamolard, and M. Le Corre. (2009). Feral cats *Felis catus* threaten the endangered endemic Barau's petrel *Pterodroma baraui* at Reunion Island (western Indian Ocean). *Waterbirds* **32**: 330–336.

Fedyk, A. (1974). Gross body composition in postnatal development of the bank vole. I. Growth under laboratory conditions. *Acta Theriologica* 19: 381–401.

Feldman, E. C. and R. W. Nelson. (1996). *Canine and Feline Endocrinology and Reproduction*, 2nd edition. Philadelphia, PA: W. B. Saunders, 785 pp.

Feldman, H. N. (1993). Maternal care and differences in the use of nests in the domestic cat. *Animal Behaviour* 45: 13–23.

Feldman, H. N. (1994). Methods of scent marking in the domestic cat. *Canadian Journal of Zoology* 72: 1093–1099.

Fendt, M. (2006). Exposure to urine of canids and felids, but not of herbivores, induces defensive behavior in laboratory rats. *Journal of Chemical Ecology* 32: 2617–2627.

Ferris, D. A., R. A. Flores, C. W. Shanklin, and M. K. Whitworth. (1995). Proximate analysis of food service wastes. *Applied Engineering in Agriculture* 11: 567–572.

Fettman, M. J., C. A. Stanton, L. L. Banks, D. W. Hamar, D. E. Johnson, and R. L. Hegstad et al. (1997). Effects of neutering on bodyweight [sic], metabolic rate and glucose tolerance of domestic cats. *Research in Veterinary Science* 62: 131–136.

Fisher, P., D. Algar, and J. Sinagra. (1999). Use of Rhodamine B as a systemic bait marker for feral cats (*Felis catus*). *Wildlife Research* 26: 281–285.

Fitzgerald, B. M. (1978). Feeding ecology of feral house cats in New Zealand forest [sic]. *Carnivore Genetics Newsletter* 4: 67–71.

Fitzgerald, B. M. (1988). Diet of domestic cats and their impact [sic] on prey populations. In *The Domestic Cat: The Biology of Its Behaviour*, D. C. Turner and P. Bateson (eds.). Cambridge: Cambridge University Press, pp. 123–144 + appendix (2 pp.)

Fitzgerald, B. M. and B. J. Karl. (1979). Foods of feral house cats (*Felis catus* L.) in a forest of the Orongorongo Valley, Wellington. *New Zealand Journal of Zoology* 6: 107–126.

Fitzgerald, B. M. and B. J. Karl. (1986). Home range of feral house cats (*Felis catus* L.) in forest of the Orongorongo Valley, Wellington, New Zealand. *New Zealand Journal of Ecology* 9: 71–81.

Fitzgerald, B. M., B. J. Karl, and C. R. Veitch. (1991). The diet of feral cats (*Felis catus*) on Raoul Island, Kermadec Group. *New Zealand Journal of Ecology* 15: 123–129.

Fitzgerald, B. M. and C. R. Veitch. (1985). The cats of Herekopare Island, New Zealand; their history, ecology and effects on birdlife. *New Zealand Journal of Zoology* 12: 319–330.

Flannelly, K. J. and R. J. Blanchard. (1981). Dominance: cause or description of social relationships. *Behavioral and Brain Sciences* 4: 438–440.

Flux, J. E. C. (2007). Seventeen years of predation by one suburban cat in New Zealand. *New Zealand Journal of Zoology* 34: 289–296.

Flynn, M. F., E. M. Hardie, and P. J. Armstrong. (1996). Effect of ovariohysterectomy on maintenance energy requirement in cats. *Journal of the American Veterinary Medical Association* 209: 1572–1581.

Foley, P., J. E. Foley, J. K. Levy, and T. Paik. (2005). Analysis of the impact of trap-neuter-return programs on populations of feral cats. *Journal of the American Veterinary Medical Association* 227: 1775–1781.

Forbush, E. H. (1916). *The Domestic Cat: Bird Killer, Mouser and Destroyer of Wild Life; Means of Utilizing and Controlling It*. Economic Biology – Bulletin No. 2. Boston, MA: Commonwealth of Massachusetts, State Board of Agriculture, 112 pp.

Foster, M. A. and F. L. Hisaw. (1935). Experimental ovulation and the resulting pseudopregnancy in anoestrous cats. *Anatomical Record* 62: 75–92.

Fox, M. W. (1969). Behavioral effects of rearing dogs with cats during the 'critical period of socialization.' *Behaviour* 35: 273–280.

Fox, M. W. (1975). The behaviour of cats. In *The Behaviour of Domestic Animals*, 2nd edn., E. S. E. Hafez (ed.). London: Baillière Tindall, pp. 107–126.

França, L. R., and C. L. Godinho. (2003). Testis morphometry, seminiferous epithelium cycle length, and the daily sperm production in domestic cats (*Felis catus*). *Biology of Reproduction* 68: 1554–1561.

Fritts, T. H. and G. H. Rodda. (1998). The role of introduced species in the degradation of island ecosystems: a case history of Guam. *Annual Review of Ecology and Systematics* **29**: 113–140.

Funaba, M., T. Tanaka, M. Kaneko, T. Iriki, Y. Hatano, and M. Abe. (2001). Fish meal vs. corn gluten meal as a protein source for dry cat food. *Journal of Veterinary Medical Science* **63**: 1355–1357.

Furet, L. (1989). Régime alimentaire et distribution du chat haret (*Felis catus*) sur l'île Amsterdam. *Revue Écologique (Terre et Vie)* **44**: 33–45.

Gage, F. H. (1981). Dominance: measure first and then define. *Behavioral and Brain Science* **4**: 440–441.

Gallo, P. V., J. Werboff, and K. Knox. (1980). Protein restriction during gestation and lactation: development of attachment behavior in cats. *Behavioral and Neural Biology* **29**: 216–223.

Gallo, P. V., J. Werboff, and K. Knox. (1984). Development of home orientation in offspring of protein-restricted cats. *Developmental Psychobiology* **17**: 437–449.

Gauthreaux, S. A. Jr., (1981). Behavioral dominance from an ecological perspective. *Behavioral and Brain Sciences* **4**: 441.

Genovesi, P., M. Besa, and S. Toso. (1995). Ecology of a feral cat *Felis catus* population in an agricultural area of northern Italy. *Wildlife Biology* **1**: 233–237.

George, W. G. (1974). Domestic cats as predators and factors in winter shortages of raptor prey. *Wilson Bulletin* **86**: 384–396.

Germain, E., S. Benhamou, and M-L. Poulle. (2008). Spatio-temporal sharing between the European wildcat, the domestic cat and their hybrids. *Journal of Zoology* **276**: 195–203.

Ghiselin, M. T. (1974). *The Economy of Nature and the Evolution of Sex*. Berkeley, CA: University of California Press, xii + 346 pp.

Giradet, S. A. B., C. R. Veitch, and J. L. Craig. (2001). Bird and rat numbers on Little Barrier Island, New Zealand, over the period of cat eradication 1976–80. *New Zealand Journal of Zoology* **28**: 13–29.

Gil-Sánchez, J. M., G. Valenzuela, and J. F. Sánchez. (1999). Iberian wild cat *Felis silvestris tartessia* predation on rabbit [sic] *Oryctolagus cuniculus*: functional response and age selection. *Acta Theriologica* **44**: 421–428.

Girardet, S. A. B., C. R. Veitch, and J. L. Craig. (2001). Bird and rat numbers on Little Barrier Island, New Zealand, over the period of cat eradication 1976–80. *New Zealand Journal of Zoology* **8**: 13–29.

Gittleman, J. L. (1989). Carnivore group living: comparative trends. In *Carnivore Behavior, Ecology, and Evolution*, Vol. 1, J. L. Gittleman (ed.). Ithaca, NY: Cornell University Press, pp. 183–207.

Gittleman, J. L. and P. H. Harvey. (1982). Carnivore home-range size, metabolic needs and ecology. *Behavioral Ecology and Sociobiology* **10**: 57–63.

Glass, E. N., J. Odle, and D. H. Baker. (1992). Urinary taurine excretion as a function of taurine intake in adult cats. *Journal of Nutrition* **122**: 1135–1142.

Glen, A. S. and C. R. Dickman. (2005). Complex interactions among mammalian carnivores in Australia, and their implications for wildlife management. *Biological Reviews* **80**: 387–401.

Glickman, S. E., C. J. Zabel, S. I. Yoerg, M. L. Weldele, C. M. Drea, and L. G. Frank. (1997). Social facilitation, affiliation, and dominance in the social life of spotted hyenas. *Annals of the New York Academy of Sciences* **807**: 175–184.

Glover, T. E., P. F. Watson, and R. C. Bonney. (1985). Observations on variability in LH release and fertility during oestrus in the domestic cat (*Felis catus*). *Journal of Reproduction and Fertility* **75**: 145–152.

Goericke-Pesch, S. (2010). New developments in non-surgical methods. *Journal of Feline Medicine and Surgery* **12**: 539–546.

González-Redondo, P., L. Velarde Gómez, L. Guerrero-Herrero, and V. M. Fernández-Cabanás. (2010). Composición química de la carne de conejo silvestre (*Oryctolagus cuniculus*) y viabilidad de su predicción mediante espectroscopía de infrarrojo cercano. *Información Técnica Económica Agraria* **106**: 184–196.

Goodrowe, K. L., P. F. Watson, and R. C. Bonney. (1985). Pituitary and gonadal response to exogenous LH-releasing hormone in the male domestic cat. *Journal of Endocrinology* 105: 175–181.

Goodrowe, K. L., J. G. Howard, P. M. Schmidt, and D. E. Wildt. (1989). Reproductive biology of the domestic cat with special reference to endocrinology, sperm function and in-vitro fertilization. *Journal of Reproduction and Fertility Supplement* 39: 73–90.

Gooneratne, S. R., C. T. Eason, C. J. Dickson, H. Fitzgerald, and G. Wright. (1995). Persistence of sodium monofluoroacetate in rabbits and risk to non-target species. *Human and Experimental Toxicology* 14: 212–216.

Gordon, J. K., C. Matthaei, and Y. van Heezik. (2010). Belled collars reduce catch of domestic cats in New Zealand. *Wildlife Research* 37: 372–378.

Gorman, S. and J. Levy. (2004). A public policy toward the management of feral cats. *Pierce Law Review* 2: 157–181.

Goszczyński, J., D. Krauze, and J. Gryz. (2009). Activity and exploration range of house cats in rural areas of central Poland. *Folia Zoologica* 58: 363–371.

Gould, S. J. and R. C. Lewontin. (1979). The spandrels of San Marco and the panglossian paradigm: a critique of the adaptationist programme. *Proceedings of the Royal Society of London* 205B: 581–598.

Green, A. S., J. J. Ramsey, C. Villaverde, D. K. Asami, A. Wei, and A. J. Fascetti. (2008). Cats are able to adapt protein oxidation to protein intake provided their requirement for dietary protein is met. *Journal of Nutrition* 138: 1053–1060.

Green, J. D., C. D. Clemente, and J. de Groot. (1957). Rhinencephalic lesions and behavior in cats. *Journal of Comparative Neurology* 108: 505–545.

Green, P. and E. Yavin. (1993). Elongation, desaturation, and esterification of essential fatty acids by fetal rat brain in vivo. *Journal of Lipid Research* 34: 2099–2107.

Greer, M. B. and M. L. Calhoun. (1966). Anal sacs of the cat (*Felis domesticus*). *American Journal of Veterinary Research* 27: 773–781.

Green, R. G., W. E. Carlson, and A. Evans. (1941). A deficiency disease of foxes produced by feeding fish: B_1 avitaminosis analogous to Wernicke's disease of man. *Journal of Nutrition* 21: 243-256.

Greulich, W. W. (1934). Artificially induced ovulation in the cat (*Felis domestica*). *Anatomical Record* 58: 217–224.

Griffiths, S., S. Dow, and O. Burman. (2010). Ultrasonic vocalizations and their associations with the non-vocalization behaviour of the endangered Turkish spiny mouse *Acomys cilicius* Spitzenberger in a captive population. *Bioacoustics* 19: 143–157.

Gruber, H. E., J. S. Girgus, and A. Banuazizi. (1971). The development of object permanence in the cat. *Developmental Psychology* 4: 9–15.

Gudermuth, D. F., L. Newton, P. Daels, and P. Concannon. (1997). Incidence of spontaneous ovulation in young, group-housed cats based on serum and faecal concentrations of progesterone. *Journal of Reproduction and Fertility Supplement* 51: 177–184.

Guitton, D., R. M. Douglas, and M. Volle. (1984). Eye-head coordination in cats. *Journal of Neurophysiology* 52: 1030–1050.

Gunther, R. (1951). The absolute threshold for vision in the cat. *Journal of Physiology* 114: 8–15.

Guttilla, D. A. and P. Stapp. (2010). Effects of sterilization on movements of feral cats at a wildland–urban interface. *Journal of Mammalogy* 91: 482–489.

Guyot, G. W., T. L. Bennett, and H. A. Cross. (1980). The effects of social isolation on the behavior of juvenile domestic cats. *Developmental Psychobiology* 13: 317–329.

Hahn, M. E. and M. J. Lavooy. (2005). A review of the methods of studies on infant ultrasound production and maternal retrieval in small rodents. *Behavior Genetics* 35: 31–52.

Hall, L. S., M. A. Kasparian, D. van Vuren, and D. A. Kelt. (2000). Spatial organization and habitat use of feral cats (*Felis catus* L.) in Mediterranean California. *Mammalia* 64: 19–28.

Hall, S. L. and J. W. S. Bradshaw. (1998). The influence of hunger on object play by adult domestic cats. *Applied Animal Behaviour Science* 58: 143–150.

Hall, S. L., J. W. S. Bradshaw, and I. H. Robinson. (2002). Object play in adult domestic cats: the roles of habituation and disinhibition. *Applied Animal Behaviour Science* 79: 263–271.

Hall, V. E. and G. N. Pierce Jr., (1934). Litter size, birth weight and growth to weaning in the cat. *Anatomical Record* 60: 111–124.

Hamilton, W. D. (1964a). The genetical evolution of social behaviour. I. *Journal of Theoretical Biology* 7: 1–16.

Hamilton, W. D. (1964b). The genetical evolution of social behaviour. II. *Journal of Theoretical Biology* 7: 17–52.

Hammer, V. A., Q. R. Rogers, and J. G. Morris. (1996). Dietary crude protein increases slightly the requirement for threonine in kittens. *Journal of Nutrition* 126: 1496–1504.

Hammond, C. (1981). Long term management of feral cat colonies. In *The Ecology and Control of Feral Cats*, O. Jackson (chmn.). South Mimms, Potters Bar, Hertfordshire (UK): Universities Federation for Animal Welfare, pp. 89–91.

Hamner, C. E., L. L. Jennings, and N. J. Sojka. (1970). Cat (*Felis catus* L.) spermatozoa require capacitation. *Journal of Reproduction and Fertility* 23: 477–480.

Hand, J. L. (1986). Resolution of social conflicts: dominance, egalitarianism, spheres of dominance, and game theory. *Quarterly Review of Biology* 61: 201–220.

Hanson, J. L. and L. M. Hurley. (2012). Female presence and estrous state influence mouse ultrasonic courtship vocalizations. *PLoS One* 7(7): e40782. (11 pp.)

Harestad, A. S. and F. L. Bunnell. (1979). Home range and body weight – a reevaluation. *Ecology* 60: 389–402.

Hargrove, D. M., Q. R. Rogers, C. C. Calvert, and J. G. Morris. (1988). Effects of dietary excesses of the branched-chain amino acids on growth, food intake and plasma amino acid concentrations of kittens. *Journal of Nutrition* 118: 311–320.

Hargrove, D. M., J. G. Morris, and Q. R. Rogers. (1994). Kittens choose a high leucine diet even when isoleucine and valine are the limiting amino acids. *Journal of Nutrition* 124: 689–693.

Harper, A. E., N. J. Benevenga, and R. Wohlheuter. (1970). Effects of ingestion of disproportionate amounts of amino acids. *Physiological Reviews* 50: 428–558.

Harper, G. A. (2004). Feral cats on Stewart Island/Rakiura: population regulation, home range size and habitat use. DOC Science Internal Series 174. Wellington: New Zealand Department of Conservation, 35 pp.

Harper, G. A. (2010). Diet of feral cats on subantarctic Auckland Island. *New Zealand Journal of Ecology* 34: 259–261.

Harrington, F. H. (1978). Ravens attracted to wolf howling. *Condor* 80: 236–237.

Hart, B. L. (1978). *Feline Behavior: A Practitioner Monograph*. Santa Barbara, CA: Veterinary Practice Publishing, 110 pp.

Hart, B. L. and L. A. Hart. (1985). *Canine and Feline Behavioral Therapy*. Philadelphia, PA: Lea and Febiger, x + 275 pp.

Hart, B. L. and M. G. Leedy. (1987). Stimulus and hormonal determinants of flehmen behavior in cats. *Hormones and Behavior* 21: 44–52.

Haspel, C. and R. E. Calhoon. (1989). Home ranges of free-ranging cats (*Felis catus*) in Brooklyn, New York. *Canadian Journal of Zoology* 67: 178–181.

Hawkins, C. C., W. E. Grant, and M. T. Longnecker. (1999). Effect of subsidized house cats on California birds and rodents. *Transactions of the Western Section of the Wildlife Society* 35: 29–33.

Hayes, K. C. and R. E. Carey. (1975). Retinal degeneration associated with taurine deficiency in the cat. *Science* 188: 949–951.

Hayes, R. D., A. M. Baer, U. Wotschikowsky, and A. S. Harestad. (2000). Kill rate by wolves on moose in the Yukon. *Canadian Journal of Zoology* 78: 49–59.

Heffner, R. S. and H. E. Heffner. (1988). Sound localization acuity in the cat: effect of azimuth, signal duration, and test procedure. *Hearing Research* 36: 221–232.

Hendriks, W. H., M. F. Tarttelin, and P. J. Moughan. (1995a). Twenty-four hour feline [sic] excretion patterns in entire and castrated cats. *Physiology and Behavior* 58: 467–469.

Hendriks, W. H., P. J. Moughan, M. F. Tarttelin, and A. D. Woolhouse. (1995b). Felinine: a urinary amino acid of Felidae. *Comparative Biochemistry and Physiology* **112B**: 581–588.

Hendriks, W. H., P. J. Moughan, and M. F. Tarttelin. (1996). Urinary excretion of endogenous nitrogen metabolites in adult domestic cats using a protein-free diet and the regression technique. *Journal of Nutrition* **127**: 623–629.

Hendriks, W. H., P. J. Moughan, and M. F. Tarttelin. (1997). Urinary excretion of endogenous nitrogen metabolites in adult domestic cats using a protein-free diet and the regression technique. *Journal of Nutrition* **127**: 623–629.

Hendriks, W. H., M. M. A. Emmens, B. Trass, and J. R. Pluske. (1999). Heat processing changes the protein quality of canned cat foods as measured with a rat bioassay. *Journal of Animal Science* **77**: 669–676.

Herbert, M. J. and C. M. Harsh. (1944). Observational learning by cats. *Journal of Comparative Psychology* **37**: 81–95.

Herron, M. A. and R. F. Sis. (1974). Ovum transport in the cat and the effect of estrogen administration. *Journal of the American Veterinary Medical Association* **10**: 1277–1279.

Heusner, A. A. (1982a). Energy metabolism and body size. I. Is the 0.75 mass exponent of Kleiber's equation a statistical artifact? *Respiration Physiology* **48**: 1–12.

Heusner, A. A. (1982b). Energy metabolism and body size. II. Dimensional analysis and energetic non-similarity. *Respiration Physiology* **48**: 13–25.

Hewson-Hughes, A. K., V. L. Hewson-Hughes, A. T. Miller, S. R. Hall, S. J. Simpson, and D. Raubenheimer. (2011a). Geometric analysis of macronutrient selection in the adult domestic cat, *Felis catus*. *Journal of Experimental Biology* **214**: 1039–1051.

Hewson-Hughes, A. K., M. S. Gilham, S. Upton, A. Colyer, R. Butterwick, and A. T. Miller. (2011b). The effect of dietary starch level on postprandial glucose and insulin concentrations in cats and dogs. *British Journal of Nutrition* **106**: S105–S109.

Heyward, R. P. and G. L. Norbury. (1998). Secondary poisoning of ferrets and cats after 1080 rabbit poisoning. *Wildlife Research* **25**: 75–80.

Hill, R. C. and K. C. Scott. (2004). Energy requirements and body surface area of cats and dogs. *Journal of the American Veterinary Medical Association* **225**: 689–694.

Hinde, R. A. (1978). Dominance and role – two concepts with dual meanings. *Journal of Social and Biological Structures* **1**: 27–38.

Hinde, R. A. and S. Datta. (1981). Dominance: an intervening variable. *Behavioral and Brain Sciences* **4**: 442.

Hirsch, E., C. Dubose, and H. L. Jacobs. (1978). Dietary control of food intake in cats. *Physiology and Behavior* **20**: 287–295.

Ho, E. (2013). World's oldest bird stumps scientists by giving birth at age 62. http://newsfeed.time.com/2013/02/12/worlds-oldest-wild-bird-stumps-scientists-by-giving-birth-at-age-62/ (accessed 25 February 2014).

Hoenig, M. and D. C. Ferguson. (2002). Effects of neutering on hormonal concentrations and energy requirements in male and female cats. *American Journal of Veterinary Research* **63**: 634–639.

Hoenig, M., K. Thomaseth, M. Waldron, and D. C. Ferguson. (2007). Insulin sensitivity, fat distribution, and adipocytokine response to different diets in lean and obese cats before and after weight loss. *American Journal of Physiology* **292**: R227–R234.

Holy, T. E. and Z. Guo. (2005). Ultrasonic songs of male mice. *PLoS Biology* **3**(12): e386. (8 pp.)

Homberger, D. G., K. Ham, T. Ogunbakin, J. A. Bonin, B. A. Hopkins, and M. C. Osborn *et al.* (2009). The structure of the cornified claw sheath in the domesticated cat (*Felis catus*): implications for the claw-shedding mechanism and the evolution of cornified digital end organs. *Journal of Anatomy* **214**: 620–643.

Hoppitt, W. J. E., G. R. Brown, R. Kendal, L. Rendell, A. Thornton, and M. M. Webster *et al.* (2008). Lessons from animal teaching. *Trends in Ecology and Evolution* **23**: 486–493.

Horn, J. A., N. Mateus-Pinilla, R. E. Warner, and E. J. Heske. (2011). Home range, habitat use, and activity patterns of free-roaming domestic cats. *Journal of Wildlife Management* 75: 1177–1185.

Howard, J. G., J. L. Brown, M. Bush, and D. E. Wildt. (1990). Teratospermic and normospermic domestic cats: ejaculate traits, pituitary-gonadal hormones, and improvement of spermatozoal motility and morphology after swim-up processing. *Journal of Andrology* 11: 204–215.

Howard, J. G., M. A. Barone, M. A. Donoghue, and D. E. Wildt. (1992). The effect of pre-ovulatory anaesthesia on ovulation in laparoscopically inseminated domestic cats. *Journal of Reproduction and Fertility* 96: 175–186.

Howard, W. E. (1957). Amount of food eaten by small carnivores. *Journal of Mammalogy* 38: 516–517.

Howard, W. E. (1960). Innate and environmental dispersal of individual vertebrates. *American Midland Naturalist* 63: 152–161.

Howell, P. G. (1984). An evaluation of the biological control of the feral cat *Felis catus* (Linnaeus, 1758). *Acta Zoologica Fennica* 172: 111–113.

Hsu, Y. and L. L. Wolf. (1999). The winner and loser effect: integrating multiple experiences. *Animal Behaviour* 57: 903–910.

Hu, D., C. Glidden, J. S. Lippert, L. Schnell, J. S. MacIvor, and J. Meisler. (2001). Habitat use and limiting factors in a population of Hawaiian dark-rumped petrels on Mauna Loa, Hawai'i. *Studies in Avian Biology* 22: 234–242.

Hubbard, A. L., S. McOrist, T. W. Jones, R. Boid, R. Scott, and N. Easterbee. (1992). Is survival of European wildcats *Felis silvestris* in Britain threatened by interbreeding with domestic cats? *Biological Conservation* 61: 203–208.

Hubbs, E. L. (1951). Food habits of feral house cats in the Sacramento Valley. *California Fish and Game* 37: 177–189.

Huberman, A. D. and C. M. Niell. (2011). What can mice tell us about how vision works? *Trends in Neurosciences* 34: 464–473.

Hughes, K. L. and M. R. Slater. (2002). Implementation of a feral cat management program on a university campus. *Journal of Applied Animal Welfare Science* 5: 15–28.

Hughes, K. L., M. R. Slater, and L. Haller. (2002). The effects of implementing a feral cat spay/neuter program in a Florida county animal control service. *Journal of Applied Animal Welfare Science* 5: 285–298.

Hurst, J. L., C. E. Payne, C. M. Nevison, A. D. Marie, R. E. Humphries, and D. H. L. Robertson et al. (2001). Individual recognition in mice mediated by major urinary proteins. *Nature* 414: 631–634.

Hutchings, S. (2003). The diet of feral house cats (*Felis catus*) at a regional rubbish tip, Victoria. *Wildlife Research* 30: 103–110.

Hyvärinen, J., L. Hyvärinen, and I. Linnankoski. (1981). Modification of parietal association cortex and functional blindness after binocular deprivation in young monkeys. *Experimental Brain Research* 42: 1–8.

Imber, M. J. (1975). Petrels and predators. *Bulletin of the International Council for Bird Preservation* 12: 260–263.

Imber, M. J., J. A. West, and W. J. Cooper. (2003). Cook's petrel (*Pterodroma cookii*): historic distribution, breeding biology and effects of predators. *Notornis* 50: 221–230.

Ishida, Y., T. Yahara, E. Kasuya, and K. Yamane. (2001). Female control of paternity during copulation: inbreeding avoidance in feral cats. *Behaviour* 138: 235–250.

Iverson, J. B. (1978). The impact of feral cats and dogs on populations of the West Indian rock iguana, *Cyclura carinata*. *Biological Conservation* 14: 63–73.

Izawa, M. and Y. Ono. (1986). Mother–offspring relationship in the feral cat population. *Journal of the Mammalogical Society of Japan* 11: 27–34.

Izawa, M., T. Doi, and Y. Ono. (1982). Grouping patterns of feral cats (*Felis catus*) living on a small island in Japan. *Japanese Journal of Ecology* 32: 373–382.

Jackson, O. F. and J. D. Tovey. (1977). Water balance studies in domestic cats. *Feline Practice* 7: 30–33.

Jackson, W. B. (1951). Food habits of Baltimore, Maryland, cats in relation to rat populations. *Journal of Mammalogy* **32**: 458–461.

Jacobsen, K. L., E. J. DePeters, Q. R. Rogers, and S. J. Taylor. (2004). Influences of stage of lactation, teat position and sequential milk sampling on the composition of domestic cat milk (*Felis catus*). *Journal of Animal Physiology and Animal Nutrition* **88**: 46–58.

Jameson, K. A. M., C. Appleby, and L. C. Freeman. (1999). Finding an appropriate order for a hierarchy based on probabilistic dominance. *Animal Behaviour* **57**: 991–998.

Jarvis, J. R. and C. M. Wathes. (2012). Mechanistic modeling of vertebrate spatial contrast sensitivity and acuity at low luminance. *Visual Neuroscience* **29**: 169–181.

Jaso-Friedmann, L., J. H. Leary III,, K. Praveen, M. Waldron, and M. Hoenig. (2008). The effects of obesity and fatty acids on the feline immune system. *Veterinary Immunology and Immunopathology* **122**: 146–152.

Jessup, D. A. (2004). The welfare of feral cats and wildlife. *Journal of the American Veterinary Medical Association* **225**: 1377–1383.

John, E. R., P. Chesler, F. Bartlett, and I. Victor. (1968). Observation learning in cats. *Science* **159**: 1489–1491.

Johnson, D. D., J. S. Eastridge, D. R. Neubauer, and C. H. McGowan. (1995). Effect of sex class on nutrient content of meat from young goat [sic]. *Journal of Animal Science* **73**: 296–301.

Johnson, J. C. Jr., P. R. Utley, R. L. Jones, and W. C. McCormick. (1975). Aerobic digested municipal garbage as feedstuff for cattle. *Journal of Animal Science* **41**: 1487–1495.

Johnson, L. M. and V. L. Gay. (1981a). Luteinizing hormone in the cat. I. Tonic secretion. *Endocrinology* **109**: 240–246.

Johnson, L. M. and V. L. Gay. (1981b). Luteinizing hormone in the cat. II. Mating-induced secretion. *Endocrinology* **109**: 247–252.

Johnson, M., D. Algar, M. O'Donoghue, and J. Morris. (2011). Field efficacy of the Curiosity feral cat bait on three Australian islands. In *Island Invasives: Eradication and Management*, C. R. Veitch, M. N. Clout, and D. R. Towns (eds.). Gland, Switzerland: International Union for the Conservation of Nature, pp. 182–187.

Johnsson, J. I. and A. Åkerman. (1998). Watch and learn: preview of the fighting ability of opponents alters contest behaviour in rainbow trout. *Animal Behaviour* **56**: 771–776.

Johnstone, I. P., B. J. Bancroft, and J. R. McFarlane. (1984). Testosterone and androstenedione profiles in the blood of domestic tom-cats. *Animal Reproduction Science* **7**: 363–375.

Jones, A. L. and C. T. Downs. (2011). Managing feral cats on a university's campuses: how many are there and is sterilization having an effect?. *Journal of Applied Animal Welfare Science* **14**: 304–320.

Jones, B. R., R. L. Sanson, and R. S. Morris. (1997). Elucidating the risk factors of feline urologic syndrome. *New Zealand Veterinary Journal* **45**: 100–108.

Jones, C. (2002). A model for the conservation management of a 'secondary' prey: sooty shearwater (*Puffinus griseus*) colonies on mainland New Zealand as a case study. *Biological Conservation* **108**: 1–12.

Jones, E. (1977). Ecology of the feral cat, *Felis catus* (L.), (Carnivora: Felidae) on Macquarie Island. *Australian Wildlife Research* **4**: 249–262.

Jones, E. (1989). Felidae. In *Fauna of Australia. Mammalia 1B*, D. W. Walton and B. J. Richardson (eds.). Canberra: Australian Government Publishing Service, 13 pp.

Jones, E. and B. J. Coman. (1981). Ecology of the feral cat *Felis catus* (L.), in south-eastern Australia. I. Diet. *Australian Wildlife Research* **8**: 437–547.

Jones, E. and B. J. Coman. (1982a). Ecology of the feral cat, *Felis catus* (L.), in south-eastern Australia. II. Reproduction. *Australian Wildlife Research* **9**: 111–119.

Jones, E. and B. J. Coman. (1982b). Ecology of the feral cat, *Felis catus* (L.), in south-eastern Australia. III. Home ranges and population ecology in semi-arid north-west Victoria. *Australian Wildlife Research* **9**: 409–420.

Jongman, E. C. (2007). Adaptation of domestic cats to confinement. *Journal of Veterinary Behavior* **2**: 193–196.

Joseph, J. P. and D. Boussaoud. (1985). Role of the cat substantia nigra pars reticulata in eye and head movements. I. Neural activity. *Experimental Brain Research* 57: 286–296.

Kalcounis-Rueppell, M. C., R. Petric, J. R. Briggs, C. Carney, M. M. Marshall, and J. T. Willse et al. (2010). Differences in ultrasonic vocalizations between wild and laboratory California mice (*Peromyscus californicus*). *PLoS One* 5(4): e9705. (10 pp.)

Kalz, B., K. M. Schiebe, I. Wegner, and J. Priemer. (2000). Gesundheitsstatus und Mortalitätsursachen verwilderter Hauskatzen in einem Untersuchungsgebiet in Berlin-Mitte. *Berliner und Munchener Tierärztliche Wochenschrift* 113: 417–422.

Kanarek, R. B. (1975). Availability and caloric density of the diet as determinants of meal patterns in cats. *Physiology and Behavior* 15: 611–618.

Kanchuk, M. L., R. C. Backus, C. C. Calvert, J. G. Morris, and Q. R. Rogers. (2003). Weight gain in gonadectomized normal and lipoprotein lipase-deficient male domestic cats results from increased food intake and not decreased energy expenditure. *Journal of Nutrition* 133: 1866–1874.

Kane, E. (1989). Feeding behaviour of the cat. In *Nutrition of the Dog and Cat: Waltham Symposium 7*, I. H. Burger and J. P. W. Rivers (eds.). New York: Cambridge University Press, pp. 147–158.

Kane, E., J. G. Morris, and Q. R. Rogers. (1981a). Acceptability and digestibility by adult cats of diets made with various sources and levels of fat. *Journal of Animal Science* 53: 1516–1523.

Kane, E., Q. R. Rogers, J. G. Morris, and P. M. B. Leung. (1981b). Feeding behavior of the cat fed laboratory and commercial diets. *Nutrition Research* 1: 499–507.

Kane, E., P. M. B. Leung, Q. R. Rogers, and J. G. Morris. (1987). Diurnal feeding and drinking patterns of adult cats as affected by changing the level of fat in the diet. *Appetite* 9: 89–98.

Kang, I., R. E. Reem, A. L. Kaczmarowski, and J. G. Malpeli. (2009). Contrast sensitivity of cats and humans in scotopic and mesopic conditions. *Journal of Neurophysiology* 102: 831–840.

Karbowski, J. (2007). Global and regional brain metabolic scaling and its functional consequences. *BMC Biology* 5: 18. (11 pp.)

Karl, B. J., and H. A. Best. (1982). Feral cats on Stewart Island; their foods and their effect on kakapo. *New Zealand Journal of Zoology* 9: 287–294.

Karlson, P. and M. Lüscher. (1959). 'Pheromones': a new term for a class of biologically active substances. *Nature* 183: 55–56.

Kaufman, G. A. and D. W. Kaufman. (1977). Body composition of the old-field mouse (*Peromyscus polionotus*). *Journal of Mammalogy* 58: 429–434.

Kaufman, J. H. (1983). On the definitions and functions of dominance and territoriality. *Biological Reviews of the Cambridge Philosophical Society* 58: 1–20.

Kaufman, L. W., G. Collins, W. L. Hill, and K. Collins. (1980). Meal cost and meal patterns in an uncaged domestic cat. *Physiology and Behavior* 25: 135–137.

Kauhala, K., K. Holmala, W. Lammers, and J. Schregel. (2006). Home ranges and densities of medium-sized carnivores in south-east Finland, with special reference to rabies spread. *Acta Theriologica* 51: 1–13.

Kean, S. (2012). Reinventing the pill: male birth control. *Science* 338: 318–320.

Keen, C. L., B. Lonnerdal, M. S. Clegg, L. S. Hurley, J. G. Morris, and Q. R. Rogers et al. (1982). Developmental changes in composition of cats' milk: trace elements, minerals, protein, carbohydrate and fat. *Journal of Nutrition* 112: 1763–1769.

Keitt, B. S. and B. R. Tershy. (2003). Cat eradication significantly decreases shearwater mortality. *Animal Conservation* 6: 307–308.

Keitt, B. S., C. Wilcox, B. R. Tershy, D. A. Croll, and C. J. Donlan. (2002). The effect of feral cats on the population viability of black-vented shearwaters (*Puffinus opisthomelas*) on Natividad Island, Mexico. *Animal Conservation* 5: 217–223.

Keller, L. F. and D. M. Waller. (2002). Inbreeding effects in wild populations. *Trends in Ecology and Evolution* 17: 230–241.

Kendall, P. T., S. E. Blaza, and P. M. Smith. (1983). Comparative digestible energy requirements of adult beagles and domestic cats for body weight maintenance. *Journal of Nutrition* 113: 1946–1955.

Kerby, G. and D. W. Macdonald. (1988). Cat society and the consequences of colony size. In *The Domestic Cat: The Biology of Its Behaviour*, D. C. Turner and P. Bateson (eds.). Cambridge: Cambridge University Press, pp. 67–81.

Kerr, K. R., A. N. Beloshapka, C. L. Morris, and K. S. Swanson. (2011). Nitrogen metabolism of four raw meat diets in domestic cats. *British Journal of Nutrition* **106**: S174–S177.

Kerr, K. R., B. M. Vester Boler, C. L. Morris, K. J. Liu, and K. S. Swanson. (2012). Apparent total tract energy and macronutrient digestibility and fecal fermentation end-product concentrations of domestic cats fed extruded, raw beef-based, and cooked-based diets. *Journal of Animal Science* **90**: 515–522.

Kettelhut, I. C., M. C. Foss, and R. H. Migliorini. (1980). Glucose homeostasis in a carnivorous animal (cat) and in rats fed a high-protein diet. *American Journal of Physiology* **239**: R437–R444.

Kienzle, E. (1993a). Carbohydrate metabolism of the cat. 1. Activity of amylase in the gastrointestinal tract of the cat. *Journal of Animal Physiology and Animal Nutrition* **69**: 92–101.

Kienzle, E. (1993b). Carbohydrate metabolism of the cat. 2. Digestion of starch. *Journal of Animal Physiology and Animal Nutrition* **69**: 102–114.

Kienzle E. (1994). Effect of carbohydrates on digestion in the cat. 1994. *Journal of Nutrition* **124**: 2568S–2571S.

Kienzle, E., G. Edstadtler-Pietsch, and R. Rudnick. (2006). Retrospective study on the energy requirements of adult colony cats. *Journal of Nutrition* **136**: 1973S–1975S.

King, W. B. (1973). Observation status of birds of central Pacific Islands. *Wilson Bulletin* **85**: 89–103.

King, W. B. (1985). Island birds: will the future repeat the past? In *Conservation of Island Birds: Case Studies for the Management of Threatened Island Species*, P. J. Moors (ed.). ICBP Technical Publication No. 3. Cambridge: International Council for Bird Preservation, pp. 3–15.

Kinney, E. C. (1973). Average or mean pH. *Progressive Fish-culturist* **35**: 93.

Kirkpatrick, J. F. (1985). Seasonal testosterone levels, testosterone clearance, and testicular weights in male domestic cats. *Canadian Journal of Zoology* **63**: 1285–1287.

Kirkpatrick, R. D. and M. J. Rauzon. (1986). Foods of feral cats *Felis catus* on Jarvis and Howland Islands, central Pacific Ocean. *Biotropica* **18**: 72–75.

Kitchener, A. C. (1999). Watch with mother: a review of social learning in the Felidae. *Symposia of the Zoological Society of London* **72**: 236–258.

Kleiman, D. G. and J. F. Eisenberg. (1973). Comparison of canid and felid social systems from an evolutionary perspective. *Animal Behaviour* **21**: 637–659.

Klopfer, P. H. (1973). *Behavioral Aspects of Ecology*, 2nd edn. Englewood Cliffs, NJ: Prentice-Hall, xxi + 200 pp.

Knutson, B., J. Burgdorf, and J. Panksepp. (1998). Anticipation of play elicits high frequency ultrasonic vocalizations in young rats. *Journal of Comparative Psychology* **112**: 65–73.

Koepke J. E. and K. H. Pribram. (1971). Effect of milk on the maintenance of sucking behavior in kittens from birth to six months. *Journal of Comparative and Physiological Psychology* **75**: 363–377.

Kokko, H. and P. Lundberg. (2001). Dispersal, migration, and offspring retention in saturated habitats. *American Naturalist* **157**: 188–202.

Kolb, B. and A. J. Nonneman. (1975). The development of social responsiveness in kittens. *Animal Behaviour* **23**: 368–374.

Kolokotrones, T., V. Savage, E. J. Deeds, and W. Fontana. (2010). Curvature in metabolic scaling. *Nature* **464**: 753–756.

Konecny, M. J. (1987a). Home range and activity patterns of feral house cats in the Galápagos Islands. *Oikos* **50**: 17–23.

Konecny, M. J. (1987b). Food habits and energetics of feral house cats in the Galápagos Islands. *Oikos* **50**: 24–32.

Korschgen, L. J. (1957). Food habits of coyotes, foxes, house cats and bobcats. P-R Series No. 15, Missouri Conservation Commission, Fish and Game Division, 63 pp.

Kowalsky, J. R., T. K. Pratt, and J. C. Simon. (2002). Prey taken by feral cats (*Felis catus*) and barn owls (*Tyto alba*) in Hanawi Natural Area Reserve, Maui, Hawai'i. *'Elepaio* **62**(5): 127–131.

Krebs, J. R. and R. Dawkins. (1984). Animal signals: information or manipulation. In *Behavioural Ecology: An Evolutionary Approach*, 2nd edn., J. R. Krebs and N. B. Davies (eds.). Sunderland, MA: Sinauer Associates, pp. 380–402.

Krauze-Gryz, D., J. B. Gryz, J. Goszczyński, P. Chylarecki, and M. Żmilhorski. (2012). The good, the bad, and the ugly: space use and intraguild interactions among three opportunistic predators – cat (*Felis catus*), dog (*Canis lupus* familiaris), and red fox (*Vulpes vulpes*) – under human pressure. *Canadian Journal of Zoology* **90**: 1402–1413.

Kuhlman, G., D. P. Laflamme, and J. M. Ballam. (1993). A simple method for estimating the metabolizable energy content of dry cat foods. *Feline Practice* **21**(2): 16–20.

Kuo, Z. Y. (1930). The genesis of the cat's responses to the rat. *Journal of Comparative Psychology* **11**: 1–35.

Kuo, Z. Y. (1938). Further study on the behavior of the cat toward the rat. *Journal of Comparative Psychology* **25**: 1–8.

Laflamme, D. P. and S. S. Hannah. (2005). Increased dietary protein promotes fat loss and reduces loss of lean body mass during weight loss in cats. *International Journal of Applied Research in Veterinary Medicine* **3**: 62–68.

Landau, H. G. (1951). On dominance relations and the structure of animal societies: I. Effect of inherent characteristics. *Bulletin of Mathematical Biophysics* **13**: 1–19.

Langham, N. P. E. (1990). The diet of feral cats (*Felis catus* L.) on Hawke's Bay farmland, New Zealand. *New Zealand Journal of Zoology* **17**: 243–255.

Langham, N. P. E. (1992). Feral cats (*Felis catus* L.) on New Zealand farmland. II. Seasonal activity. *Wildlife Research* **19**: 707–720.

Langham, N. P. E. and R. E. R. Porter. (1991). Feral cats (*Felis catus* L.) on New Zealand farmland. I. Home range. *Wildlife Research* **18**: 741–760.

Latimer, H. B. and H. L. Ibsen. (1932). The postnatal growth in body weight of the cat. *Anatomical Record* **52**: 1–5.

Laundré, J. (1977). The daytime behaviour of domestic cats in free-roaming populations. *Animal Behaviour* **25**: 990–998.

Lavigne, A. J., D. M. Bird, J. J. Negro, and D. Lacombe. (1994). Growth of hand-reared kestrels. I. The effect of two different diets and feeding frequency. *Growth, Development and Aging* **58**: 191–201.

Le Galliard, J-F., R. Ferrière, and U. Dieckmann. (2005). Adaptive evolution of social traits: origin, trajectories, and correlations. *American Naturalist* **165**: 206–224.

Lekcharoensuk, C., C. A. Osborne, and J. P. Lulich. (2001). Epidemiologic study of risk factors for lower urinary tract diseases in cats. *Journal of the American Veterinary Medical Association* **218**: 1429–1435.

Lepczyk, C. A., A. G. Mertig, and J. Liu. (2003). Landowners and cat predation across rural-to-urban landscapes. *Biological Conservation* **115**: 191–201.

Lesel, R. (1971). Rapport sur l'etat de developpement de la population de chat feral (*Felis lybica* L.) aux Iles Kerguelen au 1" Janvier 1968. *Terres australes et antarctiques françaises (TAAF)* **55/56**: 55–63.

Levi, T. and C. Wilmers. (2012). Wolves–coyotes–foxes: a cascade among carnivores. *Ecology* **93**: 921–929.

Levy, J. K. and P. C. Crawford. (2004). Humane strategies for controlling feral cat populations. *Journal of the American Veterinary Medical Association* **225**: 1354–1360.

Levy, J. K., D. W. Gale, and L. A. Gale. (2003). Evaluation of the effect of a long-term trap–neuter–return and adoption program on a free-roaming cat population. *Journal of the American Veterinary Medical Association* **222**: 42–46.

Levy, J. K., J. A. Friary, L. A. Miller, S. J. Tucker, and K. A. Fagerstone. (2011). Long-term fertility control in female cats with GonaCon™, a GnRH immunocontraceptive. *Theriogenology* **76**: 1517–1525.

Levy, J. K., L. A. Miller, P. C. Crawford, J. W. Ritchey, M. K. Ross, and K. A. Fagerstone. (2004). GnRH immunocontraception of male cats. *Theriogenology* **62**: 1116–1130.

Leyhausen, P. (1965). The communal organization of solitary mammals. *Symposium of the Zoological Society of London* **14**: 249–263.

Leyhausen, P. (1973). Social organization and density tolerance in mammals (1965). In *Motivation and Animal Behavior: An Ethological View*, K. Lorenz and P. Leyhausen (eds.). New York: D. Van Nostrand, pp. 120–143.

Leyhausen, P. (1979). *Cat Behavior: The Predatory and Social Behavior of Domestic and Wild Cats*. New York: Garland STPM Press, xv + 340 pp.

Leyva, H., T. Madley, and G. H. Stabenfeldt. (1989a). Effect of light manipulation on ovarian activity and melatonin and prolactin secretion in the domestic cat. *Journal of Reproduction and Fertility Supplement* **39**: 125–133.

Leyva, H, T. Madley, and G. H. Stabenfeldt. (1989b). Effect of melatonin on photoperiod responses, ovarian secretion of oestrogen, and coital responses in the domestic cat. *Journal of Reproduction and Fertility Supplement* **39**: 135–142.

Li, G., J. E. Janecka, and W. J. Murphy. (2010). Accelerated evolution of *CES7*, a gene encoding a novel major urinary protein in the cat family. *Molecular and Biological Evolution* **28**: 911–920.

Li, X., W. Li, H. Wang, J. Cao, K. Maehashi, and L. Huang *et al.* (2005). Pseudogenization of a sweet-receptor gene accounts for cats' indifference toward sugar. *PLoS Genetics* **1**: 27–35.

Liberg, O. (1980). Spacing patterns in a population of rural free roaming domestic cats. *Oikos* **35**: 336–349.

Liberg, O. (1982). Correction factors for important prey categories in the diet of domestic cats. *Acta Theriologica* **27**: 115–122.

Liberg, O. (1983). Courtship behaviour and sexual selection in the domestic cat. *Applied Animal Ethology* **10**: 117–132.

Liberg, O. (1984a). Home range and territoriality in free ranging house cats. *Acta Zoologica Fennica Supplement* **171**: 283–285.

Liberg, O. (1984b). Food habits and prey impact by feral and house-based domestic cats in a rural area in southern Sweden. *Journal of Mammalogy* **65**: 424–432.

Liberg, O. and M. Sandell. (1988). Spatial organisation and reproduction tactics in the domestic cat and other felids. In *The Domestic Cat: The Biology of Its Behaviour*, D. C. Turner and P. Bateson (eds.). Cambridge: Cambridge University Press, pp. 83–98.

Liberg, O., M. Sandell, D. Pontier, and E. Natoli. (2000). Density, spatial organisation and reproductive tactics in the domestic cat and other felids. In *The Domestic Cat: The Biology of its Behaviour*, 2nd edn., D. C. Turner and P. Bateson (eds.). Cambridge: Cambridge University Press, pp. 119–147.

Liche, H. (1939). Œstrous cycle in the cat. *Nature* **143**: 100.

Lima, S. L. (1987). Clutch size in birds: a predation perspective. *Ecology* **68**: 1062–1070.

Lima, S. L. (1998). Stress and decision making under the risk of predation: recent developments from behavioral, reproductive, and ecological perspectives. *Advances in the Study of Behavior* **27**: 215–290.

Litvaitis, J. A. and W. W. Mautz. (1980). Food and energy use by captive coyotes. *Journal of Wildlife Management* **44**: 56–61.

Loizos, C. (1966). Play in mammals. *Symposium of the Zoological Society of London* **18**: 1–9.

Long, J. A., D. W. Wildt, B. A. Wolfe, J. K. Critser, R. V. DeRossi, and J. Howard. (1996). Sperm capacitation and the acrosome reaction are compromised in teratospermic domestic cats. *Biology of Reproduction* **54**: 638–646.

Longcore, T., C. Rich, and L. M. Sullivan. (2009). Critical assessment of claims regarding management of feral cats by trap-neuter-return. *Conservation Biology* **23**: 887–894.

Longley, W. H. (1911). The maturation of the egg and ovulation in the domestic cat. *American Journal of Anatomy* **12**: 139–172.

López, K. P., M. W. Shilling, and A. Corzo. (2011). Broiler genetic strain and sex effects on meat characteristics. *Poultry Science* **90**: 1105–1111.

Lorenz, K. and P. Leyhausen. (1973). *Motivation of Human and Animal Behavior: An Ethological View*. New York: Van Nostrand Reinhold, xix + 423 pp.

Loss, S. R., T. Will, and P. P. Marra. (2013). The impact of free-ranging domestic cats on wildlife of the United States. *Nature Communications* **4**: 1396, 7 pp.

Loveridge, G. G. (1986). Bodyweight [sic] changes and energy intake of cats during gestation and lactation. *Animal Technology* **37**: 7–15.

Loveridge, G. G. and J. P. W. Rivers. (1989). Bodyweight [sic] changes and energy intakes of cats during pregnancy and lactation. In *Nutrition of the Dog and Cat: Waltham Symposium 7*, I. H. Burger and J. P. W. Rivers (eds.). New York: Cambridge University Press, pp. 113–132.

Ludwig, J. A. and J. F. Reynolds. (1988). *Statistical Ecology: A Primer on Methods and Computing*. New York: John Wiley & Sons, Inc. xviii + 337 pp.

Lukas, D. and T. H. Clutton-Brock. 2013. The evolution of social monogamy in mammals. *Science* **341**: 526–530.

Lundie-Jenkins, G., L. K. Corbett, and C. M. Phillips. (1993). Ecology of the rufous hare-wallaby, *Lagorchestes hirsutus* Gould (Marsupialia: Macropodidae), in the Tanami Desert, Northern Territory. III. Interactions with introduced mammal species. *Wildlife Research* **20**: 495–511.

Lynch, M., J. Conery, and R. Bürger. (1995). Mutation accumulation and the extinction of small populations. *American Naturalist* **146**: 489–518.

Macaluso, C., S. Onoe, and G. Niemeyer. (1992). Changes in glucose level affect rod function more than cone function in the isolated, perfused cat eye. *Investigative Ophthalmology and Visual Science* **33**: 2798–2808.

MacCracken, J. G. (1982). Coyote food in a southern California suburb. *Wildlife Society Bulletin* **10**: 280–281.

Macdonald, D. (1981). The behaviour and ecology of farm cats. In *The Ecology and Control of Feral Cats*, O. Jackson (chmn.). South Mimms, Potters Bar, Hertfordshire (UK): Universities Federation for Animal Welfare, pp. 23–29.

Macdonald, D. W. (1983). The ecology of carnivore social behaviour. *Nature* **301**: 379–384.

Macdonald, D. W. and P. J. Apps. (1978). The social behaviour of a group of semi-dependent farm cats, *Felis catus*: a progress report. *Carnivore Genetics Newsletter* **3**: 256–269.

Macdonald, D. W., P. J. Apps, G. M. Carr, and G. Kerby. (1987). Social dynamics, nursing coalitions and infanticide among farm cats *Felis catus*. *Advances in Ethology* **28**: 1–66.

MacDonald, M. L., Q. R. Rogers, and J. G. Morris. (1983). Role of linoleate as an essential fatty acid for the cat independent of arachidonate synthesis. *Journal of Nutrition* **113**: 1422–1433.

MacDonald, M. L. and Q. R. Rogers, and J. G. Morris. (1984a). Nutrition of the domestic cat, a mammalian carnivore. *Annual Review of Nutrition* **4**: 521–562.

MacDonald, M. L., B. C. Anderson, Q. R. Rogers, C. A. Buffington, and J. G. Morris. (1984b). Essential fatty acid requirements of cats: pathology of essential fatty acid deficiency. *American Journal of veterinary Research* **45**: 1310–1317.

Maher, C. R. and D. F. Lott. (1995). Definitions of territoriality used in the study of variation in vertebrate spacing systems. *Animal Behaviour* **49**: 1581–1597.

Mahlow, J. C. and M. R. Slater. (1996). Current issues in the control of stray and feral cats. *Journal of the American Veterinary Medical Association* **209**: 2016–2020.

Mahon, P. S., P. B. Banks, and C. R. Dickman. (1998). Population indices for wild carnivores: a critical study in sand-dune habitat, south-western Queensland. *Wildlife Research* **25**: 11–22.

Mandeville, S., V. Yaylayan, and B. K. Simpson. (1992). Proximate analysis, isolation and identification of amino acids and sugars from raw and cooked commercial shrimp waste. *Food Biotechnology* **6**: 51–64.

Marquet, P. A., R. A. Quiñones, S. Abades, F. Labra, M. Tognelli, and M. Arim et al. (2005). Scaling and power-laws in ecological systems. *Journal of Experimental Biology* **208**: 1749–1769.

Marshall, W. H. (1961). A note on the food habits of feral cats on Little Barrier Island, New Zealand. *New Zealand Journal of Science* **4**: 822–824.

Martin, G. J. W. and J. S. Rand. (1999). Food intake and blood glucose in normal and diabetic cats fed ad libitum. *Journal of Feline Medicine and Surgery* **1**: 241–251.

Martin, P. (1982). The energy cost of play: definition and estimation. *Animal Behaviour* **30**: 294–295.

Martin, P. (1984a). The meaning of weaning. *Animal Behaviour* **32**: 1257–1259.

Martin, P. (1984b). The (four) whys and wherefores of play in cats: a review of functional, evolutionary, developmental and causal issues. In *Play in Animals and Humans*, P. K. Smith (ed.). Oxford: Blackwell Scientific, pp. 71–94.

Martin, P. (1984c). The time and energy costs of play behaviour in the cat. *Zeitschrift für Tierpsychologie* **64**: 298–312.

Martin, P. (1986). An experimental study of weaning in the domestic cat. *Behaviour* **99**: 221–249.

Martin, P. and P. Bateson. (1985a). The ontogeny of locomotor play behaviour in the domestic cat. *Animal Behaviour* **33**: 502–510.

Martin, P. and P. Bateson. (1985b). The influence of experimentally manipulating a component of weaning on the development of play in domestic cats. *Animal Behaviour* **33**: 511–518.

Martin, P. and T. M. Caro. (1985). On the functions of play and its role in behavioral development. *Advances in the Study of Behavior* **15**: 59–103.

Masserman, J. H. and P. W. Siever. (1944). Dominance, neurosis, and aggression. *Psychosomatic Medicine* **6**: 7–16.

Matias, R. and P. Catry. (2008). The diet of feral cats at New Island, Falkland Islands, and impact on breeding birds. *Polar Biology* **31**: 609–616.

Maxim, P. E. (1981). Dominance: a useful dimension of social communication. *Behavioral and Brain Sciences* **4**: 444–445.

May, B. J. and A. Y. Huang. (1996). Sound orientation behavior in cats. I. Localization of broadband noise. *Journal of the Acoustical Society of America* **100**: 1059–1069.

May, R. M. (1988). Control of feline delinquency. *Nature* **332**: 392–393.

Mayntz, D., V. H. Nielsen, A. Sørensen, S. Toft, D. Raubenheimer, and C. Hejlesen et al. (2009). Balancing of protein and lipid intake by a mammalian carnivore, the mink, *Mustela vison*. *Animal Behaviour* **77**: 349–355.

Mazaki-Tovi, M., S. K. Abood, and P. A. Schenck. (2011). Effect of omega-3 fatty acids on serum concentrations of adipokines in healthy cats. *American Journal of Veterinary Research* **72**: 1259–1265.

McDonald, D. L. (1977). Play and exercise in the California ground squirrel (*Spermophilus beecheyi*). *Animal Behaviour* **25**: 782–784.

McIlroy, J. C. (1981). The sensitivity of Australian animals to 1080 poison II. Marsupial and eutherian Carnivores. *Australian Wildlife Research* **8**: 385–399.

McIlroy, J. C. and E. J. Gifford, (1992). Secondary poisoning hazards associated with 1000-treated carrot- baiting campaigns against rabbits, *Oryctolagus cuniculus*. *Wildlife Research* **19**: 629–641.

McIlroy, J. C., E. J. Gifford, and S. M. Carpenter. (1988). The effect of rainfall on '1080'-treated meat baits used in poisoning campaigns against wild dogs. *Australian Wildlife Research* **15**: 473–483.

McLean, J. G. and E. A. Monger. (1989). Factors determining the essential fatty acid requirements of the cat. In *Nutrition of the Dog and Cat: Waltham Symposium 7*, I. H. Burger and J. P. W. Rivers (eds.). New York: Cambridge University Press, pp. 329–342.

McLean, L., J. L. Hurst, C. J. Gaskell, J. C. M. Lewis, and R. J. Beynon. (2007). Characterization of cauxin in the urine of domestic and big cats. *Journal of Chemical Ecology* **33**: 1997–2009.

McMurry, F. B. and C. C. Sperry. (1941). Food of feral house cats in Oklahoma, a progress report. *Journal of Mammalogy* **22**: 185–190.

Meckstroth, A. M., A. K. Miles, and S. Chandra. (2007). Diets of introduced predators using stable isotopes and stomach contents. *Journal of Wildlife Management* **71**: 2387–2392.

Medina, F. M. and R. García. (2007). Predation of insects by feral cats *(Felis silvestris catus* L., 1758) on an oceanic island (La Palma, Canary island. *Journal of Insect Conservation* **11**: 203–207.

Medina, F. M. and M. Nogales. (2007). Habitat use of feral cats in the main environments of an Atlantic island (La Palma, Canary Islands). *Folia Zoologica* **56**: 277–283.

Medina, F. M. and M. Nogales. (2009). A review on the impacts of feral cats *(Felis silvestris catus)* in the Canary Islands: implications for the conservation of its endangered fauna. *Biodiversity and Conservation* **18**: 829–846.

Medina, F. M., R. García, and M. Nogales. (2006). Feeding ecology of feral cats on a heterogeneous subtropical oceanic island (La Palma, Canarian Archipelago). *Acta Theriologica* **51**: 75–83.

Medina, F. M., E. Bonnaud, E. Vidal, B. R. Tershy, E. S. Zavaleta, and C. J. Donlan *et al.* (2011). A global review of the impacts of invasive cats on island endangered vertebrates. *Global Change Biology* **17**: 3503–3510 + appendices and additional references (25 pp.).

Mellen, J. D. (1992). Effects of early rearing experience on subsequent adult sexual behavior using domestic cats *(Felis catus)* as a model for exotic small felids. *Zoo Biology* **11**: 17–32.

Mellen, J. D. (1993). A comparative analysis of scent-marking, social and reproductive behavior in 20 species of small cats *(Felis). American Zoologist* **33**: 151–166.

Mendelssohn, H. (1982). Wolves in Israel. In *Wolves of the World: Perspectives of Behavior, Ecology, and Conservation*, F. H. Harrington and P. C. Paquet (eds.). Park Ridge, NJ: Noyes Publications, pp. 173–195.

Mendes-de-Almeida, F., M. C. Ferreira Faria, A. Serricella Branco, M. L. Serrão, A. Moreira Souza, and N. Almosny *et al.* (2004). Sanitary conditions of a colony of urban feral cats *(Felis catus* Linnaeus, 1758) in a zoological garden of Rio de Janeiro, Brazil. *Revista do Instituto de Medicina Tropical de São Paulo* **46**: 269–274.

Mendes-de-Almeida, F., G. L. Remy, L. C. Gershony, D. P. Rodriguez, M. Chame, and N. V. Labarthe. (2011). Reduction of feral cat *(Felis catus* Linnaeus 1758) colony size following hysterectomy of adult female cats. *Journal of Feline Medicine and Surgery* **13**: 436–440.

Mendl, M. (1988). The effects of litter-size variation on the development of play behaviour in the domestic cat: litters of one and two. *Animal Behaviour* **36**: 20–34.

Mendoza, D. L. and J. M. Ramirez. (1987). Play in kittens *(Felis domesticus)* and its association with cohesion and aggression. *Bulletin of the Psychonomic Society* **25**: 27–30.

Mercer, J. R. and S.V.P. S. Silva. (1989). Tryptophan metabolism in the cat. In *Nutrition of the Dog and Cat: Waltham Symposium 7*, I. H. Burger and J. P. W. Rivers (eds.). New York: Cambridge University Press, pp. 207–227.

Meredith, M. A. and B. E. Stein. (1996). Spatial determinants of multisensory integration in cat superior colliculus neurons. *Journal of Neurophysiology* **75**: 1843–1857.

Meredith, M. A., H. R. Clemo, and B. E. Stein. (1991). Somatotopic component of the multisensory map in the deep laminae of the cat superior colliculus. *Journal of Comparative Neurology* **312**: 353–370.

Meredith, M. A., M. T. Wallace, and B. E. Stein. (1992). Visual, auditory and somatosensory convergence in output neurons of the cat superior colliculus: multisensory properties of the tecto-reticulo-spinal projection. *Experimental Brain Research* **188**: 181–186.

Meredith, M. A., H. R. H. Clemo, and L. R. Dehneri. (2000). Responses to innocuous, but not noxious, somatosensory stimulation by neurons in the ferret superior colliculus. *Somatosensory and Motor Research* **17**: 297–308.

Michael, R. P. (1958). Sexual behaviour and the vaginal cycle in the cat. *Nature* **181**: 567–568.

Michael, R. P. (1961). Observations upon the sexual behaviour of the domestic cat *(Felis catus* L.) under laboratory conditions. *Behaviour* **18**: 1–24.

Miles, R. C. (1958). Learning in kittens with manipulatory, exploratory, and food incentives. *Journal of Comparative and Physiological Psychology* **51**: 39–42.

Miller, J. D., C. S. Watson, and W. P. Covell. (1963). Deafening effects of noise on cat [sic]. *Acta Oto-laryngologica* **56** (Suppl. 176): 1–91.

Millward, D. J. (1995). A protein-stat mechanism for regulation of growth and maintenance of the lean body mass. *Nutrition Research Reviews* **8**: 93–120.

Mirmovitch, V. (1995). Spatial organisation of urban feral cats (*Felis catus*) in Jerusalem. *Wildlife Research* **22**: 299–310.

Miyazaki, M., K. Kamiie, S. Soeta, H. Taira, and T. Yamashita. (2003). Molecular cloning and characterization of a novel carboxylesterase-like protein that is physiologically present at high concentrations in the urine of domestic cats (*Felis catus*). *Biochemical Journal* **370**: 101–110.

Miyazaki, M., T. Yamashita, Y. Suzuki, Y. Saito, S. Soeta, and H. Taira et al. (2006a). A major urinary protein of the domestic cat regulates the production of felinine, a putative pheromone precursor. *Chemistry and Biology* **13**: 1071–1079.

Miyazaki, M., T. Yamashita, M. Hosokawa, H. Taira, and A. Suzuki. (2006b). Species-, sex-, and age-dependent urinary excretion of cauxin, a mammalian carboxylesterase. *Comparative Biochemistry and Physiology* **145B**: 270–277.

Miyazaki, M., T. Yamashita, H. Taira, and A. Suzuki. (2008). The biological function of cauxin, a major urinary protein of the domestic cat (*Felis catus*). In *Chemical Signals in Vertebrates 11*, J. L. Hurst, R. J. Beynon, S. C. Roberts, and T. Wyatt (eds.). New York: Springer Verlag, pp. 51–60.

Moelk, M. (1979). The development of friendly approach behavior in the cat: a study of kitten–mother relations and the cognitive development of the kitten from birth to eight weeks. *Advances in the Study of Behavior* **10**: 163–224.

Molsher, R. L. (2001). Trapping and demographics of feral cats (*Felis catus*) in central New South Wales. *Wildlife Research* **28**: 631–636.

Molsher, R., A. Newsome, and C. Dickman. (1999). Feeding ecology and population dynamics of the feral cat (*Felis catus*) in relation to the availability of prey in central-eastern New South Wales. *Wildlife Research* **26**: 593–607.

Molsher, R., C. Dickman, A. Newsome, and W. Mueller. (2005). Home ranges of feral cats (*Felis catus*) in central-western New South Wales, Australia. *Wildlife Research* **32**: 587–595.

Molteno, A J., A. Sliwa, and P. R. K. Richardson. (1998). The role of scent marking in a free-ranging, female black-footed cat (*Felis nigripes*). *Journal of Zoology (London)* **245**: 35–41.

Moore, J. and R. Ali. (1984). Are dispersal and inbreeding avoidance related? *Animal Behaviour* **32**: 94–112.

Moors, P. J. and I. A. E. Atkinson. (1984). Predation on seabirds by introduced animals, and factors affecting its severity. In *Status and Conservation of the World's Seabirds*, J. P. Croxall, P. G. H. Evans, and R. W. Schreiber (eds.). ICBP Technical Publication No. 2. Cambridge: International Council for Bird Preservation, pp. 667–690.

Morris, J. G. (1999). Ineffective vitamin D synthesis in cats is reversed by an inhibitor of 7-dehydrocholestrol [sic]-Δ^7-reductase. *Journal of Nutrition* **129**: 903–908.

Morris, J. G. (2001). Unique nutrient requirements of cats appear to be diet-induced evolutionary adaptations. *Recent Advances in Animal Nutrition in Australia* **13**: 187–194.

Morris, J. G. (2002). Idiosyncratic nutrient requirements of cats appear to be diet-induced evolutionary adaptations. *Nutrition Research Reviews* **15**: 153–168.

Morris, J. G. and Q. R. Rogers. (1978a). Ammonia intoxication in the near-adult cat as a result of a dietary deficiency of arginine. *Science* **199**: 431–432.

Morris, J. G. and Q. R. Rogers. (1978b). Arginine: an essential amino acid for the cat. *Journal of Nutrition* **108**: 1944–1953.

Morris, J. G., J. Trudell, and T. Pencovic. (1977). Carbohydrate digestion by the domestic cat (*Felis catus*). *British Journal of Nutrition* **37**: 365–373.

Morris, J. G., Q. R. Rogers, D. L. Winterrowd, and E. M. Kamikawa. (1979). The utilization of ornithine and citrulline by the growing kitten. *Journal of Nutrition* **109**: 724–729.

Morris, J. G., Q. R. Rogers, S. W. Kim, and R. C. Backus. (1994). Dietary taurine requirement of cats is determined by microbial degradation of taurine in the gut. *Advances in Experimental Medicine and Biology* **359**: 59–70.

Morris, J. G., K. E. Earle, and P. A. Anderson. (1999). Plasma 25-hydroxyvitamin D in growing kittens is related to dietary intake of colecalciferol. *Journal of Nutrition* **129**: 909–912.

Moseby, K. E., H. Neilly, J. L. Read, and H. A. Crisp. (2012). Interactions between a top order predator and exotic mesopredators in the Australian rangelands. *International Journal of Ecology*. doi:10.1155/2012/250352. (15 pp.)

Moundras, C., C. Remesy, and C. Demigne. (1993). Dietary protein paradox: decrease of amino acid availability induced by high-protein diets. *American Journal of Physiology* **264**: G1057–G1065.

Mugford, R. A. (1977). External influences on the feeding of carnivores. In *The Chemical Senses and Nutrition*, M. R. Kare and O. Maller (eds.). New York: Academic Press, pp. 25–50.

Mugford, R. A. and C. J. Thorne. (1980). Comparative studies of meal patterns in pet and laboratory housed dogs and cats. In *Nutrition of the Dog and Cat*, R. S. Anderson (ed.). Oxford: Pergamon Press, pp. 3–14.

Munson, L. (2006). Contraception in felids. *Theriogenology* **66**: 126–134.

Murphy, E. C., L. Robbins, J. B. Young, and J. E. Dowding. (1999). Secondary poisoning of stoats after an aerial 1080 poison operation in Pureora Forest, New Zealand. *New Zealand Journal of Ecology* **23**: 175–182.

Murphy, W. J., S. Sun, Z-q. Chen, N. Yuhki, D. Hirschmann, and M. Menotti-Raymond *et al.* (2000). A radiation hybrid map of the cat genome: implications for comparative mapping. *Genome Research* **10**: 691–702.

Musolf, K., F. Hoffmann, and D. J. Penn. (2010). Ultrasonic courtship vocalizations in wild house mice, *Mus musculus musculus*. *Animal Behaviour* **79**: 757–764.

Mykytowycz, R. (1960). Social behaviour of an experimental colony of wild rabbits, *Oryctolagus cuniculus* (L.). III. Second breeding season. *Wildlife Research* **5**: 1–20.

Myrcha, A. and W. Walkowa. (1968). Changes in the caloric value of the body during the postnatal development of white mice. *Acta Theriologica* **13**: 391–400.

Nagy, K. A. (1987). Field metabolic rate and food requirement scaling in mammals and birds. *Ecological Monographs* **57**: 111–128.

Nagy, K. A. (2001). Food requirements of wild animals: predictive equations for free-living mammals, reptiles, and birds. *Nutrition Abstracts and Reviews* **71B**: 21R–31R.

Nagy, K. A., I. A. Girard, and T. K. Brown. (1999). Energetics of free-ranging mammals, reptiles, and birds. *Annual Review of Nutrition* **19**: 247–277.

Natoli, E. (1985a). Behavioural responses of urban feral cats to different types of urine marks. *Behaviour* **94**: 234–243.

Natoli, E. (1985b). Spacing patterns in a colony of urban stray cats (*Felis catus* L.) in the historic centre of Rome. *Applied Animal Behaviour Science* **14**: 289–304.

Natoli, E. and E. De Vito. (1991). Agonistic behaviour, dominance rank and copulatory success in a large multi-male feral cat, *Felis cautus* L., colony in central Rome. *Animal Behaviour* **42**: 227–241.

Natoli, E., A. Baggio, and D. Pontier. (2001). Male and female agonistic and affiliative relationships in a social group of farm cats (*Felis catus* L.). *Behavioural Processes* **53**: 137–143.

Natoli, E., L. Maragliano, G. Cariola, A. Faini, R. Bonanni, and S. Cafazzo *et al.* (2006). Management of feral domestic cats in the urban environment of Rome (Italy). *Preventive Veterinary Medicine* **77**: 180–185.

Nelson, N. S., E. Berman, and J. F. Stara. (1969). Litter size and sex distribution in an outdoor feline colony. *Carnivore Genetics Newsletter* **1**: 181–191.

Nelson, S. H., A. D. Evans, and R. B. Bradbury. (2005). The efficacy of collar-mounted devices in reducing the rate of predation of wildlife by domestic cats. *Applied Animal Behaviour Science* **94**: 273–285.

Neubauer, K., K. Jewgenow, S. Blottner, D. E. Wildt, and B. S. Pukazhenthi. (2004). Quantity rather than quality in teratospermic males: a histomorphometric and flow cytometric evaluation of spermatogenesis in the domestic cat (*Felis catus*). *Biology of Reproduction* **71**: 1517–1524.

Neville, P. F. and J. Remfry. (1984). Effect of neutering on two groups of feral cats. *Veterinary Record* **114**: 447–450.

Nguyen, P. G., H. J. Dumon, B. S. Siliart, L. J. Martin, R. Sergheraert, and V. C. Biourge. (2004). Effects of dietary fat and energy on body weight and composition after gonadectomy in cats. *American Journal of Veterinary Research* **65**: 1708–1713.

Noble, G. K. (1939). The rôle of dominance in the social life of birds. *Auk* **56**: 263–273.

Nogales, M., J. L. Rodriguez, G. Delgado, V. Quilis, and O. Truijillo. (1992). The diet of feral cats (*Felis catus*) on Alegranza Island (north of Lanzarote, Canary Islands. *Folia Zoologica* **41**: 209–212.

Nogales, M. and F. M. Medina. (1996). A review of the diet of feral domestic cats (*Felis silvestris* f. *catus*) on the Canary Islands, with new data from the laurel forest of La Gomera. *Zeitschrift für Säugetierkunde* **61**: 1–6.

Nogales, M., F. M. Medina, and A. Valido. (1996). Indirect seed dispersal by the feral cats *Felis catus* in island ecosystems (Canary Islands). *Ecography* **19**: 3–6.

Nogales, M., A. Martín, B. R. Tershy, C. J. Donlan, D. Veitch, and N. Puerta et al. (2004). A review of feral cat eradication on islands. *Conservation Biology* **18**: 310–319.

Norbury, G. L., D. C. Norbury, and R. P. Heyward. (1998a). Behavioral responses of two predator species to sudden declines in primary prey. *Journal of Wildlife Management* **62**: 45–68.

Norbury, G. L., D. C. Norbury, and R. P. Heyward. (1998b). Space use and denning behaviour of wild ferrets (*Mustela furo*) and cats (*Felis catus*). *New Zealand Journal of Ecology* **22**: 149–159.

Nowell, K. and P. Jackson (eds.). (1996). *Wild Cats: Status Survey and Conservation Action Plan*. Gland, Switzerland: International Union for the Conservation of Nature, xxiv + 382 pp.

NRC (National Research Council) of the National Academies. (2006). *Nutrient Requirements of Dogs and Cats*. Washington, DC: National Academies Press, xii + 398 pp.

Nutter, F. B., J. F. Levine, and M. K. Stoskopf. (2004). Reproductive capacity of free-roaming domestic cats and kitten survival rate. *Journal of the American Veterinary Medical Association* **225**: 1399–1402.

O'Brien, S. J. and N. Yuhki. (1999). Comparative genome organization of the major histocompatibility complex: lessons from the Felidae. *Immunological Reviews* **167**: 133–144.

Oliveira, N. and F. M. Hilker. (2010). Modelling disease introduction as biological control of invasive predators to preserve endangered prey. *Bulletin of Mathematical Biology* **72**: 444–468.

Olmstead, C. E., J. R. Villablanca, M. Torbiner, and D. Rhodes. (1979). Development of thermoregulation in the kitten. *Physiology and Behavior* **23**: 489–495.

Oppenheimer, E. C. (1980). *Felis catus*: population densities in an urban area. *Carnivore Genetics Newsletter* **4**: 72–80.

Oro, D., J. Salvador Aguilar, J. Manuel Igual, and M. Louzao. (2004). Modelling demography and extinction risk in the endangered Balearic shearwater. *Biological Conservation* **116**: 93–102.

Paape, S. R., V. M. Shille, H. Seto, and G. H. Stabenfeldt. (1975). Luteal activity in the pseudopregnant cat. *Biology of Reproduction* **13**: 470–474.

Packer, C. (1986). The ecology of sociality in felids. In *Ecological Aspects of Social Evolution: Birds and Mammals*, D. I. Rubenstein and R. W. Wrangham (eds.). Princeton, NJ: Princeton University Press, pp. 429–451.

Packer, C. and L. Ruttan. (1988). The evolution of cooperative hunting. *American Naturalist* **132**: 159–198.

Page, R. A., T. Schnelle, E. K. V. Kalko, T. Bunge, and X. E. Bernal. (2012). Sequential assessment of prey through the use of multiple sensory cues by an eavesdropping bat. *Naturwissenschaften* **99**: 505–509.

Page, R. J. C., J. Ross, and D. H. Bennett. (1992). A study of the home ranges, movements and behaviour of the feral cat population at Avonmouth Docks. *Wildlife Research* **19**: 263–277.

Panaman, R. (1981). Behaviour and ecology of free-ranging female farm cats (*Felis catus* L.). *Zeitschrift für Tierpsychologie* **56**: 59–73.

Papes, F., D. W. Logan, and L. Stowers. (2010). The vomeronasal organ mediates interspecies defensive behaviors through detection of protein pheromone homologs. *Cell* **141**: 692–703.

Park, H. J., J. S. Park, M. G. Hayek, G. A. Reinhart, and B. P. Chew. (2011). Dietary fish oil and flaxseed oil suppress inflammation and immunity in cats. *Veterinary Immunology and Immunopathology* **141**: 301–306.

Parmalee, P. W. (1953). Food habits of the feral house cat in east-central Texas. *Journal of Wildlife Management* **17**: 375–376.

Parr, S. J., M. J. Hill, J. Nevill, D. V. Merton, and N. J. Shah. (2000). Alien species case-study: eradication of introduced mammals in Seychelles in 2000. Gland, Switzerland: World Conservation Union, 18 pp. + Annex 1 (4 pp.).

Pascal, M. (1980). Structure et dynamique de la population de chats harets de l'archipel des Kerguelen. *Mammalia* **44**: 161–182.

Pasternak, T. and K. Horn. (1991). Spatial vision of the cat: variation with eccentricity. *Visual Neuroscience* **6**: 151–158.

Pasternak, T. and W. H. Merigan. (1981). The luminance dependence of spatial vision in the cat. *Vision Research* **21**: 1333–1339.

Pawlosky, R., A. Barnes, and N. Salem Jr., (1994). Essential fatty acid metabolism in the feline: relationship between liver and brain production of long-chain polyunsaturated fatty acids. *Journal of Lipid Research* **35**: 2032–2040.

Pawlosky, R. J. and N. Salem Jr., (1996). Is dietary arachidonic acid necessary for feline reproduction? *Journal of Nutrition* **126**: 1081S–1085S.

Peacock, D. E., B. D. Williams, and P. E. Christensen. (2007). 'Total fluorine' analysis of seed of Australian *Gastrolobium* spp. showing temporal, spatial and morphological variation. *Journal of Fluorine Chemistry* **128**: 631–635.

Peacock, D. E., P. E. Christensen, and B. D. Williams. (2011). Historical accounts of toxicity to introduced carnivores consuming bronzewing pigeons (*Phaps chalcoptera* and *P. elegans*) and other vertebrate fauna in south-west Western Australia. *Australian Zoologist* **35**: 826–842.

Pearre Jr., S. and R. Maass. (1998). Trends in the prey size-based trophic niches of feral and house cats *Felis catus* L. *Mammal Review* **28**: 125–139.

Pearson, O. P. (1964). Carnivore–mouse predation: an example of its intensity and bioenergetics. *Journal of Mammalogy* **45**: 177–188.

Pech, R. P., A. R. E. Sinclair, A. E. Newsome, and P. C. Catling. (1992). Limits to predator regulation of rabbits in Australia: evidence from predator-removal experiments. *Oecologia* **89**: 102–112.

Pech, R. P., A. R. E. Sinclair, and A. E. Newsome. (1995). Predation models for primary and secondary prey species. *Wildlife Research* **22**: 55–64.

Peck, D. R., L. Faulquier, P. Pinet, S. Jaquemet, and M. Le Corre. (2008). Feral cat diet and impact on sooty terns at Juan de Nova Island, Mozambique Channel. *Animal Conservation* **11**: 65–74.

Pet Food Institute. (2012). http://www.petfoodinstitute.org/?page=PetPopulation (accessed 25 February 2014).

Petraitis, P. S. (1981). Dominance rankings and problems of intransitive relationships. *Behavioral and Brain Sciences* **4**: 445–446.

Phillips, D. P., M. B. Calford, J. D. Pettigrew, L. M. Aitkin, and M. N. Semple. (1982). Directionality of sound pressure transformation at the cat's pinna. *Hearing Research* **8**: 13–28.

Phillips, R. B., B. D. Cooke, K. Campbell, V. Carrion, C. Marquez, and H. L. Snell. (2005). Eradicating feral cats to protect Galapagos land iguanas: methods and strategies. *Pacific Conservation Biology* **11**: 257–267.

Phillips, R. B., C. S. Winchell, and R. H. Schmidt. (2007). Dietary overlap of an alien and native carnivore on San Clemente Island, California. *Journal of Mammalogy* **88**: 173–180.

Pierpaoli, M., Z. S. Birò, M. Herrmann, K. Hupe, M. Fernandes, and B. Ragni et al. (2003). Genetic distinction of wildcat (*Felis silvestris*) populations in Europe, and hybridization with domestic cats in Hungary. *Molecular Ecology* **12**: 2585–2598.

Pion, P. D., M. D. Kittleson, Q. R. Rogers, and J. G. Morris. (1987). Myocardial failure in cats associated with low plasma taurine: a reversible cardiomyopathy. *Science* **237**: 764–768.

Plantinga, E. A., G. Bosch, and W. H. Hendriks. (2011). Estimation of the dietary nutrient profile of free-roaming feral cats: possible implications for nutrition of domestic cats. *British Journal of Nutrition* **106**: S35–S48.

Podberscek, A. L., J. K. Blackshaw, and A. W. Beattie. (1991). The behaviour of laboratory colony cats and their reactions to a familiar and unfamiliar person. *Applied Animal Behaviour Science* **31**: 119–130.

Polsky, R. H. (1975). Hunger, prey feeding, and predatory aggression. *Behavioral Biology* **13**: 81–93.

Pomianowski, J. F., D. Mikulski, K. Pudyszak, R. G. Cooper, M. Angowski, and A. Józwik *et al.* (2008). Chemical composition, cholesterol content, and fatty acid profile of pigeon meat as influenced by meat-type breeds. *Poultry Science* **88**: 1306–1309.

Pontier, D. and E. Natoli. (1996). Male reproductive success in the domestic cat (*Felis catus* L.): a case history. *Behavioural Processes* **37**: 85–88.

Pontier, D. and E. Natoli. (1999). Infanticide in rural male cats (*Felis catus* L.) as a reproductive mating tactic. *Aggressive Behavior* **25**: 445–449.

Pontier, D., L. Say, F. Debias, J. Bried, J. Thiouloyse, and T. Micol *et al.* (2002). The diet of feral cats (*Felis catus* L.) at five sites on the Grande Terre, Kerguelen archipelago. *Polar Biology* **25**: 833–837.

Pontier, D., L. Say, S. Devillard, and F. Bonhomme. (2005). Genetic structure of the feral cat (*Felis catus* L.) introduced 50 years ago to a sub-Antarctic island. *Polar Biology* **28**: 268–275.

Poole, W. E. (1960). Breeding of the wild rabbit, *Oryctolagus cuniculus* (L.), in relation to the environment. *Wildlife Research* **5**: 21–43.

Popper, K. (1968). *The Logic of Scientific Discovery*. New York: Harper Torchbooks, 480 pp.

Populin, L. C. and T. C. T. Yin. (1998a). Behavioral studies of sound localization in the cat. *Journal of Neuroscience* **18**: 2147–2160.

Populin, L. C. and T. C. T. Yin. (1998b). Pinna movements of the cat during sound localization. *Journal of Neuroscience* **18**: 4233–4243.

Port, M., P. Kappeler, and R. A. Johnstone. (2011). Communal defense of territories and the evolution of sociality. *American Naturalist* **178**: 787–800.

Powell, R. A. (2012). Diverse perspectives on mammal home ranges or a home range is more than location densities. *Journal of Mammalogy* **93**: 887–889.

Powell, R. A. and M. S. Mitchell. (2012). What is a home range? *Journal of Mammalogy* **93**: 948–958.

Powers, S. T., A. S. Penn, and R. A. Watson. (2011). The concurrent evolution of cooperation and the population structures that support it. *Evolution* **65**: 1527–1543.

Preisser, E. L., D. I. Bolnick, and M. F. Benard. (2005). Scared to death? The effects of intimidation and consumption in predator–prey interactions. *Ecology* **86**: 501–509.

Prentiss, P. G., A. V. Wolf, and H. A. Eddy. (1959). Hydropenia in cat and dog. Ability of the cat to meet its water requirements solely from a diet of fish or meat. *American Journal of Physiology* **196**: 625–632.

Prince, P. A., P. Rothery, J. P. Croxall, and A. G. Wood. (1994). Population dynamics of black-browed and grey-headed albatrosses *Diomedea melanophris* and *D. chrysostoma* at Bird Island, South Georgia. *Ibis* **136**: 50–71.

Pukazhenthi, B. S., D. E. Wildt, and J. G. Howard. (2001). The phenomenon and significance of teratospermia in felids. *Journal of Reproduction and Fertility Supplement* **57**: 423–433.

Pusey, A., and M. Wolf. (1996). Inbreeding avoidance in animals. *Trends in Ecology and Evolution* **11**: 201–206.

Queller, D. C. (1994). Genetic relatedness in viscous populations. *Evolutionary Ecology* **8**: 70–73.

Quillfeldt, P., I. Schenk, R. A. R. McGill, I. J. Strange, J. F. Masello, and A. Gladbach *et al.* (2008). Introduced mammals coexist with seabirds at New Island, Falkland Islands: abundance, habitat preferences, and stable isotope analysis of diet. *Polar Biology* **31**: 333–349.

Quinn, T. (1997). Coyote (*Canis latrans*) food habits in three urban habitat types of western Washington. *Northwest Science* **71**: 1–5.

Rand, J. S., L. M. Fleeman, H. A. Farrow, D. J. Appleton, and R. Lederer. (2004). Canine and feline diabetes mellitus: nature or nurture? *Journal of Nutrition* 134: 2072S–2080S.

Randall, J. A., B. McCowan, K. C. Collins, S. L. Hooper, and K. Rogovin. (2005). Alarm signals of the great gerbil: acoustic variation by predator context, sex, age, individual, and family group. *Journal of the Acoustical Society of America* 118: 2706–2714.

Randi, E., M. Pierpaoli, M. Beaumont, B. Ragni, and A. Sforzi. (2001). Genetic identification of wild and domestic cats (*Felis silvestris*) and their hybrids using Bayesian clustering methods. *Molecular Biology and Evolution* 18: 1679–1693.

Ratcliffe, N., M. Bell, T. Pelembe, D. Boyle, R. Benjamin, and R. White et al. (2009). The eradication of feral cats from Ascension Island and its subsequent recolonization by seabirds. *Oryx* 44: 20–29.

Rauzon, M. J. (1985). Feral cats on Jarvis Island: their effects and their eradication. *Atoll Research Bulletin* (282): 1–32.

Rauzon, M. J., D. J. Forsell, E. N. Flint, and J. M. Gove. (2011). Howland, Baker and Jarvis Islands 25 years after cat eradication: the recovery of seabirds in a biogeographical context. In *Island Invasives: Eradication and Management*, C. R. Veitch, M. N. Clout, and D. R. Towns (eds.). Gland, Switzerland: International Union for the Conservation of Nature, pp. 345–349.

Rayner, M. J., M. E. Hauber, M. J. Imber, R. K. Stamp, and M. N. Clout. (2007). Spatial heterogeneity of mesopredator release within an oceanic island system. *Proceedings of the National Academy of Sciences* 104: 20862–20865.

Read, J. and Z. Bowen. (2001). Population dynamics, diet and aspects of the biology of feral cats and foxes in arid South Australia. *Wildlife Research* 28: 195–203.

Recio, M. R., R. Mathieu, R. Maloney, and P. J. Seddon. (2010). First results of feral cats (*Felis catus*) monitored with GPS collars in New Zealand. *New Zealand Journal of Ecology* 34: 288–296.

Rees, P. (1981). The ecological distribution of feral cats and the effects of neutering a hospital colony. In *The Ecology and Control of Feral Cats*, O. Jackson (chmn.). South Mimms, Potters Bar, Hertfordshire (UK): Universities Federation for Animal Welfare, pp. 12–22.

Reeve, H. K., S. T. Emlen, and L. Keller. (1998). Reproductive sharing in animal societies: reproductive incentives or incomplete control by dominant breeders? *Behavioral Ecology* 9: 267–278.

Remfry, J. (1981). Strategies for control. In *The Ecology and Control of Feral Cats*, O. Jackson (chmn.). South Mimms, Potters Bar, Hertfordshire (UK): Universities Federation for Animal Welfare, pp. 73–80.

Remfry, J. (1996). Feral cats in the United Kingdom. *Journal of the American Veterinary Medical Association* 208: 520–523.

Remtulla, S. and P. E. Hallett. (1985). A schematic eye for the mouse, and comparisons with the rat. *Vision Research* 25: 21–31.

Rice, F. L. and B. L. Munger. (1986). A comparative light microscopic analysis of the sensory innervation of the mystacial pad. II. The common fur between the vibrissae. *Journal of Comparative Neurology* 252: 186–205.

Rice, F. L, A. Mance, and B. L. Munger. (1986). A comparative light microscopic analysis of the sensory innervation of the mystacial pad. I. Innervation of vibrissal follicle-sinus complexes. *Journal of Comparative Neurology* 252: 154–174.

Rice, J. J., B. J. May, G. A. Spirou, and E. D. Young. (1992). Pinna-based spectral cues for sound localization in cat [sic]. *Hearing Research* 58: 132–152.

Richard, M. J., J. T. Holck, and D. C. Beitz. (1989). Lipogenesis in liver and adipose tissue of the domestic cat (*Felis domestica*). *Comparative Biochemistry and Physiology* 93B: 561–564.

Richards, S. M. (1974). The concept of dominance and methods of assessment. *Animal Behaviour* 22: 914–930.

Richdale, L. E. (1963). Biology of the sooty shearwater (*Puffinus griseus*). *Proceedings of the Zoological Society of London* 141: 1–117.

Risbey, D. A., M. Calver, and J. Short. (1997). Control of feral cats for nature conservation: 1. Field tests of four baiting methods. *Wildlife Research* **24**: 319–326.

Ritchie, E. G. and C. N. Johnson. (2009). Predator interactions, mesopredator release and biodiversity conservation. *Ecology Letters* **12**: 982–998.

Rivera-Parra, J. L., K. M. Levenstein, J. C. Bednarz, F. Hernan Vargas, V. Carrion, and P. G. Parker. (2012). Implications of goat eradication on the survivorship of the Galapagos Hawk. *Journal of Wildlife Management* **76**: 1197–1204.

Rivers, J. P. W. and T. L. Frankel. (1980). Fat in the diet of cats and dogs. In *Nutrition of the Dog and Cat*, R. S. Anderson (ed.). Oxford: Pergamon Press, 67–99.

Rivers, J. P. W., A. J. Sinclair, and M. A. Crawford. (1975). Inability of the cat to desaturate essential fatty acids. *Nature* **258**: 171–173.

Robert, K., D. Garant, and F. Pelletier. (2012). Keep in touch: does spatial overlap correlate with contact rate frequency? *Journal of Wildlife Management* **76**: 1670–1678.

Roberts, S. A., A. J. Davidson, L. McLean, R. J. Beynon, and J. L. Hurst. (2012). Pheromonal induction of spatial learning in mice. *Science* **338**: 1462–1465.

Robertson, C. J. R. (1993). Survival and longevity of the northern royal albatross *Diomedea epomophora sanfordi* at Taiaroa Head, 1937–93. *Emu* **93**: 269–276.

Robertson, I. D. (1998). Survey of predation by domestic cats. *Australian Veterinary Journal* **76**: 551–554.

Robertson, S. A. (2008). A review of feral cat control. *Journal of Feline Medicine and Surgery* **10**: 366–375.

Robinson, I. (1992). Social behaviour of the cat. In *The Waltham Book of Dog and Cat Behaviour*, C. Thorne (ed.). Oxford: Pergamon Press, pp. 79–95.

Robinson, R. and H. W. Cox. (1970). Reproductive performance in a cat colony over a 10-year period. *Laboratory Animals* **4**: 99–112.

Rockström, J., W. Steffen, K. Noone, Å. Persson, F. S. Chapin III,, and E. F. Lambin *et al.* (2009). A safe operating space for humanity. *Nature* **461**: 472–475.

Rodríguez, C., R. Torres, and H. Drummond. (2006). Eradicating introduced mammals from a forested tropical island. *Biological Conservation* **130**: 98–105.

Roemer, G. W., C. J. Donlan, and F. Courchamp. (2002). Golden eagles, feral pigs, and insular carnivores: how exotic species turn native predators into prey. *Proceedings of the National Academy of Sciences* **99**: 791–796.

Rogers, Q. R. and J. G. Morris. (1979). Essentiality of amino acids for the growing kitten. *Journal of Nutrition* **109**: 718–723.

Rogers, Q. R. and J. G. Morris. (1980). Why does the cat require a high protein diet? In *Nutrition of the Dog and Cat*, R. Anderson (ed.). Oxford: Pergamon Press, pp. 45–66.

Rogers, Q. R. and J. G. Morris. (1982). Do cats really need more crude protein? *Journal of Small Animal Practice* **23**: 521–532.

Rogers, Q. R., J. G. Morris, and R. A. Freedland. (1977). Lack of hepatic enzymatic adaptation to low and high levels of dietary protein in the adult cat. *Enzyme* **22**: 348–356.

Rogers, Q. R., A. R. Wigle, A. Laufer, V. H. Castellanos, and J. G. Morris. (2004). Cats select for adequate methionine but not threonine. *Journal of Nutrition* **134**: 2046S–2049S.

Rogers, W. W. (1932). Controlled observations on behavior of kittens toward the rat from birth to five months of age. *Journal of Comparative Psychology* **13**: 107–135.

Rolls, E. C. (1969). *They All Ran Wild: The Story of Pests on the Land in Australia*. Sydney: Angus and Robertson xx + 444 pp.

Romanowski, J. (1988). Abundance and activity of the domestic cat (*Felis silvestris* F. *catus* L.) in the suburban zone. *Polish Ecological Studies* **14**: 213–221.

Rosenblatt, J. S. (1971). Suckling and home orientation in the kitten: a comparative developmental study. In *The Biopsychology of Development*, E. Tobach, L. R. Aronson, and E. Shaw (eds.). New York: Academic Press, pp. 345–410.

Rosenblatt, J. S. and T. C. Schneirla. (1962). The behaviour of cats. In *The Behaviour of Domestic Animals*, E. S. E. Hafez (ed.). London: Baillière, Tindall and Cox, pp. 453–488.

Rosenblatt, J. S., G. Turkewitz, and T. C. Schneirla. (1961). Early socialization in the domestic cat as based on feeding and other relationships between female and young. In *Determinants of Infant Behaviour*, B. M. Foss (ed.). London: Methuen, pp. 51–74.

Rosenstein, L. and E. Berman. (1973). Postnatal body weight changes of domestic cats maintained in an outdoor colony. *American Journal of Veterinary Research* **34**: 575–577.

Rowland, N. (1981). Glucoregulatory feeding in cats. *Physiology and Behavior* **26**: 901–903.

Royle, S. A. (2004). Human interference on Ascension Island. *Environmental Archaeology* **9**: 127–134.

Russell, K. (2002). Reply to Rogers and Morris. *Journal of Nutrition* **132**: 2821–2822.

Russell, K., G. E. Lobley, J. Rawlings, D. J. Millward, and E. J. Harper. (2000). Urea kinetics of a carnivore, *Felis silvestris catus*. *British Journal of Nutrition* **84**: 597–604.

Russell, K., P. R. Murgatroyd, and R. M. Batt. (2002). Net protein oxidation is adapted to dietary protein intake in domestic cats (*Felis silvestris catus*). *Journal of Nutrition* **132**: 456–460.

Russell, K., G. E. Lobley, and D. J. Millward. (2003). Whole-body protein turnover of a carnivore, *Felis silvestris catus*. *British Journal of Nutrition* **89**: 29–37.

Russell, R. W. (1999). Comparative demography and life history tactics of seabirds: implications for conservation and marine monitoring. *American Fisheries Society Symposium* **23**: 51–76.

Rutherfurd, S. M., T. M. Kitson, A. D. Woolhouse, M. C. McGrath, and W. H. Hendriks. (2007). Felinine stability in the presence of selected urine compounds. *Amino Acids* **32**: 235–242.

Ruxton, G. D., S. Thomas, and J. W. Wright. (2002). Bells reduce predation of wildlife by domestic cats (*Felis catus*). *Journal of Zoology (London)* **256B**: 81–83.

Safford, R. J. and C. G. Jones. (1998). Strategies for land-bird conservation on Mauritius. *Conservation Biology* **12**: 169–176.

Salazar, I. and P. Sánchez-Quinteiro. (2011). A detailed morphological study of the vomeronasal organ and the accessory olfactory bulb of cats. *Microscopy Research and Technique* **74**: 1109–1120.

Salazar, I., P. Sánchez Quinteiro, J. Manuel Cifuentes, and T. Garcia Caballero. (1996). The vomeronasal organ of the cat. *Journal of Anatomy* **188**: 445–454.

Sales, G. D. (1972a). Ultrasound and aggressive behaviour in rats and other small mammals. *Animal Behaviour* **20**: 88–100.

Sales, G. D. (1972b). Ultrasound and mating behaviour in rodents with some observations on other behavioural situations. *Journal of Zoology (London)* **168**: 149–164.

Sales, G. D. and J. C. Smith. (1978). Comparative studies of the ultrasonic calls of infant murid rodents. *Developmental Psychobiology* **11**: 595–619.

Sandell, M. (1989). The mating tactics and spacing patterns of solitary carnivores. In *Carnivore Behavior, Ecology, and Evolution*, J. L. Gittleman (ed.). Ithaca, NY: Comstock/Cornell University Press, pp. 164–182.

Sanders, M. D. and R. F. Maloney. (2002). Causes of mortality at nests of ground-nesting birds in the Upper Waitaki Basin, South Island, New Zealand: a 5-year video study. *Biological Conservation* **106**: 225–236.

Sawicka-Kapusta, K. (1970). Changes in gross body composition and the caloric value of the common voles [sic] during their postnatal development. *Acta Theriologica* **15**: 67–79.

Say, L. and D. Pontier. (2004). Spacing pattern in a social group of stray cats: effects on male reproductive success. *Animal Behaviour* **68**: 175–180.

Say, L., D. Pontier, and E. Natoli. (1999). High variation in multiple paternity of domestic cats (*Felis catus* L.) in relation to environmental conditions. *Proceedings of the Royal Society of London* **266B**: 2071–2074.

Say, L., D. Pontier, and E. Natoli. (2001). Influence of oestrus synchronization on male reproductive success in the domestic cat (*Felis catus* L.). *Proceedings of the Royal Society of London* **268B**: 1049–1053.

Say, L., J-M. Gaillard, and D. Pontier. (2002a). Spatio-temporal variation in cat population density in a sub-Antarctic environment. *Polar Biology* **25**: 90–95.

Say, L., S. Devillard, E. Natoli, and D. Pontier. (2002b). The mating system of feral cats (*Felis catus* L.) in a sub-Antarctic environment. *Polar Biology* 25: 838–842.

Scattoni, M. L., L. Ricceri, and J. N. Crawley. (2011). Unusual repertoire of vocalizations in adult BTBR T+tf/J mice during three types of social encounters. *Genes, Brains and Behavior* 10: 44–56.

Schaeffer, M. C., Q. R. Rogers, and J. G. Morris. (1982). Methionine requirement of the growing kitten, in the absence of dietary cystine. *Journal of Nutrition* 112: 962–971.

Schaller, G. B. (1972). *The Serengeti Lion: A Study of Predator–Prey Relations*. Chicago, IL: University of Chicago Press, xiii + 480 pp.

Schjelderup-Ebbe, T. (1922). Beiträge zur Sozialpsychologie des Haushuhns. *Zeitschrift für Psychologie* 88: 226–252.

Schjelderup-Ebbe, T. (1935). Social behavior of birds. In *A Handbook of Social Psychology*, C. Murchison (ed.). Worcester, MA: Clark University Press, pp. 947–972.

Schmidt, P. M., R. R. Lopez, and B. A. Collier. (2007). Survival, fecundity, and movements of free-roaming cats. *Journal of Wildlife Management* 71: 915–919.

Schmidt, P. M., P. K. Chakraborty, and D. W. Wildt. (1983). Ovarian activity, circulating hormones and sexual behavior in the cat. II. Relationships during pregnancy, parturition, lactation and the postpartum estrus. *Biology of Reproduction* 28: 657–671.

Schneirla, T. C., J. S. Rosenblatt, and E. Tobach. (1963). Maternal behavior in the cat. In *Maternal Behavior in Mammals*, H. L. Rheingold (ed.). New York: John Wiley & Sons, Inc., pp. 122–168.

Schummer, A., H. Wilkens, B. Vollmerhaus, and K-H. Habermehl. (1981). *The Circulatory System, the Skin, and the Cutaneous Organs of the Domestic Mammals*. In *The Anatomy of the Domestic Animals*, 2nd edn., Vol. 3, R. Nickel, A. Schummer, and E. Seiferle (eds.). Berlin: Verlag Paul Parey, xv + 610 pp. [Transl. by W. G. Siller, and P. A. L. Wight]

Schwagmeyer, P. L. and C. H. Brown. (1983). Factors affecting male–male-competition in thirteen-lined ground squirrels. *Behavioral Ecology and Sociobiology* 13: 1–6.

Schwagmeyer, P. L. and G. A. Parker (1987). Queuing for mates in thirteen-lined ground squirrels. *Animal Behaviour* 35: 1015–1025.

Schwagmeyer, P. L. and S. J. Woontner. (1986). Scramble competition polygyny in thirteen-lined ground squirrels: the relative contributions of overt conflict and competitive mate searching. *Behavioral Ecology and Sociobiology* 19: 359–364.

Scott, J. P. (1950). The social behavior of dogs and wolves: an illustration of sociobiological systematics. *Annals of the New York Academy of Sciences* 51: 1009–1021.

Scott, J. P. and J. L. Fuller. (1965). *Genetics and the Social Behavior of the Dog*. Chicago, IL: University of Chicago Press, xviii + 468 pp.

Scott, K. C., J. K. Levy, and S. P. Gorman. (2002). Body condition of feral cats and the effect of neutering. *Journal of Applied Animal Welfare Science* 5: 203–213.

Scott, M. G. and P. P. Scott. (1957). Post-natal development of the testes and epididymis in the cat. *Journal of Physiology* 136: 40P–41P.

Scott, P. P. (1955). The domestic cat as a laboratory animal for the study of reproduction. *Journal of Physiology* 130: 47P.

Scott, P. P. and M. A. Lloyd-Jacob. (1955). Some interesting features in the reproductive cycle of the cat. *Studies on Fertility* 7: 123–129.

Scott, P. P. and M. A. Lloyd-Jacob. (1959). Reduction in the anœstrus period of laboratory cats by increased illumination. *Nature* 184: 2022.

Seefeldt, S. L. and T. E. Chapman. (1979). Body water content and turnover in cats fed dry and canned rations. *American Journal of Veterinary Research* 40: 183–185.

Segev, G., H. Livne, E. Ranen, and E. Lavy. (2011). Urethral obstruction in cats: predisposing factors, clinical, clinicopathological characteristics and prognosis. *Journal of Feline Medicine and Surgery* 13: 101–108.

Sehgal, H. S. and J. Thomas. (1987). Efficacy of two newly formulated supplementary diets for carp, *Cyprinus carpio* var. *communis* (Linn.): effects on flesh composition. *Biological Wastes* 21: 179–187.

Seitz, P. F. D. (1959). Infantile experience and adult behavior in animal subjects. II: Age of separation from the mother and adult behavior in the cat. *Psychosomatic Medicine* **21**: 353–378.

Seligman, M. E. P. (1970). On the generality of the laws of learning. *Psychological Review* **77**: 406–418.

Seyfarth, R. M. (1981). Do monkeys rank each other? *Behavioral and Brain Sciences* **4**: 447–448.

Shah, N. J. (2001). Eradication of alien predators in the Seychelles: an example of conservation action on tropical islands. *Biodiversity and Conservation* **10**: 1219–1220.

Shea, K. and P. Chesson. (2002). Community ecology theory as a framework for biological invasions. *Trends in Ecology and Evolution* **17**: 170–176.

Sherman, P. W., E. A. Lacey, H. K. Reeve, and L. Keller. (1995). The eusociality continuum. *Behavioral Ecology* **6**: 102–108.

Shille, V. M., C. Munro, S. Walker-Farmer, H. Papkoff, and G. H. Stabenfeldt. (1983). Ovarian and endocrine responses in the cat after coitus. *Journal of Reproduction and Fertility* **69**: 29–39.

Shille, V. M. and G. H. Stabenfeldt. (1979). Luteal function in the domestic cat during pseudopregnancy and after treatment with prostaglandin $F_{2\alpha}$. *Biology of Reproduction* **21**: 1217–1223.

Shille, V. M., K. E. Lundström, and G. H. Stabenfeldt. (1979). Follicular function in the domestic cat as determined by estradiol-17β concentrations in plasma: relation to estrous behavior and cornification of exfoliated vaginal epithelium. *Biology of Reproduction* **21**: 953–963.

Shimizu, M. (2001). Vocalizations of feral cats: sexual differences in the breeding season. *Mammal Study* **26**: 85–92.

Short, J., B. Turner, D. A. Risbey, and R. Carnamah. (1997). Control of feral cats for nature conservation. II. Population reduction by poisoning. *Wildlife Research* **24**: 703–714.

Siemers, B. M., E. Kriner, I. Kaipf, M. Simon, and S. Grief. (2012). Bats eavesdrop on the sound of copulating flies. *Current Biology* **22**(14): R563–R564.

Sih, T. R., J. G. Morris, and M. A. Hickman. (2002). Chronic ingestion of high levels of cholecalciferol in cats. *American Journal of Veterinary Research* **62**: 1500–1506.

Silk, J. B. (1999). Male bonnet macaques use information about third-party rank relationships to recruit allies. *Animal Behaviour* **58**: 45–51.

Silva, S. V. P. S. and J. R. Mercer. (1985). Effect of protein intake on amino acid catabolism and gluconeogenesis by isolated hepatocytes from the cat (*Felis domestica*). *Comparative Biochemistry and Physiology* **80B**: 603–607.

Silva, S. V. P. S. and J. R. Mercer. (1991). The effect of protein intake on the potential activity of the lysosomal vacuolar system in the cat. *Comparative Biochemistry and Physiology* **98A**: 551–558.

Simberloff, D. (2003). How much information on population biology is needed to manage introduced species? *Conservation Biology* **17**: 83–92.

Simonson, M., H. M. Hanson, and D. A. Brodie. (1972). Replication of effects of maternal malnourishment in another specie [sic]. *Federation Proceedings* **31**: A688. [Abstract]

Sims, V., K. L. Evans, S. E. Newson, J. A. Tratalos, and K. J. Gaston. (2008). Avian assemblage structure and domestic cat densities in urban environments. *Diversity and Distributions* **14**: 387–399.

Sinclair, A. J., J. G. McLean, and E. A. Monger. (1979). Metabolism of linoleic acid in the cat. *Lipids* **14**: 932–936.

Slater, M. R. (2001). Understanding and controlling of feral cat populations. In *Consultations in Feline Internal Medicine* 4, J. R. August (ed.). Philadelphia, PA: Saunders, pp. 561–570.

Slater, M. R. (2004). Understanding issues and solutions for unowned, free-roaming cat populations. *Journal of the American Veterinary Medical Association* **225**: 1350–1354.

Smith, A. P. and D. G. Quin. (1996). Patterns and causes of extinction and decline in Australian conilurine rodents. *Biological Conservation* **77**: 243–267.

Smith, B. A. and G. R. Jansen. (1977a). Maternal undernutrition in the feline: brain composition of offspring. *Nutrition Reports International* **16**: 497–512.

Smith, B. A. and G. R. Jansen. (1977b). Maternal undernutrition in the feline: behavioral sequelae. *Nutrition Reports International* **16**: 513–526.

Smith, D. P., J. K. Northcutt, and E. L. Steinberg. (2012). Meat quality and sensory attributes of a conventional and a label Rouge-type broiler strain obtained at retail. *Poultry Science* **91**: 1489–1495.

Smith, W., A. J. L. Butler, L. A. Hazell, M. D. Chapman, A. Pomés, and D. G. Nickels *et al.* (2004). Fel d 4, a cat lipocalin allergen. *Clinical and Experimental Allergy* **34**: 1732–1738.

Smucker, T. D., G. D. Lindsey, and S. M. Mosher. (2000). Home range and diet of feral cats in Hawaii forests. *Pacific Conservation Biology* **6**: 229–237.

Smuts, B. (1981). Dominance: an alternative view. *Behavioral and Brain Sciences* **4**: 448–449.

Snetsinger, T. J., S. G. Fancy, J. C. Simon, and J. D. Jacobi. (1994). Diets of owls and feral cats in Hawaii. *'Elepaio* **54**(8): 47–50.

Sojka, N. J., L. L. Jennings, and C. E. Hamner. (1970). Artificial insemination in the cat (*Felis catus* L. [sic]). *Laboratory Animal Care* **20**: 198–204.

Soulé, M. E., D. T. Bolger, A. C. Alberts, J. Wright, M. Sorice, and S. Hill. (1988). Reconstructed dynamics of rapid extinctions of chaparral-requiring birds in urban habitat islands. *Conservation Biology* **2**: 75–92.

Spindler, R. E. and D. E. Wildt. (1999). Circannual variations in intraovarian oocyte but not epididymal sperm quality in the domestic cat. *Biology of Reproduction* **61**: 188–194.

Spotte, S. (1992). *Captive Seawater Fishes: Science and Technology*. New York: John Wiley & Sons, Inc., xxii + 942 pp.

Spotte, S. (2007). *Bluegills: Biology and Behavior*. Bethesda, MD: American Fisheries Society, vii + 214 pp.

Spotte, S. (2012). *Societies of Wolves and Free-ranging Dogs*. Cambridge: Cambridge University Press, xii + 377 pp.

Stabenfeldt, G. H. and V. M. Shille. (1977). Reproduction in the dog and cat. In *Reproduction in Domestic Animals*, H. H. Cole and P. T. Cupps (eds.). New York: Academic Press, pp. 499–527.

Stains, H. J. (1961). Comparison of temperatures inside and outside two tree dens used by raccoons. *Ecology* **42**: 410–413.

Stein, B. E. and H. P. Clamann. (1981). Control of pinna movements and sensorimotor register in cat superior colliculus. *Brain, Behavior and Evolution* **19**: 180–192.

Stephan, Z. F. and K. C. Hayes. (1978). Vitamin E deficiency and essential fatty acid (EFA) status of cats. *Federation Proceedings* **37**: 706. [Abstract]

Stephens, C. (1996). Modeling reciprocal altruism. *British Journal for the Philosophy of Science* **47**: 533–551.

Stewart, D. R. and G. H. Stabenfeldt. (1985). Relaxin activity in the pregnant cat. *Biology of Reproduction* **32**: 848–854.

Stone, P. A., H. L. Snell, and H. M. Snell. (1994). Behavioral diversity as biological diversity: introduced cats and lava lizard wariness. *Conservation Biology* **8**: 569–573.

Stoskopf, M. K. and F. B. Nutter. (2004). Analyzing approaches to feral cat management – one size does not fit all. *Journal of the American Veterinary Medical Association* **225**: 1361–1364.

Stracey, C. M. (2011). Resolving the urban nest predator paradox: the role of alternative foods for nest predators. *Biological Conservation* **144**: 1545–1552.

Strong, B. W. and W. A. Low. (1983). Some observations of feral cats *Felis catus* in the southern Northern Territory. Technical Report 9. Alice Springs, NT (Australia): Conservation Commission of the Northern Territory, 12 pp.

Sunvold, G. D., G. C. Fahey Jr., N. R. Merchen, L. D. Bourquin, E. C. Titgemeyer, and L. L. Bauer *et al.* (1995). Dietary fiber for cats: in vitro fermentation of selected fiber sources by cat fecal inoculum and in vivo utilization of diets containing selected fiber sources and their blends. *Journal of Animal Science* **73**: 2329–2339.

Syme, G. J. (1974). Competitive orders as measures of social dominance. *Animal Behaviour* **22**: 931–940.

Takahashi, N., M. Kashino, and N. Hironaka. (2001). Structure of rat ultrasonic vocalizations and its relevance to behavior. *PLoS One* **5**(11): e14115 (7 pp.).

Takahashi, L. K., B. R. Nakashima, H. Hong, and K. Watanabe. (2005). The smell of danger: a behavioral and neural analysis of predator odor-induced fear. *Neuroscience and Biobehavioral Reviews* **29**: 1157–1167.

Tan, P. L. and J. J. Counsilman. (1985). The influence of weaning on prey-catching behaviour in kittens. *Zeitschrift für Tierpsychologie* **70**: 148–164.

Tanaka, A., A. Inoue, A. Takeguchi, T. Washizu, M. Bonkobara, and T. Aral. (2005). Comparison of expression of glucokinase gene and activities of enzymes related to glucose metabolism in livers between dog and cat. *Veterinary Research Communications* **29**: 477–485.

Tantillo, J. A. (2006). Killing cats and killing birds: philosophical issues pertaining to feral cats. In *Consultations in Feline Internal Medicine*, Vol. 5, J. R. August (ed.). St. Louis, MO: Saunders Elsevier, pp. 701–708.

Tarttelin, M. F., W. H. Hendriks, and P. J. Moughan. (1998). Relationship between plasma testosterone and urinary felinine in the growing kitten. *Physiology and Behavior* **65**: 83–87.

Taylor, P. D. (1992). Altruism in viscous populations – an inclusive fitness model. *Evolutionary Ecology* **6**: 352–356.

Taylor, R. H. (1979). How the Macquarie Island parakeet became extinct. *New Zealand Journal of Ecology* **2**: 42–45.

Taylor, R. H. (1985). Status, habits and conservation of *Cyanoramphus* parakeets in the New Zealand region. In *Conservation of Island Birds*, P. J. Moors (ed.). Technical Publication No. 3. Cambridge: International Council for Bird Preservation, pp. 195–211.

Tennent, J. and C. T. Downs. (2008). Abundance and home ranges of feral cats in an urban conservancy where there is supplemental feeding: a case study from South Africa. *African Zoology* **43**: 218–229.

Tennie, C., J. Call, and M. Tomasello. (2009). Ratcheting up the ratchet: on the evolution of cumulative culture. *Philosophical Transactions of the Royal Society* **364B**: 2405–2415.

Tershy, B. R., C. J. Donlan, B. S. Keitt, D. A. Croll, J. A. Sanchez, and B. Wood et al. (2002). Island conservation in north-west Mexico: a conservation model integrating research, education, and exotic mammal eradication. In *Turning the Tide: The Eradication of Invasive Species*. Occasional Paper of the IUCN, Species Survival Commission No. 27, C. R. Veitch and M. N. Clout (eds.). Gland, Switzerland: International Union for the Conservation of Nature, pp. 293–300.

Theraulaz, G., E. Bonabeau, and J-L. Deneubourg. (1995). Self-organization of hierarchies in animal societies: the case of the primitively eusocial wasp *Polistes dominulus* Christ. *Journal of Theoretical Biology* **174**: 313–323.

Thrall, B. E. and L. G. Miller. (1976). Water turnover in cats fed dry rations. *Feline Practice* **6**: 10–17.

Tickell, W. L. N. (2000). *Albatrosses*. New Haven, CT: Yale University Press, unnumbered front matter + 448 pp.

Tidemann, C. R., H. D. Yorkston, and A. J. Russack. (1994). The diet of cats, *Felis catus*, on Christmas Island, Indian Ocean. *Wildlife Research* **21**: 279–286.

Tiedemann, K. and E. Henschel. (1973). Early radiographic diagnosis of pregnancy in the cat. *Journal of Small Animal Practice* **14**: 567–572.

Tirindelli, R., M. Dibattista, S. Pifferi, and A. Menini. (2009). From pheromones to behavior. *Physiological Reviews* **89**: 921–956.

Tomasello, M. (1996). Do apes ape? In *Social Learning in Animals: The Roots of Culture*, C. M. Heyes and B. G. Galef Jr, (eds.). San Diego, CA: Academic Press, pp. 319–346.

Toner, G. C. (1956). House cat predation on small animals. *Journal of Mammalogy* **37**: 119.

Trevizan, L., A. de Mello Kessler, J. T. Brenna, P. Lawrence, M. K. Waldron, and J. E. Bauer. (2012). Maintenance of arachidonic acid as evidence of $\Delta 5$ desaturation in cats fed γ-linolenic and linoleic acid enriched diets. *Lipids* **47**: 413–423.

Trivers, R. L. (1971). The evolution of reciprocal altruism. *Quarterly Review of Biology* **46**: 35–57.

Trivers, R. L. (1972). Parental investment and sexual selection. In *Sexual Selection and the Descent of Man 1871–1971*, B. Campbell (ed.). Chicago, IL: Aldine Press, pp. 136–179.

Tsutsui, T., I. Murao, E. Kawakami, A. Ogasa, and G. H. Stabenfeldt. (1990). Androgen concentration in the blood and spermatogenic function of tom cats during the breeding season. *Japanese Journal of Veterinary Science* **52**: 801–806.

Tufto, J., E. J. Solberg, and T-H. Ringsby. (1998). Statistical models of transitive and intransitive dominance structures. *Animal Behaviour* **55**: 1489–1498.

Turner, D. C. and P. Bateson. (2000). Why the cat? In *The Domestic Cat: The Biology of Its Behaviour*, 2nd edn., D. C. Turner and P. Bateson (eds.). Cambridge: Cambridge University Press, pp. 3–6.

Turner, D. C. and O. Meister. (1988). Hunting behaviour of the domestic cat. In *The Domestic Cat: The Biology of Its Behaviour*, D. C. Turner and P. Bateson (eds.). Cambridge: Cambridge University Press, pp. 111–121.

Turner, D. C. and C. Mertens. (1986). Home range size, overlap and exploitation in domestic farm cats (*Felis catus*). *Behaviour* **99**: 22–45.

Twyford, K. L., P. G. Humphrey, R. P. Nunn, and L. Willoughby. (2000). Eradication of feral cats (*Felis catus*) from Gabo Island, south-east Victoria. *Ecological Management and Restoration* **1**: 42–49.

Tyndale-Biscoe, C. H. (1993). Use of viruses as vectors for immunosterilisation of feral pest mammals. In *Vaccines in Agriculture: Immunological Applications to Animal Health and Production*, P. R. Wood, P. Willadsen, J. Vercoe, J. E. Hoskinson, and D. Demeyer (eds.). Melbourne, VIC (Australia): Commonwealth Scientific and Industrial Research Organisation [CSIRO], pp. 137–143.

van Aarde, R. J. (1978). Reproduction and population ecology in the feral house cat *Felis catus* on Marion Island. *Carnivore Genetics Newsletter* **3**: 288–316.

van Aarde, R. J. (1979). Distribution and density of the feral house cat *Felis catus* on Marion Island. *South African Journal of Antarctic Research* **9**: 14–19.

van Aarde, R. J. (1980). The diet and feeding behaviour of feral cats, *Felis catus* at Marion Island. *South African Journal of Wildlife Research* **10**: 123–128.

van Aarde, R. J. (1984). Population biology and control of feral cats on Marion Island. *Acta Zoologica Fennica* **172**: 107–110.

van den Bos, R. and T. de Cock Buning. (1994). Social behaviour of domestic cats (*Felis lybica* f. *catus* L.): study of dominance in a group of female laboratory cats. *Ethology* **98**: 14–37.

van de Waal, E., C. Borgeaud, and A. Whiten. (2013). Potent social learning and conformity shape a wild primate's foraging decisions. *Science* **340**: 483–485.

Van Pelt, T. I., J. F. Piatt, B. K. Lance, and D. D. Roby. (1997). Proximate composition and energy density of some North Pacific forage fishes. *Comparative Biochemistry and Physiology* **118A**: 1393–1398.

van Rensburg, P. J. J. (1985). The feeding ecology of a decreasing feral house cat, *Felis catus*, population at Marion Island. In *Antarctic Nutrient Cycles and Food Webs*, W. R. Siegfried, P. R. Condy, and R. M. Laws (eds.). Berlin and Heidelberg: Springer-Verlag, pp. 620–624.

van Rensburg, P. J. J. and M. N. Bester. (1988). The effect of cat *Felis catus* predation on three breeding Procellariidae species on Marion Island. *South African Journal of Zoology* **23**: 301–305.

van Rensburg, P. J. J., J. D. Skinner, and R. J. van Aarde. (1987). Effects of feline panleucopaenia on the population characteristics of feral cats on Marion Island. *Journal of Applied Ecology* **24**: 63–73.

van Veelen, M., J. García, and L. Avilés. (2010). It takes grouping and cooperation to get sociality. *Journal of Theoretical Biology* **264**: 1240–1253.

Vázquez-Domínguez, E., G. Ceballos, and J. Cruzado. (2004). Extirpation of an insular subspecies by a single introduced cat: the case of the endemic deer mouse *Peromyscus guardia* on Estanque Island, Mexico. *Oryx* **38**: 347–350.

Vaughan, J. A. and T. Adams. (1967). Surface area of the cat. *Journal of Applied Physiology* **22**: 956–958.

Veitch, C. R. (1985). Methods of eradicating feral cats from offshore islands in New Zealand. In *Conservation of Island Birds: Case studies for the Management of Threatened Island Species*, P. J. Moors (ed.). ICBP Technical Publication No. 3. Cambridge: International Council for Bird Preservation, pp. 125–141.

Verberne, G. and J. de Boer. (1976). Chemocommunication among domestic cats, mediated by the olfactory and vomeronasal senses. I. Chemocommunication. *Zeitschrift für Tierpsychologie* **42**: 86–109.

Verhage, H. G., N. B. Beamer, and R. M. Brenner. (1976). Plasma levels of estradiol and progesterone in the cat during polyestrus, pregnancy and pseudopregnancy. *Biology of Reproduction* **14**: 579–585.

Vessey, S. H. (1981). Dominance as control. *Behavioral and Brain Sciences* **4**: 449.

Vester, B. M., S. L. Burke, K. J. Liu, C. L. Dikeman, L. G. Simmons, and K. S. Swanson. (2010). Influence of feeding raw or extruded feline diets on nutrient digestibility and nitrogen metabolism of African wildcats. *Zoo Biology* **29**: 676–686.

Villablanca, J. R. and C. E. Olmstead. (1979). Neurological development of kittens. *Developmental Psychobiology* **12**: 101–127.

von Borkenhagen, P. (1979). Zur Nahrungsökologie streunender Hauskatzen (*Felis sylvestris* f. *catus* Linné, 1758) aus dem Stadtbereich Kiel. *Zeitschrift für Säugetierkunde* **44**: 375–383.

Walker, C., C. J. Vierck Jr., and L. A. Ritz. (1998). Balance in the cat: role of the tail and effect of sacrocaudal transection. *Behavioural Brain Research* **91**: 41–47.

Wallace, J. L. and J. K. Levy. (2006). Population characteristics of feral cats admitted to seven trap–neuter–return programs in the United States. *Journal of Feline Medicine and Surgery* **8**: 279–284.

Wang, C-S., M. E. Martindale, M. M. King, and J. Tang. (1989). Bile-salt-activated lipase: effect on kitten growth rate. *American Journal of Clinical Nutrition* **49**: 457–463.

Ward, A. J. W. (2012). Social facilitation of exploration in mosquitofish (*Gambusia holbrooki*). *Behavioral Ecology and Sociobiology* **66**: 223–230.

Warner, R. E. (1985). Demography and movements of free-ranging domestic cats in rural Illinois. *Journal of Wildlife Management* **49**: 340–346.

Waser, P. M. (1981). Sociality or territorial defense? The influence of resource renewal. *Behavioral Ecology and Sociobiology* **8**: 231–237.

Watkins, B. P. and J. Cooper. (1986). Introduction, present status and control of alien species at the Prince Edward islands [sic], sub-Antarctic. *South African Journal of Antarctic Research* **16**: 86–94.

West, G. B. and J. H. Brown. (2005). The origin of allometric scaling laws in biology from genomes to ecosystems: towards a quantitative unifying theory of biological structure and organization. *Journal of Experimental Biology* **208**: 1575–1592.

West, M. (1974). Social play in the domestic cat. *American Zoologist* **14**: 427–436.

West, M. J. (1977). Exploration and play with objects in domestic kittens. *Developmental Psychobiology* **10**: 53–57.

West, M. J. (1979). Play in domestic kittens. In *The Analysis of Social Interactions*, R. B. Cairns (ed.). Hillsdale, NJ: Lawrence Erlbaum, pp. 179–193.

West, S. A., I. Pen, and A. S. Griffin. (2002). Cooperation and competition between relatives. *Science* **296**: 72–75.

Whalen, R. E. (1963a). Sexual behavior of cats. *Behaviour* **20**: 321–342.

Whalen, R. E. (1963b). The initiation of mating in naive female cats. *Animal Behaviour* **11**: 461–463.

White, C. R. (2010). There is no single *p*. *Nature* **464**: 691–692.

White, C. R. and R. S. Seymour. (2005). Allometric scaling of mammalian metabolism. *Journal of Experimental Biology* **208**: 1611–1619.

White, T. D. and J. C. Boudreau. (1975). Taste preferences of the cat for neurophysiologically active compounds. *Physiological Psychology* **3**: 405–410.

Wichert, B., L. Schade, S. Gebert, B. Bucher, B. Zottmaier, and C. Wenk *et al.* (2009). Energy and protein needs of cats for maintenance, gestation and lactation. *Journal of Feline Medicine and Surgery* **11**: 808–815.

Wichert, B., M. Signer, and D. Uebelhart. (2011). Cats during gestation and lactation fed with canned food *ad libitum*: energy and protein intake, development of body weight and body composition. *Journal of Animal Physiology and Animal Nutrition*: doi:10.1111/j.1439.0396.2011.01214.x. (8 pp.)

Wichert, B., J. Trossen, D. Uebelhart, M. Wanner, and S. Hartnack. (2012). Energy requirement and food intake behaviour in young adult intact male cats with and without predisposition to overweight. *Scientific World Journal 2012* : doi:10.1100/2012/509854. (6 pp.)

Wildt, D. E., S. C. Guthrie, and S. W. J. Seager. (1978). Ovarian and behavioral cyclicity of the laboratory maintained cat. *Hormones and Behavior* **10**: 251–257.

Wildt, D. E., S. W. J. Seager, and P. K. Chakraborty. (1980). Effect of copulatory stimuli on incidence of ovulation and on serum luteinizing hormone in the cat. *Endocrinology* **107**: 1212–1217.

Wildt, D. E., S. Y. W. Chan, S. W. J. Seager, and P. K. Chakraborty. (1981). Ovarian activity, circulating hormones, and sexual behavior in the cat. I. Relationships during the coitus-induced luteal phase and the estrous period without mating. *Biology of Reproduction* **25**: 15–28.

Wildt, D. E., M. Bush, J. G. Howard, S. J. O'Brien, M. Meltzer, and A. van Dyk *et al.* (1983). Unique seminal quality in the South African cheetah and a comparative evaluation in the domestic cat. *Biology of Reproduction* **29**: 1019–1025.

Wildt, D. E., J. L. Brown, and W. F. Swanson. (1998). Cats. In *Encyclopedia of Reproduction*, Vol. 1, E. Knobil and J. D. Neill (eds.). San Diego, CA: Academic Press, pp. 497–510.

Wilkins, C., R. C. Long Jr., M. Waldron, D. C. Ferguson, and M. Hoenig. (2004). Assessment of the influence of fatty acids on indices of insulin sensitivity and myocellular lipid content by use of magnetic resonance spectroscopy in cats. *American Journal of Veterinary Research* **65**: 1090–1099.

Williams, W. O., D. L. Riskin, and K. M. Mott. (2008). Ultrasonic sound as an indicator of acute pain in laboratory mice. *Journal of the American Association for Laboratory Animal Science* **47**: 8–10.

Wilson, D. R. and J. F. Hare. (2004). Ground squirrel uses ultrasonic alarms. *Nature* **430**: 523.

Wilson, D. S., G. B. Pollock, and L. A. Dugatkin. (1992). Can altruism evolve in purely viscous populations? *Evolutionary Ecology* **6**: 331–341.

Winslow, C. N. (1938). Observations of dominance-subordination in cats. *Journal of Genetic Psychology* **52**: 425–428.

Winslow, C. N. (1944a). The social behavior of cats. I. Competitive and aggressive behavior in an experimental runway situation. *Journal of Comparative Psychology* **37**: 297–313.

Winslow, C. N. (1944b). The social behavior of cats. II. Competitive, aggressive, and food-sharing behavior when both competitors have access to the goal. *Journal of Comparative Psychology* **37**: 315–326.

Winter, L. (2004). Trap–neuter–release programs: the reality and the impacts. *Journal of the American Veterinary Medical Association* **225**: 1369–1376.

Wolf, A. V., P. G. Prentiss, L. G. Douglas, and R. J. Swett. (1959). Potability of sea water with special reference to the cat. *American Journal of Physiology* **196**: 633–641.

Wood, B., B. R. Tershy, M. A. Hermosillo, C. J. Donlan, J. A. Sanchez, and B. S. Keitt *et al.* (2002). Removing cats from islands in north-west Mexico. In *Turning the Tide: The Eradication of Invasive Species*. Occasional Paper of the IUCN, Species Survival Commission No. 27, C. R. Veitch and M. N. Clout (eds.). Gland, Switzerland: International Union for the Conservation of Nature, pp. 374–380.

Wood, T. C., R. J. Montali, and D. E. Wildt. (1997). Follicle-oocyte atresia and temporal taphonomy in cold-stored domestic cat ovaries. *Molecular Reproduction and Development* **46**: 190–200.

Woods, M., R. A. McDonald, and S. Harris. (2003). Predation of wildlife by domestic cats *Felis catus* in Great Britain. *Mammal Review* 33: 174–188.

Wooller, R. D., J. S. Bradley, and J. P. Croxall. (1992). Long-term population studies of seabirds. *Trends in Ecology and Evolution* 7: 111–114.

Wyrwicka, W. (1978). Imitation of mother's inappropriate food preference in weanling kittens. *Pavlonian Journal of Biological Science* 13: 55–72.

Wyrwicka, W. (1979). Imitation of mother by weanling kittens in eating odorless and tasteless jellied agar. *Federation Proceedings* 38: 1309. [Abstract]

Wyrwicka, W. and A. M. Long. (1980). Observations on the initiation of eating of new kitten food by weanling kittens. *Pavlonian Journal of Biological Science* 15: 115–122.

Xiccato, G., M. Bernardini, C. Castellini, A. Dalle Zotte, P. I. Queaque, and A. Trocino. (1999). Effect of postweaning feeding on the performance and energy balance of female rabbits at different physiological states. *Journal of Animal Science* 77: 416–426.

Yamane, A. (1998). Male reproductive tactics and reproductive success of a group-living feral cat (*Felis catus*). *Behavioural Processes* 43: 239–249.

Yamane, A. (1999). Male homosexual mounting in the group-living feral cat (*Felis catus*). *Ethology Ecology and Evolution* 11: 399–406.

Yamane, A., Y. Ono, and T. Doi. (1994). Home range size and spacing pattern of a feral cat population on a small island. *Journal of the Mammalogical Society of Japan* 19: 9–20.

Yamane, A., T. Doi, and Y. Ono. (1996). Mating behaviors, courtship rank and mating success of male feral cat [sic] (*Felis catus*). *Journal of Ethology* 14: 35–44.

Yamane, A., J. Emoto, and N. Ota. (1997). Factors affecting feeding order and social tolerance to kittens in the group-living feral cat (*Felis catus*). *Applied Animal Behaviour Science* 52: 119–127.

Yerkes, R. M. and D. Bloomfield. (1910). Do kittens instinctively kill mice? *Psychological Bulletin* 7: 253–263.

Young, W. C. (1941). Observations and experiments on mating behavior in female mammals. *Quarterly Review of Biology* 16: 135–156. [pp. 149–150]

YoungLai, E. V., L. W. Belbeck, P. Dimond, and P. Singh. (1976). Testosterone production by ovarian follicles of the domestic cat (*Felis catus*). *Hormone Research* 7: 91–98.

Yu, S., Q. R. Rogers, and J. G. Morris. (1997). Absence of a salt (NaCl) preference or appetite in sodium-replete or depleted kittens. *Appetite* 29: 1–10.

Zaghini, G. and G. Biagi. (2005). Nutritional peculiarities and diet palatability in the cat. *Veterinary Research Communications* 29: 39–44.

Zaher, U. U., P. P. Buffiere, J. P. Steyer, and S. S. Chen. (2009). A procedure to estimate proximate analysis of mixed organic wastes. *Water Environment Research* 81: 407–415.

Zaunbrecher, K. I. and R. E. Smith. (1993). Neutering of feral cats as an alternative to eradication programs. *Journal of the American Veterinary Medical Association* 203: 449–452.

Zavaleta, E. S., R. J. Hobbs, and H. A. Mooney. (2001). Viewing invasive species removal in a whole-ecosystem context. *Trends in Ecology and Evolution* 16: 454–459.

Index

Note: page numbers followed with f refer to figures or photographs

aggregation 62, 232–3
agonistic behavior
    breeding behavior and 94–6
    defined 2
    dominance/submission and 3
    dominant-submissive behavior and 10–15
    early weaning and separation and 114
    territoriality and 9–10
allogrooming 21, 61
alloparental care 59
altruism 63–70. *See also* cooperation
amino acids 140–141, 145–6, 168–70
anal sacs 46–7
arch 13
arch-back 13, 14f, 130
arch-tail 12
audition 190–196

basal metabolic rate (BMR) 37–40
biological control 233–7
birds 214–24, 244–5, 247–50
birth process 100–104
blink 12
breeding. *See* reproduction

carbohydrates 150–155
cat food. *See* nutrition
chin-down 12
chomp 12
clawing 45–6
clumped patterns 26
colonies 62–3
communication 2–3
competition 16–17

competitive exclusion 20
cooperation 54–5, 58–61, 64–5, 68–9.
    See also altruism
cooperative breeders 59
crouching behavior 3, 12
cultural transmission 120

daily activity 23–5, 189
dens 100
density-mediated interactions (DMIs) 220
development
    dens 100
    early maturation 104–8
    early predatory behavior 113–15
    early weaning and separation 113–14, 126–7
    intrauterine 98–100
    nursing 108–9
    parturition 100–104
    survival 111–13
    weaning 109–11, 128
diel activity 23–5, 189
dispersal 26–30
displays, defined 11
dominance 1–6, 15–18, 27
dominance hierarchies 6–10, 16–18
dominance rank 6, 7
dominance reversal 4
dominant-submissive behavior 10–15
dyadic asymmetry 2–3, 5

ears-flat 12, 13f
ears-folded 12
emulative learning 116–21, 210–213

Free-ranging Cats: Behavior, Ecology, Management, First Edition. Stephen Spotte.
© 2014 John Wiley & Sons, Ltd. Published 2014 by John Wiley & Sons, Ltd.
Companion Website: www.wiley.com/go/spotte/cats

energy
    costs of pregnancy and lactation 172–8
    needs for 166–72
    obesity and 178–80
    overview 162–6
eradication programs 245–7. *See also* management
exploration 12

family unit 62
felinine 44
female choice 96–7
female reproductive biology 72–84
feral, defined 19
fiber 155–6
flehmen response 44
food. *See also* foraging; nutrition; predatory behavior
    home-range and 20–21, 31, 33, 37
    interactions and 9–10, 51–3, 55–6
    play and 128
    preferences 120–121
    survival and 111–13
foraging
    effects on wildlife 215–24, 227
    hunting methods 200–205
    learned *vs.* instinctive 210–213
    predation and 182–5
    prey detection 190–200
    prey selection 207–10
    prey types 205–7
    scavenging and 185–9
    timing of 189–90
forepaw-raised 13
freeloaders 60–61

glands 45–8
gluconeogenesis 168–72, 178
group breeders 59

habitat selection 41–3
half-sit 12
halloween 13, 14f
head-avert 13
head-high 12
head-low 12
head-rub 11, 12f
hiss 12, 13f
home-range
    boundaries 31–3
    defined 20
    size 33–41
    territory *vs.*, 20–23

hunting. *See* foraging
hydropenia 159–60
hyperthermia 161

imitative learning 116
inbreeding avoidance 27–31
integrated control 241–5
interaction
    of asocial domestic cats 49–52
    cooperation 58–61
    kinship 61–6
    social *vs.* solitary 52–8, 66–71
interdigital glands 45
intransitive hierarchies 7, 9
intrauterine development 98–100
island ecology 214–24, 241–5, 246–7

kinship 61–6
kittens. *See also* play
    early maturation 104–8
    early predatory behavior 113–15
    early weaning and maturation 113–14, 126–7
    emulative learning 116–21
    intrauterine development 98–100
    nursing 108–9
    survival 111–13
    weaning 109–11

lactation 174–8
learning 116–21
lick-lips 12
litter size 102–4
locomotor play 123, 126, 132
look-away 13
lures and traps 239–41

male-male interactions 11–12
male reproductive biology 84–8
management
    biological control 233–7
    challenges 214–15
    effects on wildlife 215–24, 227
    eradication programs 245–7
    integrated control 241–5
    lures and traps 239–41
    poisoning 237–9
    secondary prey management 247–50
    trap-neuter-release 224–33
mating. *See* reproduction
monogamy hypothesis 64
mutualism 69

neck-flex 13, 14f, 130
neighborhoods 31
neutering 164–5. *See also* trap-neuter-release
nitrogen balance 143, 168, 170–171, 183, 199
noncompetitive exclusion 20
noncooperative breeders 59
nose-sniff 11
nursing behavior 108–9. *See also* lactation
nutrition. *See also* food; foraging; obesity
    carbohydrates 150–155
    cat milk 101–2, 103, 173–4
    fiber 155–6
    overview 137–8
    play and 128
    in pregnancy 106–8
    proteins 139–50, 162–3, 165, 167–72, 178–9
    proximate composition 138–9, 151–2
    vitamins 156–7

obesity 178–80
object permanence 113
object play 123, 124f, 126–9, 132, 135
object-rub 12, 14f
observational learning 116
olfaction 190–191, 199–200

pair-bonding 50–51
parturition 100–104, 174. *See also* pregnancy
paw-strike 12
perianal sniff 11
pheromones 199–200
philopatry 20, 58
piloerection 3, 12–13, 14f
play behavior
    analysis of 130–136
    defined 122
    function and structure of 121–5
    ontogenesis of 125–30
poisoning 237–9
polygyny 88–91, 96–7
population density 9–10
pouncing 131
predatory behavior. *See also* foraging; management
    early 113–15
    effects on wildlife 215–24, 227
    emulative learning and 116–21
    foraging and 182–5
    prey detection 190–200

prey selection 207–10
prey types 205–7
pregnancy 172–4. *See also* parturition
prior attributes 3–5, 7, 15–16
promiscuity 63–4, 88–91
proteins 139–50, 162–3, 165, 167–72, 178–9
proteinuria 44
proximate composition. *See under* nutrition
pupils-small 12

reciprocal reinforcement 8
relative dominance 1
reproduction
    cooperation and 59–60
    female biology 72–84
    female choice 96–7
    female mating behavior 91–2
    male biology 84–8
    male mating behavior 93–6
    promiscuity and 63–4, 88–91
Resource Dispersion Hypothesis (RDH) 70

scavenging 185–9
scent-making 43–8
seabirds 214–24, 244–5, 247–50
seawater 160, 203, 205
self-lick 12
shelter. *See* habitat selection
side-step 13, 14f
sinus hairs 47f
sit 12
smack-lips 12
sniff 12
social dynamics hypothesis 7–9
social facilitation 17
sociality 56–7. *See also* interaction
social play 123, 126–9, 132
solitary animals 56. *See also* interaction
somatosensory functions 197–9
space
    diel activity 23–5
    dispersal 26–7
    habitat selection 41–3
    home-range boundaries 31–3
    home-range size 33–41
    inbreeding avoidance 27–31
    scent-making 43–8
sphinx 13
spit 12
spray 12
stand-straight 12

status 6–10
stiff-legged 12
strays, defined 19
submissive behaviors 2
supraindividual characteristics 3
survival 111–13

tail-lash 12
tail-up 11
territoriality 9–10, 20–23
trait-mediated interactions (TMIs) 220
transitive hierarchies 7, 9
trap-neuter-release (TNR) 224–33
traps and lures 239–41

uniform patterns 26
urine 44, 47

ventrum-pressed 12
vibrissae 47f
vision 190–195
vitamins 156–7
vomeronasal organ (VNO) 199–200

water balance 158–62
weaning 109–11, 113–14, 126–7, 128
wildlife effects 215–24, 227

yowls 12